EVOLUTIONARY THEORY AND VICTORIAN CULTURE

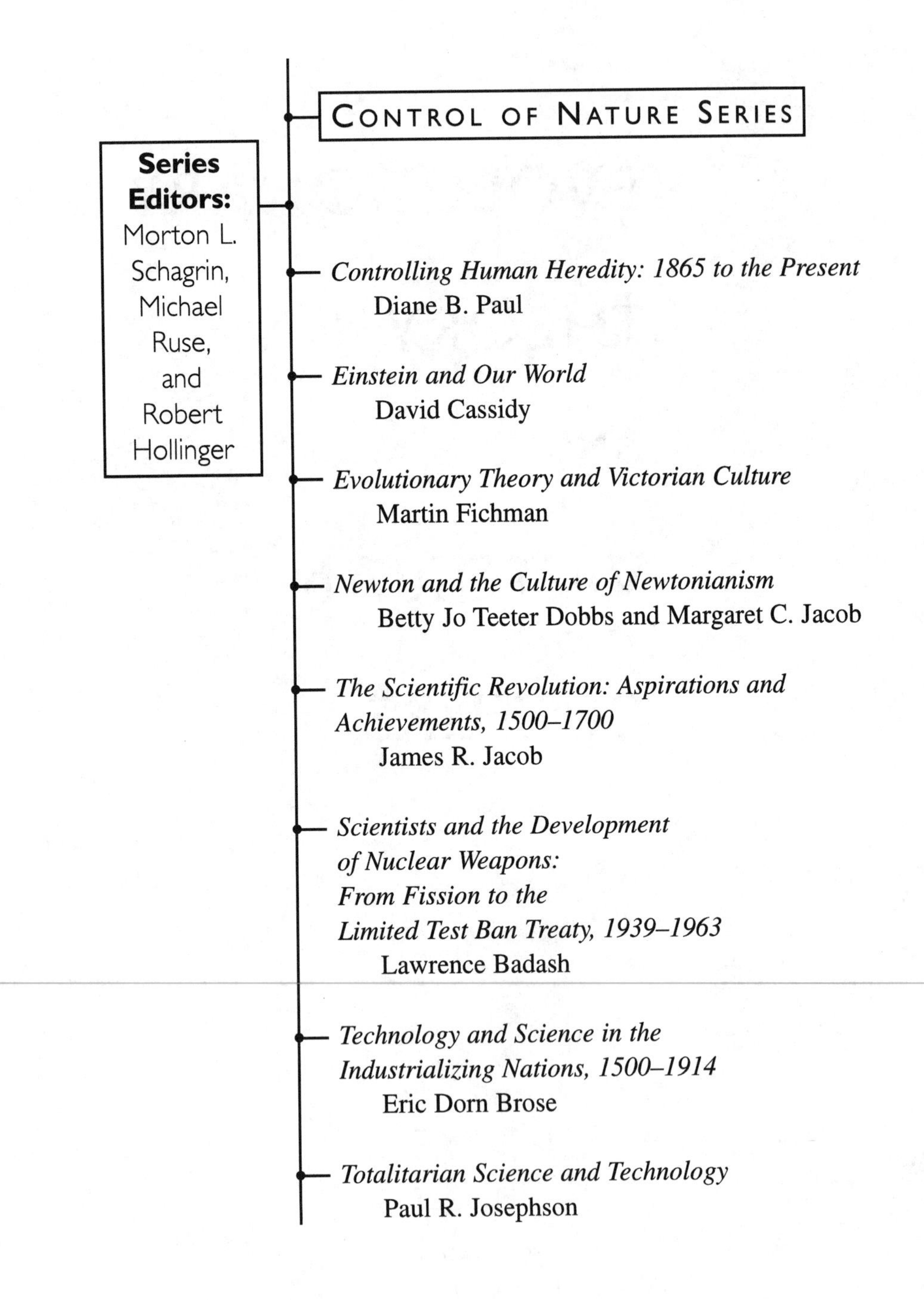
CONTROL OF NATURE SERIES
Series Editors:
Morton L. Schagrin, Michael Ruse, and Robert Hollinger
Controlling Human Heredity: 1865 to the Present
Diane B. Paul
Einstein and Our World
David Cassidy
Evolutionary Theory and Victorian Culture
Martin Fichman
Newton and the Culture of Newtonianism
Betty Jo Teeter Dobbs and Margaret C. Jacob
The Scientific Revolution: Aspirations and Achievements, 1500–1700
James R. Jacob
Scientists and the Development of Nuclear Weapons: From Fission to the Limited Test Ban Treaty, 1939–1963
Lawrence Badash
Technology and Science in the Industrializing Nations, 1500–1914
Eric Dorn Brose
Totalitarian Science and Technology
Paul R. Josephson

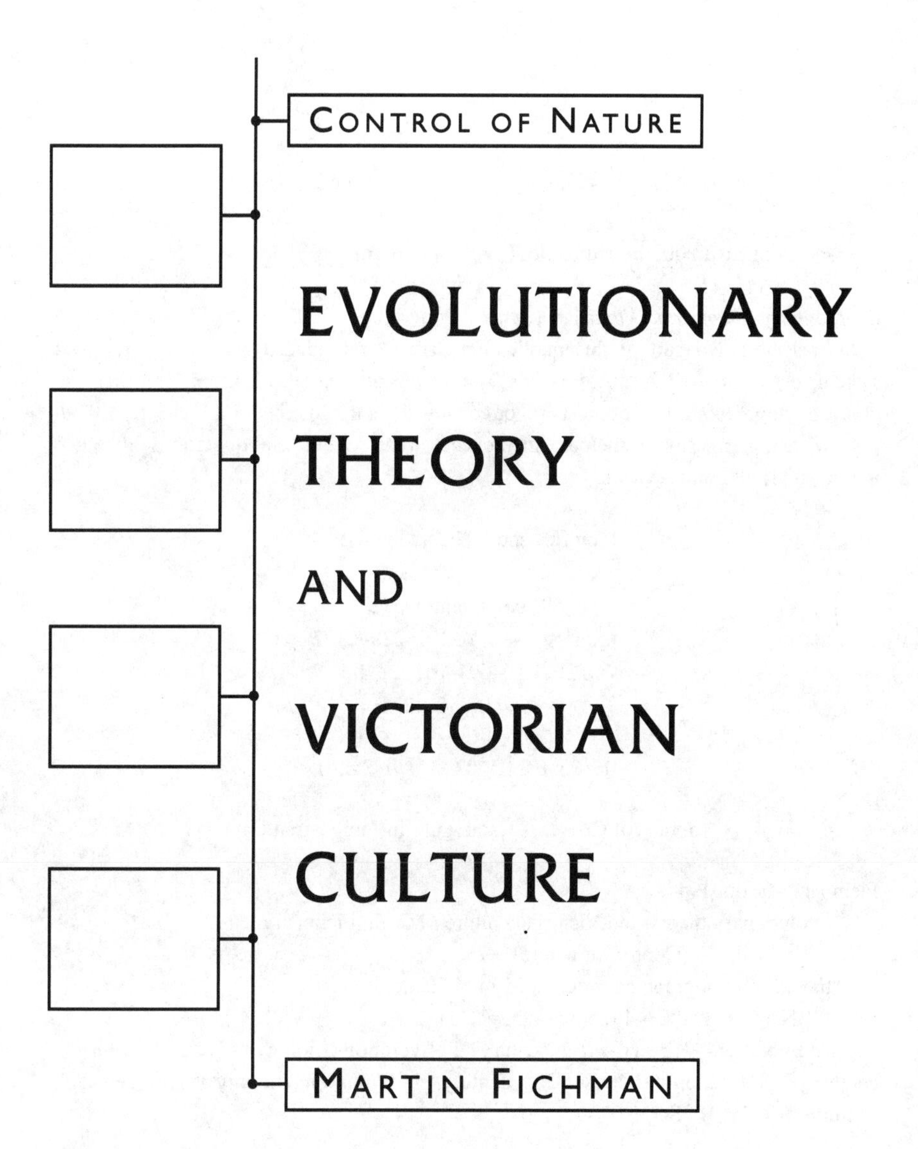

CONTROL OF NATURE

EVOLUTIONARY THEORY AND VICTORIAN CULTURE

MARTIN FICHMAN

Humanity Books
an imprint of Prometheus Books
59 John Glenn Drive, Amherst, New York 14228-2197

Published 2002 by Humanity Books, an imprint of Prometheus Books

Inquiries should be addressed to
Humanity Books
59 John Glenn Drive
Amherst, New York 14228–2197
VOICE: 716–691–0133, ext. 207
FAX: 716–564–2711

06 05 04 03 02 5 4 3 2 1

Library of Congress Cataloging-in-Publication Data

Fichman, Martin, 1944–.
Evolutionary theory and Victorian culture / Martin Fichman.
p. cm. — (Control of nature)
Includes bibliographical references and index.
ISBN 1–59102–003–4 (paper : alk. paper)
1. Evolution—History—19th century. 2. Evolution (Biology)—History—19th century. 3. Evolution—History—20th century. 4. Evolution (Biology)—History—20th century. I. Title. II. Series.

B818 .F43 2002
146'.7'09034—dc21

2002068813

Printed in the United States of America on acid-free paper

To Ken and Harriet

CONTENTS

AUTHOR'S PREFACE

Evolutionary theory has transformed the way humans view their past, present, and future. In domains as diverse as genetic engineering, sociobiology, race, gender, and the environmental balance of our planet, evolution affects the relationship between human beings and nature in fundamental and controversial ways. However, the controversial nature of evolutionary biology is not a recent phenomenon. The most well-known landmark in the development of evolutionary theory is the publication of Charles Darwin's *Origin of Species* in 1859. But even in the decades leading up to *Origin*, the implications of evolutionary theory were already subjects of heated argument. Moreover, Darwin was only one figure in the development of evolutionary theory. This book broadens the historical perspective by including the equally important role played in Victorian debates on the cultural meaning(s) of evolution by such figures as Alfred Russel Wallace, Herbert Spencer, and Thomas Henry Huxley.

Victorian culture was marked by intense interest in questions concerning the appropriate role of evolutionary biology in fundamental aspects of social, political, and religious life. This study focuses on the reception of evolutionary thought in the Victorian age from two perspectives. First, the nineteenth century saw the triumph of the idea of societal "improvement"—what has been termed the "cult of progress." Evolutionism was one of the key factors in the emergence of science as a political and cultural force. The Victorian debates concerning evolution are significant elements of that historical process that ushered in the modern era of technoscientific society. An investigation of the reception of evolutionary thought in the Victorian period provides today's reader with an invaluable historical record of how science came to be inseparable from the broader culture. The second perspective deals with how the analysis of the debates on evolution in the Victorian era affords a rich source of insight and historical data concerning questions about humanity's past, present, and future—questions that have become even more demanding and urgent today than they were a century ago.

This study also relates the Victorian controversies to cultural debates in the twentieth and very early twenty-first centuries. An account of the Scopes trial (1925) demonstrates that evolutionary theory continued to have highly charged implications, particularly for the relationship between science and religion. The book concludes with an overview of the current debates about scientific creationism and shows that the use and abuse of evolutionary science is an enduring legacy from the Victorian era. Evolutionary theory was, and remains, one of the most ideologically charged fields of science. A major goal of this study is to acquaint the contemporary reader with the most recent historiographic trends that have revolutionized our very understanding of the evolution revolution. As will become clear, the historical debates on evolutionism did not resolve the profound issues raised by the concept of human origins from nonhuman ancestors. The Victorian efforts to come to terms with the implications of human evolution did pose crucial questions relating to humanity's place in nature. The answers to such questions were and still are diverse. There are facts about human evolution, but there are also numerous opinions about its cultural meaning(s). This historical perspective on evolutionary debates will, hopefully, enable the modern reader to understand why evolution has been so inflammatory a science. Such historical contextualization provides the tools for a more sophisticated assessment of the controversies surrounding the meaning of evolution as we embark upon a new century.

1 EVOLUTIONISM IN CULTURAL CONTEXT

Science and Society in the Victorian Age

Science and society are interwoven profoundly at the start of the twenty-first century. The varied impacts of scientific and technological developments—such as genetic engineering, computers, the Human Genome Project, nuclear weapons, biological and chemical warfare, global telecommunications, environmental changes—are today obvious and pervasive. Quite literally, contemporary societies are inconceivable without their massive scientific infrastructure. It is therefore sometimes easy to forget how recent this interconnection is in terms of the long span of human history. Science has its origins deep in the human past. But it was in the nineteenth century that the mutual interaction between science and the broader culture first became a—perhaps *the*—dominant feature of human societies. One of the clearest examples of this intimate link between scientific thought and human aspirations and actions is evolutionism. That the human implications of evolution are paramount for most people should come as no surprise. Indeed, both Charles Darwin (1809–1882) and Alfred Russel Wallace (1823–1913), the codiscoverers of natural selection theory, were from the outset of their careers deeply concerned with the question of human evolution and its bearing on philosophical, religious, and sociopolitical matters. Surely, Darwin's closing statement in *Origin of Species* (1859)—a lengthy treatise conspicuous in its omission of any significant mention of the species *Homo sapiens*—ranks as one of the greater understatements in history: "Light," Darwin suggested, "will be thrown on the origin of man and his history" (Darwin 1859, 488).

To understand the full impact of Darwin's and Wallace's ideas, it is necessary to grasp the temper of the age in which they lived. The world of the mid-nineteenth century witnessed dramatic advances in science. But "science" meant only partly the empirical approach to nature. More immediately, more tangibly, science meant to most people the secondary results of that empirical method: the products of technology. During the long reign of Queen Victoria, science and technology transformed many of the conditions of people's lives. The first railroad was built in England in 1825, when Victoria was a little girl. Before that, the maximum speed of land travel was, for up-to-date Englishmen as it had been for Caesars and Pharaohs, the speed of the horse. But before the queen died, at the century's close, almost all of Great Britain's railroads had been built. Science had begun that liberation of humans from animal muscle, that acceleration toward (then) inconceivable velocities, which became characteristic of the twentieth century.

Science was also impressive. It was doing things, making things *work*. The practical, empirical British (and European and North American) mind was fascinated (Briggs 1989). While Victoria occupied the throne, transatlantic steamship service was begun, power-driven machines revolutionized industry, the telegraph and telephone were developed, and the electric lamp and the automobile were invented. Eight years before *Origin*, in 1851, the Victorians celebrated progress at the first world's fair in the dazzling Crystal Palace Exhibition held in London's Hyde Park. When Darwin and his family, wealthy even by comfortable Victorian standards, visited the Crystal Palace, they were representative of those who had benefited handsomely from Britain's global industrial supremacy (Desmond and Moore 1991, 391–397). Science seemed to many (but not all) to banish doubt. In this cultural milieu, evolutionary theory was new, revolutionary, even heretical, yet it was persuasive. It was persuasive because science was persuasive. Evolution became a watchword to the late Victorians. By the end of the nineteenth century, hardly a field of thought remained untouched by the concept. Historians had begun looking at the past as a living organism. Legal theorists studied the law as a developing social institution. Psychologists and philosophers explored the implications of evolutionary theory for their fields (R. Richards 1987). Writers explored evolutionary themes in their fiction to show how their characters developed in an empirical way (Beer 1983). Anthropologists and sociologists invoked evolutionary analogies in their analyses of social forms. Apologists for wealth showed how the poor were "unfit," and how progress, under the leadership of the "fit," was inevitable. Half a century after the publication of *Origin*, the word *evolution*, which in 1800 had been used in mostly a narrow and technical scientific sense, seemed capable of explaining just about anything. The titles of major books of that period reflect this: Charles S. Wake's *The Evolution of Morality* (1878), Edwin Caird's *The Evolution of Religion* (1893), William Carlile's *The Evolution of Modern Money* (1901), and, finally, Thomson Hudson's *The Evolution of the Soul* (1904).

The application of Darwin's and Wallace's ideas to nonbiological realms, whether legitimate or not, was encouraged by the very language of natural selection theory (R. Young 1985). Although the controversies surrounding evolution and religion are perhaps the most well-known, there was scarcely any subject that remained immune to evolutionary incursion. Darwin and Wallace themselves often probed the cultural implications of evolution. Darwin's extensive correspondence reveals his abiding concern with a vast range of topics, from anthropology to psychology to language to politics to theology (F. H. Burkhardt 1983). Wallace wrote frequently on subjects as diverse as anthropology, education, spiritualism, phrenology, vaccination, and land nationalization and other sociopolitical issues (Charles Smith 1991). The vocabulary they employed in first presenting their theories drew heavily on the terminology of the social sciences and political philosophy. It was Thomas Malthus's (1766–1834) *An Essay on the Principle of Population* (1798) that contributed one crucial spark to both Darwin's and Wallace's earliest thoughts on natural selection. Malthus's vivid depiction of the most morbid aspects of struggle and conflict in human as well as animal societies—war, cannibalism, infanticide, epidemics, famine—affected Darwin and Wallace profoundly. And Malthus was only one of the numerous authors who seemed obsessed with the idea of struggle. Victorian Britain, with its unchecked industrial expansion, seemed to demand the metaphor of struggle in politics, religion, and economics as well as science. It was, moreover, Herbert Spencer's phrase "survival of the fittest" that Wallace preferred to his and Darwin's expression "natural selection." Wallace thought that survival of the fittest was a direct rather than metaphorical expression of the process of evolution. In one crucial sense, Wallace was right. The scientific community debated natural selection. For the general public, however, evolutionism was synonymous with survival of the fittest and struggle for existence.

The Challenge of Evolutionism

It was, of course, not so much the process of evolution that caused such cultural upheaval in the Victorian era as what people thought evolutionary theory implied or predicted. Evolutionism not only opened a new framework for biology, it became a popular and intellectual sensation. The debates about evolution challenged and transformed long-held ideas about humanity and its relationship to God, nature, and the course of history. Darwin's *Origin* may be taken as a major symbol of evolutionary thought. Indeed, Darwin was once described as the most dangerous man in England. But as this book will demonstrate, evolutionism was a far broader force than *Origin* itself. Darwin's impact was crucial, but the full cultural landscape of Victorian evolutionism requires that we look at numerous individuals and many versions of evolutionary theory in addition to Darwin and Darwinism. Thomas Henry Huxley, Spencer, Wallace, and a veritable army of scientists, philosophers, religious and social thinkers, as well as broad sections of the general public, were all essential actors in the evolutionary drama of the nineteenth century. Their beliefs, theories, and activities will form the basis for an assessment of the cultural context of evolutionism.

Until the last two decades, many scholars tended to simplify Victorian cultural history as a battle between the forces of tradition, notably religion, and secular forces, notably evolutionism. Although the words *secular* and *religious* remain useful, these terms and the often unexamined assumptions that lie behind them no longer provide an adequate analytic framework for probing the Victorian age. The secularization and modernization of British, European, and North American culture *did* accelerate during the course of the nineteenth century. The consequences and effects of that transformation are felt throughout the world today. But the transformation was anything but inevitable, unproblematic, or systematically steady. Victorian science, religion, politics, and societal and institutional structures were complex and dynamically interacting and changing aspects of the human condition. Only when this is recognized does it become possible to understand Victorian culture—and the crucial role of evolutionism within that culture—as a pluralistic and shifting landscape. Evolutionism, and the allied powers of science, technology, and human invention, did succeed in replacing—in important respects—God and eternity with natural forces and human history as the new backdrop for humankind and its social development. By the close of the nineteenth century, human domination of the world—other species, the environment, and health and disease—seemed capable of unlimited potential. Evolutionism was a key ingredient in the cult of progress that the Victorian era bequeathed to the twentieth century (Turner 1993, 35–37, 71–72, 119–120, 126–127). But the control of nature was part fact and part illusion.

Events of the late twentieth and early twenty-first centuries have shown how fragile humanity's conquest of the natural world actually is. The presumed triumph of secular science over traditional religious and other cultural authorities is also questionable. The resurgence of religious and other nonmaterialist modes for comprehending the human condition in our own time testifies to the inadequacy and inaccuracy of viewing the struggles, perplexities, and debates of the Victorian age as leading to the smooth path of scientific certitude. We shall look closely at the significance of contemporary controversies over scientific creationism as one example of the ongoing dialectic between evolutionary biology and the broader culture. To understand the Victorian world in its rich complexity, we need to recognize nineteenth-century evolutionary science not as a clear-cut set of guidelines for finding out what the natural world is really like, but rather as being deeply embedded in the context of culture and society. Evolutionary theories, and evolutionary forces, are inseparable from human interventions in continuing political, social, economic, and religious struggles and controversies. As will become clear, evolutionism—like science generally—is no monolithic entity. Its boundaries in the Victorian era, as in our own, are never absolute, its definition never certain or uncontested. The contents and methods of evolutionary science are never "innocent" or ideologically neutral, never without influence from other nonscientific enterprises. The relationship between evolutionism and Victorian culture is a two-way street. We will explore various facets of that traffic in ideas and practices in this book (Lightman 1997a, 9, 15). Our goal is to understand the nineteenth-century evolutionary debates in their full richness and ambiguity. Those debates were puzzling, often traumatic, to the society of that age. The questions asked and the answers offered tell us much not only about the Victorian era but about our own hopes and anxieties in an age of profound biological promise and uncertainty.

Pre-Nineteenth-Century Cultural Contexts of Science

In discussing the implications of evolutionism for Victorian culture, it is important to be reminded that science has had an extremely ancient relationship to human social and individual concerns. When our early ancestors first began the attempt to comprehend the natural world in a certain orderly way, we may say that the scientific imagination began to stir (Worthen 1991). We possess few records of these earliest approaches to scientific thinking in prehistoric times. We possess more evidence from the first great civilizations—Egypt, Mesopotamia, Phoenicia, Israel, and India—which, while still primarily imbued with magical and utilitarian concerns, made the first momentous approaches to scientific reasoning. The crucial point is that these early cultures perceived, however tenuously, an intellectual link between the forces of nature and the practical activities of humanity. It was still a long time before the idea emerged that by comprehending the universe in a certain orderly, logical manner, human societies could gain some degree of effective influence over the forces of nature (M. Greene 1992). But the stage was set, before the nineteenth century, for the powerful belief that science might afford humans the power to modify, possibly control, nature.

The history of science in the classical world of Greece and Rome and in China, India, and Islam during the Middle Ages is a record of remarkable scientific achievements (Lindberg 1992; Bodde 1991; Butt 1991). But none of these cultures gave rise to what we term a modern scientific outlook or worldview. It was only in a fairly localized region of the globe, Europe, during the period roughly from 1500 to 1700, that a complex network of specific cultural, sociopolitical, economic, and religious forces permitted science—both as a theoretical and institutionalized structure—to emerge as a defining feature of society (Seitz 1992). During the Renaissance, from the late fourteenth to the late sixteenth centuries, a new spirit of geographical exploration and economic, political, and religious innovation permeated all aspects of culture, everywhere uprooting accepted ideas (DaCosta Kaufmann 1992). By the seventeenth century, certain crucial aspects of European thought and society began to take on those forms that historians call early modern culture. Central to these developments was the Scientific Revolution (1500–1700)—associated with individuals like Galileo Galilei, Francis Bacon, René Descartes, and Isaac Newton—during which the conceptual and institutional foundations of modern science were erected. Although the rise of modern science owes a great debt to classical and medieval achievements, a new image of science was constructed. This image emphasized a mechanistic philosophy of nature, with important but increasingly marginalized residues of magical and mystical traditions (Martin 1992; Goodman and Russell 1991; Cohen 1985). The Scientific Revolution had broad repercussions in general culture: scientific ideas and methodology permeated the thought of the seventeenth and eighteenth centuries (Jacob 1988). One of the clearest indications of the impact of scientific thought and practice occurred in the cultural context of the European Enlightenment.

During the eighteenth century, the twin symbols of reason and nature dominated Enlightenment attempts to establish what were grandly hoped to be objective sciences of humanity and society, as well as physics, astronomy, chemistry, and biology. Led by a group of brilliant individuals known as the *philosophes*—including Denis Diderot, Jean Lerond D'Alembert, Francois Marie Arouet de Voltaire, David Hume, and the Marquis de Condorcet—the Enlightenment epitomized the explicit effort to propagandize on behalf of the scientific outlook. In books and pamphlets, in vast encyclopedias and in newspapers and journals, in poetry and prose, by serious argument and by witty mockery, the disciples of Enlightenment created a potent, albeit controversial, polemic for social and cultural change. Fundamental to these efforts was the *philosophes'* conviction that humanity's foremost task consisted in the establishment, via the scientific method, of greater and greater control over the environment, both social and material (Golinski 1992; Hankins 1985; Rousseau 1980; Maniquis 1992).

The various uses and interpretations of science by different thinkers warn us against assuming that there was a single, unified approach to the Enlightenment. Yet virtually all the *philosophes*—Jean Jacques Rousseau is the most notable exception (Charlton 1984)—were united in their conviction that the benefits resulting from the application of a scientific method would include social reform and progress. They believed that with science as a guide, humanity would at last be able to control desires rationally and pursue mutual concerns more justly and humanely. The Enlightenment was predicated on the idea that social progress would automatically issue from an increased satisfaction of material needs *coupled with* a general liberation of the intellect from what was deemed traditional superstition and irrationality. And, indeed, since the start of the nineteenth century there is no doubt that the human species has been able to use science and technology to transform both physical and social environments radically. Whether or not this transformation has been on the whole a beneficial one was a matter of increasing debate in the nineteenth century, and continues to be so today. That nineteenth-century society was engaged in such a momentous debate is testimony to the power, if not the correctness, of the Enlightenment's vision (Hulme and Jordanova 1990). This broader debate was played out with great drama and consequence in the Victorian cultural controversies that greeted the theory of evolution by natural selection.

Setting the Stage for Evolutionism

Evolution was a controversial subject for nineteenth-century audiences for a number of reasons. The concept of evolution itself was hardly novel. There had been vague evolutionary hypotheses at least as early as the period of classical Greece. By the start of the eighteenth century, however, naturalists generally agreed that the explanation for the origin of the wide variety of species of animals and plants lay in some version of the doctrine of *special creation*. Evolutionary speculations were not absent. Indeed, in the work of Georges Louis Leclerc, Comte de Buffon (1707–1788), and Jean Baptiste de Lamarck (1744–1829), evolutionary ideas found two influential exponents long before Darwin and Wallace. But the doctrine of special creation, which held that all living organisms existing at the present day had been created in precisely their present forms by God, retained its scientific and ideological dominance. The creationist hypothesis argued the plausible and agreeable notion that all species had been perfectly preadapted to suit their particular environments and geographical locales by a rationally ordained system of divine means and ends. This *argument from design* was the basis of the creationist position, which flourished well into the early nineteenth century. Creationism maintained that the adaptive features of each characteristic structure of animals and plants had been purposely created by a divine intelligence. It was argued that the order and complexity of the natural world could not be the product of natural forces alone. Only God, it was asserted, could impose such a system of design on nature. Moreover, creationists held that the structure of living beings illustrated the Creator's power, wisdom, and benevolence. Humans, of course, were the supreme example of God's design. The perfect structure of the human hand and eye, for example, seemed the most obvious illustration of how our bodies had been specially created to serve our needs. *Homo sapiens*, it was further believed, was a unique species with no animal (or prehuman) ancestors (Bowler 1989).

The scientific and popular adherence to the doctrine of special creation during the eighteenth century was symbolized in the person of the great Swedish botanist and taxonomist Carl Linnaeus (1707–1778). His *Systema Natura* (*The System of Nature*; 1735) assigned to every plant and animal species its particular place in one grand hierarchical chain of being. Linnaeus's taxonomic system introduced the modern binary system of biological nomenclature by which each species is known by its generic plus specific name. Linnaeus lent his great authority to the notion that the scientific classification of animals and plants was rationally comprehensible, because the biologist's classification mirrored the natural system, which itself was the product of God's wisdom and plan. Later in his career, Linnaeus gradually developed certain doubts that nature was a completely divinely ordained scale of unchangeable organic forms. But his public position was clearly on the side of creationism. Like Newton, Linnaeus was popularly perceived as providing definitive scientific support for a theological worldview of a harmonious, unchanging nature. His influence was spread not only through his masterly texts but also through his many students. A great many foreign students came to Linneaus in Uppsala, Sweden, to learn the foundations of systematic natural science.

All of them later helped to spread the Linnean doctrine in their own countries. In addition to its compatibility with dominant religious systems, Linnaeus's theory also seemed the ideal guide to the practical and economic demands placed upon eighteenth-century naturalists. They had to fit into some meaningful framework the vast number of new types of animals and plants rapidly being discovered during the commercial and scientific voyages of European travelers and explorers. Linnaeus's biological theories at once explicated and reflected a complex network of scientific, cultural, economic, and religious forces (Lindroth 1973). Although there were occasional disruptions, the natural world, as well as human societies, seemed stable and orderly, part of a divine plan. Or so most people were conditioned to believe.

In Britain, the creationist position was given a classic expression in William Paley's *Natural Theology* (1802). The subtitle of Paley's work, "Evidences of the Existence and Attributes of the Deity, Collected from the Appearances of Nature," epitomized the comforting late–eighteenth-century "official" portrait of nature and society. Although the reality for many people, especially the masses of laboring poor, was anything but comfortable, most political and economic elites endorsed the view of a well-ordered natural and social universe. This image of a cozy world would soon be challenged by the Darwin-Wallace theory of evolution. Additional factors during the nineteenth century contributed to the Victorians' sense that an old order was being rapidly and radically transformed into a new cultural reality, at once alluring and frightening. The continued advance of industrialization, both in Europe and globally, dramatically altered the physical as well as sociopolitical landscape during the course of the century (Mathias and Davis 1991). The rise of the industrial working classes, the greater power won by the middle and professional classes, and the (relative) decline of the landed aristocracy combined to produce a powerful but confusing transformation at all levels of society (D. Coleman 1992). The dramatic growth of industrial towns and ever larger cities—abetted by the striking spread of the railways—created a social, political, and economic environment unprecedented in world history (Bagwell 1988; Goodman and Honeyman 1988). In such an exhilarating yet intimidating complex of novel developments, it was inevitable that the Darwin-Wallace theory of evolution would provide a touchstone for debates on the vexing question of humanity's place in nature, and of science's appropriate role in elucidating that question.

Some fifteen years prior to the publication of Darwin's *Origin*, Victorian society was given a preview of just how heated debates about evolution could become. In 1844, *Vestiges of the Natural History of Creation* was published anonymously by the Scottish author, publisher, and amateur scientist Robert Chambers. *Vestiges* was a fascinating if flawed synthesis that attempted to pull together the most recent findings and speculations concerning geology, natural history, and the moral sciences. The key to Chambers's synthesis was his belief that everything in nature is progressing toward a higher state. In biology, he interpreted progression as transmutation, what would soon be more commonly known as evolutionism. Chambers's book promoted the basic idea of evolution, but his scientific speculation about the mechanism of evolution was defective and surely did not anticipate the Darwin-Wallace theory of natural selection. But *Vestiges* caused a sensation among the scientific community as well as the general public. It sold extremely well and succeeded in spreading the highly controversial view of the materialistic and evolutionary connections between humans and the rest of the animal kingdom. Despite or because of highly critical reviews and attacks from both the scientific and religious communities—that the book would undermine religion and the social order itself—*Vestiges* remained at the center of an enduring controversy. (Bowler 1989, 141–147). Chambers's book was "an evolutionary epic that ranged from the formation of the solar system to reflections on the destiny of the human race." As one of the most controversial works of science and philosophy of the early Victorian era, *Vestiges* proved how explosive a concept evolution was destined to be. Its readership numbered over 100,000 and spanned all sectors of Victorian society. Aristocrats and handloom weavers, evangelicals and militant freethinkers, science journalists and the wives of cotton manufacturers devoured it. So, too, did Queen Victoria, Alfred Tennyson, Florence Nightingale, Thomas Carlyle—and Darwin and Wallace (Secord 2000, 1–2, 506–514).

The Darwin-Wallace Theory of Evolution

The joint discovery of the principle of natural selection by Wallace and Darwin is among the most celebrated episodes in the history of science. And although their paths to discovery displayed common features, there is no doubt that the two naturalists arrived independently at strikingly similar, though not identical, conclusions concerning the origin of species (Kottler 1985). For both men, it was the impact of extensive firsthand observations of the flora and fauna of exotic lands that sharpened their desire to provide a rigorous description of the evolutionary process. Wallace and Darwin were determined to transfer evolutionary theory from the realm of speculation to the domain of demonstrable scientific evidence. Significantly, certain ideas of the English political economist Thomas Malthus, found in his *Essay on the Principle of Population*, provided a direct clue to the mechanism of evolution (Winch 1987). In particular, Malthus's ideas on the tendency of rapid human population growth to outstrip increases in available food supply were critical to both Darwin's and Wallace's formulation of natural selection (McKinney 1972, 80–96). Malthus's book unleashed a storm of controversy that resonates powerfully in contemporary debates on global population growth.

For Darwin, it was the famous voyage aboard the *Beagle* to South America—in particular, to the Galapagos Islands—from 1831 to 1836, which began the transformation of the youthful natural theologian into a convinced evolutionist (Browne 1995; Desmond and Moore 1991). His collecting successes and observations of the vast panorama of animal and plant life, including his astute and extensive thinking about the human inhabitants with whom he came in contact, had a profound impact upon him. Modeling his activities on the comprehensive intellectual range of Alexander von Humboldt, Darwin eventually succeeded in mastering an equivalent breadth of scientific and personal concerns. He recorded everything he saw in the natural history domain and much else from fields such as ethnology and political economy. Most crucially, Darwin gathered a vast assemblage of specimens from most of the places he visited. The specimens were sent back to England where upon his return Darwin was able to utilize them in support of his developing theory of natural selection (Browne 1995, 230). Wallace's travels to South America (1848–1852) and to the Malay Archipelago (1854–1862) were equally decisive (Brooks 1984; McKinney 1972; Camerini 1996; Fichman 1981, 29).

It is crucial to distinguish at the outset between the general theory of evolution and the specific theory of natural selection. Both Darwin and Wallace convincingly demonstrated the *fact* of evolution. Their various writings provide a vast and well-chosen body of evidence showing that animals and plants cannot have been separately created in their present forms but must have evolved from earlier forms by slow transformation. Their unique innovation, however, was to provide a *mechanism* by which such transformations could, and would, be brought about: natural selection. Natural selection rendered the idea of evolution scientifically intelligible. It therefore exerted a decisive impact on subsequent intellectual and cultural history. Yet while many endorsed the first part of Darwin's and Wallace's theory, evolution, the principle of natural selection elicited considerable criticism from the start. Wallace himself later came to question its complete sufficiency, especially in the case of human evolution (Kottler 1974; Fichman 2001). In the crucial period of the late 1850s and early 1860s, however, both Darwin and Wallace insisted that both parts of their theory were inseparable.

Biogeography

Many factors stimulated the first evolutionary ideas of the young Darwin and the young Wallace. Biogeographical questions were central to the growing dissatisfaction each felt with regard to creationist explanations of the still puzzling facts of animal and plant distribution. Biogeography is the science that studies the geographical occurrence, or range, of each animal and plant species over the globe (Browne 1983). Wallace was to devote much of his long scientific career toward establishing biogeography on a secure evolutionary foundation. For Darwin, the Galapagos islands represented isolated ecosystems that served as different natural "laboratories" to compare and contrast species and varieties. The islands of the Malay Archipelago served a similar purpose for Wallace. Thus, travel to exotic lands was absolutely central to the discovery of the principle of natural selection. Three types of observations in South America especially struck Darwin, and although they did not provoke his first theorization on speciation, they did provide him later with excellent examples of speciation, once natural selection had been identified as the probable cause. Darwin's conversion to the theory of evolution—once thought to have been a typical "eureka" experience stemming from his famous visit to the Galapagos Archipelago—is now generally seen as a slow and largely postvoyage development in his scientific thinking. The *Beagle* voyage, therefore, served primarily as a catalyst in shaking Darwin's adherence to special creationism. Once that bond was shattered, he was able to turn to evolutionary hypotheses as alternative explanatory tools (Sulloway 1982).

Darwin's first observation was that species of Galapagos finches differed slightly from island to island. But the species showed general resemblance to one another and also to other finches on the section of the South American mainland nearest to the islands. If, he queried, each species had been specially created, why was there such an inordinate amount of creative activity here? Should not the identity of conditions and location of the several volcanic islands of the Galapagos have caused these creatures to be identical? For, the theory of special creation held that each species of animal and plant had been created perfectly adapted to its particular environmental niche. Moreover, why did the Galapagos species so closely resemble the finches of the nearest coast of lush, tropical South America, where environmental conditions were so dissimilar? Shouldn't they resemble those species of birds on the Cape Verdes Islands (off the northwestern coast of Africa) whose harsh volcanic character and midoceanic position afforded a climate and ecology so similar to that of the Galapagos?

As Darwin traveled up and down the mainland of South America, he made his second finding. He saw that species occupying a particular ecological niche in one region were replaced in neighboring regions—even those with dissimilar environments—by other species that while different appeared closely related in structure. Darwin wondered why the rabbitlike animals on the savanna of La Plata resembled the peculiar South

American rodent that lived near the savannas. Should they not resemble the savanna-dwelling rabbits of North America or the Old World? Darwin generalized these findings, which were inconsistent with special creation, to conclude that the fauna of the different climate zones of South America are *historically* related to each other. They showed little or no relationship to the fauna of equivalent climate and ecological zones on other continents, precisely because they were not related to them (in any direct evolutionary sense).

As his third observation, Darwin found in the pampas of central Argentina fossil remains of large mammals (Glyptodon) covered with bony armor like that of the armadillos now living in the same region. When Darwin returned to London, these spectacular fossils would be the young naturalist's entrée into the world of professional science (Desmond and Moore 1991, 205). Why were these extinct animals built on the same plan as those now living? According to special creation theory, fossils represent extinct species from earlier epochs separated in time by universal catastrophes (like the biblical deluge). Since new, postcatastrophe species were supposedly created by God to fit entirely new environmental conditions, there should be no relation between fossils and living forms. Moreover, fossils should display similarities, if any existed, to contemporary fossils found elsewhere on the globe, since all would have been created for the same (earlier) geological epoch. On the hypothesis that species were immutable and had not changed since they were originally created, there was no scientifically consistent creationist answer to any of Darwin's questions (Barrett 1987).

Common Ancestry

On the other hand, such observations were wholly consistent with an evolutionary explanation. If species *were* subject to modification through time and to divergence into different lines (new species), then all of Darwin's queries could be resolved simply. The finches of the Galapagos resemble each other and those of South America *because* they are descended from a common ancestor. They differ slightly because each has adapted to its own particular mode of life and, in some instances, is restricted to one island only. Further, although the volcanic nature and ecological conditions of the Galapagos and Cape Verdes Islands are so similar, their species are *historically* unrelated to each other. Instead, they are related to those of their (respective) adjacent continents. The Galapagos species share a common ancestor with the South American species. The Cape Verdes species share a common African ancestor. They do not, despite their similar environments, share a common ancestor with each other. In like manner, the hares of South American grasslands are related to neighboring rodent species because all share some common (South American) ancestor. Finally, the fossil Glyptodon resembles the armadillo now living because both share a common ancestor (De Beer 1963).

The example of the armadillo/Glyptodon relationship is particularly important. If living animals and plants display affinities with extinct species, then there is no reason to believe, as catastrophist biologists assert, that extinct organisms have no living descendants. They may in fact have representatives alive today. This meant that the entire fossil record was available to Darwin and Wallace as paleontological evidence for evolution (Desmond and Moore 1991, 210). To be sure, nineteenth-century paleontology was an undeveloped science. Darwin admitted (in chapter 10 of *Origin*) the difficulty posed by the incompleteness of the geological record for the acceptance of evolutionary theory. There was an absence of any graded sequence of fossils illustrating the probable course of evolution of even a single species. It is to Darwin's and Wallace's great credit that they were able to formulate the principle of natural selection despite such obstacles.

The vast accumulation of data by Darwin and Wallace during their respective travels armed them with a powerful working hypothesis: Species have undergone change and have originated one after the other by *descent with modification* from common ancestors. Groups of such descendent species share the same common ancestor. The general theory of evolution, when supplemented with natural selection as the mechanism of species change, was soon recognized as bringing together a broad and diverse range of biological phenomena under the explanatory potential of a single, unified scientific framework (D. Young 1993). A relationship was disclosed between the data of hitherto separate fields of study. Such data could be drawn from embryology, comparative behavior, comparative anatomy, mimicry (the fact that certain species so closely resemble another unrelated species as to be mistaken for that species), animal and plant breeding (domestic variation), and, of course, paleontology. The data from these different fields was boldly—and correctly—reinterpreted by Darwin and Wallace as the consequence of descent with modification, geographical or reproductive isolation, and divergence of species owing to their occupying different ecological niches (Brooks 1984).

But what caused evolutionary change? For Darwin and Wallace, at first almost alone among their contemporaries in this belief, evolution was, scientifically, a two-step process: first, the appearance of variations (later called mutations) in nature and second, the sorting of this variation by natural selection. Darwin and Wallace argued that the existence of *heritable* variations within a species coupled with the production of more offspring than could possibly survive constituted the conditions under which "favorable variations" tended to be preserved and "injurious variations" eliminated. Over long periods of time—that is, over many generations—and under the continued selective influence of the environment (the "struggle for existence"), a group of organisms would eventually have accumulated many new favorable variations. They would then differ sufficiently from their ancestors to constitute a new taxonomic status; thus, the "origin of species." Both Darwin and Wallace realized that evolution must be a universal phenomenon. If different species of finches could evolve from a common ancestor, then, given enough time, entire genera, families, even classes and orders of plants and animals could also be formed in a similar process. Ultimately, all living organisms must be related through their common descent from some simple, original form or forms. Furthermore, since all organisms vary and all reproduce themselves in greater numbers than can survive, there must always be competition, whether direct or indirect, between variants for scarce resources. Thus, the principle of natural selection, too, is universally valid (D. Young 1993; De Beer 1963).

Natural Selection: A Case of Simultaneous Discovery

The simultaneous discovery of the principle of natural selection by Wallace and Darwin has given birth to an enormous, often extremely controversial if not contradictory literature. Most recent scholars, as well as many of Darwin's and Wallace's contemporaries, have analyzed the relationship between the two codiscoverers as a competition for scientific priority. Such a focus, however, tends to reconstruct the actual history of the joint discovery in terms of the palm of victory being awarded to either Darwin (more frequently) or to Wallace (less so). Historian of biology Barbara G. Beddall, among others, has offered a more realistic and less partisan analytical framework. Instead of invoking the analogy of a "zero-sum game" with winners and losers, Beddall suggests the non–zero-sum game metaphor. Rather than pitting Darwin against Wallace for the spoils of victory, historians can appreciate and record the enormous contributions made by both Wallace and Darwin. By focusing on the interaction rather than competition between the two formulators of natural selection theory, the manner in which Wallace and Darwin benefited from each other's work becomes the central object of historical reconstruction (Beddall 1988).

Although the exact nature of Thomas Malthus's influence continues to be the subject of debate, it is clear that *An Essay on the Principle of Population* provided both Wallace and Darwin with a critical insight into solving the question of how species originate (R. Young 1985). Wallace's autobiographical rendition of his moment of discovery provides a dramatic, if remote (it was written nearly half a century after the event), statement of scientific creativity. According to his autobiography, it was during an illness in the Malay Archipelago in late February 1858 that Wallace recalled the work of Malthus, which he had read some twelve years before. Wallace's illness proved a blessing in disguise. "Suffering from a sharp attack of intermittent fever, and every day during the cold and succeeding hot fits [having] to lie down for several hours," Wallace had "nothing to do but think over any subjects then particularly interesting me." Specifically, it was Malthus's vivid demonstration of "the positive checks to increase" disease, war, accidents, and famine, which keep the population of savage races down to a much lower average than civilized races that sparked Wallace's chain of reasoning:

> It then occurred to me that these causes or their equivalents are continually acting in the case of animals also; and as animals usually breed much more rapidly than does mankind, the destruction every year from these causes must be enormous in order to keep down the numbers of each species, since they evidently do not increase regularly from year to year, as otherwise the world would long ago have been densely crowded with those that breed most quickly. Vaguely thinking over the enormous and constant destruction which this implied, it occurred to me to ask the question, Why do some die and some live? And the answer was clearly, that on the whole the best fitted live. From the effects of disease the most healthy escaped, from enemies, the strongest, the swiftest, or the most cunning. . . . Then it suddenly flashed upon me that this self-acting process would necessarily *improve the race*, because in every generation the inferior would inevitably be killed off and the superior would remain—that is, the *fittest would survive*. (Wallace 1905, vol. I, 361–362)

It at once became clear to Wallace that natural selection (though he did not yet use that term) was the mechanism he had been seeking.

Combining the brilliant geologist Sir Charles Lyell's (1797–1875) description of the gradual fluctuations of land and sea, of climate, of food supply, and of predators with his own field experience of organic variation in nature, Wallace realized that given sufficient time new species would evolve in response to altered environmental conditions. The exquisite and often complex adaptations of animals were now explicable, not as the product of planned design but as the outcome of evolutionary change (Wallace 1905, vol. I, 360–363).

Upon recovery, Wallace wrote out his theory as an essay entitled "On the Tendency of Varieties to Depart Indefinitely from the Original Type" (1858). He mailed it to Darwin with the request that Darwin show it to Lyell, "should he think it sufficiently novel and interesting" (Beddall 1988). It is clear that Wallace, although aware that Darwin was preparing for publication a massive work on species and varieties, did not know that the latter, too, had discovered natural selection but had not yet published on it. Darwin, on the contrary, had likely discerned Wallace's progress on the species question from his letters as well as his pre-1858 articles on evolution. A recent biographer of Darwin suggests that although Darwin was obviously aware of the evolutionary overtones in Wallace's essays and letters, he "blindly stared straight past the implications in Wallace's words. . . . He was not prepared to see the possibility that someone else might be hesitantly circling around before arriving at the same theory. His own work, not Wallace's, was primary" (Browne 1995, 537–538). Having labored for so long in preparing *Origin*, Darwin may well have been lulled into a sense of security and limitless time for its completion. He was, thus, dismayed—but not entirely surprised—to receive Wallace's 1858 sketch whose "terms now stand as Heads of my Chapters. . . . I never saw a more striking coincidence. If Wallace had my [manuscript] sketch written out in 1842 he could not have made a better short abstract!" (Darwin to Lyell, 18 June 1858, Burkhardt, Smith, et al. 1983, vol. 7, 107).

As Lyell had warned, Darwin was forestalled and tormented over the proper course of action to follow. Wallace had not specifically instructed Darwin to publish the essay, but the latter realized that publication was the only honorable step (Burkhardt, Smith, et al. 1983, vol. 7, 107). Fortunately for Darwin, his close friends Lyell and Joseph Dalton Hooker arranged a compromise by which both he and Wallace were accorded priority. On 1 July 1858, Wallace's essay was read before the Linnean Society, preceded by extracts from an unpublished essay on natural selection written by Darwin in 1844 and from a copy of a letter from Darwin to Asa Gray (dated 5 September 1857), which discussed the *principle of divergence*, an important part of the theory not discussed in the 1844 manuscript. The title of the joint papers and letter to Gray is "On the Tendency of Species to Form Varieties; and on the Perpetuation of Varieties and Species by Natural Means of Selection." Neither of the two principals was at that meeting. Wallace was in the Dutch East Indies; Darwin was at his rural estate in Down, grieving over the death of his infant son Charles Waring Darwin on June 28 from scarlet fever, and worrying about his other children.

Wallace, in distant Malaysia, was ignorant of the distress his essay had caused Darwin and of the skillful manner in which Lyell and Hooker had extricated Darwin from his dilemma (Desmond and Moore 1991, 467–472). Wallace never (at least publicly) questioned the propriety of the joint publication. He wrote home that Darwin had shown his essay to "Dr. Hooker and Sir C. Lyell, who thought so highly of it that they immediately read it before the Linnean Society," thus ensuring Wallace "the acquaintance and assistance of these eminent men on [his] return home," and went so far as to assure Darwin that he considered the theory of natural selection to be Darwin's and Darwin's only (Marchant 1916, 57, 131). Wallace's persistent deference—public and private—to Darwin was generous, but curious in the extreme. Darwin's higher social status may have awed the younger man in the late 1850s. And Wallace did, in later years, issue statements establishing the independence of his discovery and emphasizing that his essay had been printed without his knowledge "and of course without any correction of proofs" (Beddall 1988). Yet, it is primarily by Wallace's own efforts that the theory of evolution by natural selection is frequently known as Darwinism. In fact, there were significant differences between his and Darwin's formulations of the theory—differences that intensified through the years (Kottler 1985). Some of those differences were apparent as early as 1860, when Wallace annotated the copy of the first edition of *Origin,* which Darwin had his publisher send to him in the East Indies (Beddall 1988). But in 1858 the relatively unknown Wallace could well be satisfied with having his name indelibly associated with that of a member of Britain's scientific elite. Whatever criticisms Wallace may have had regarding the joint publication, they remained publicly muted.

Criticism of the Darwin-Wallace Theory

The joint Darwin-Wallace announcement of the theory of evolution by natural selection at the meeting of the Linnean Society of London on July 1, 1858, marks a turning point both in the history of biology and the his-

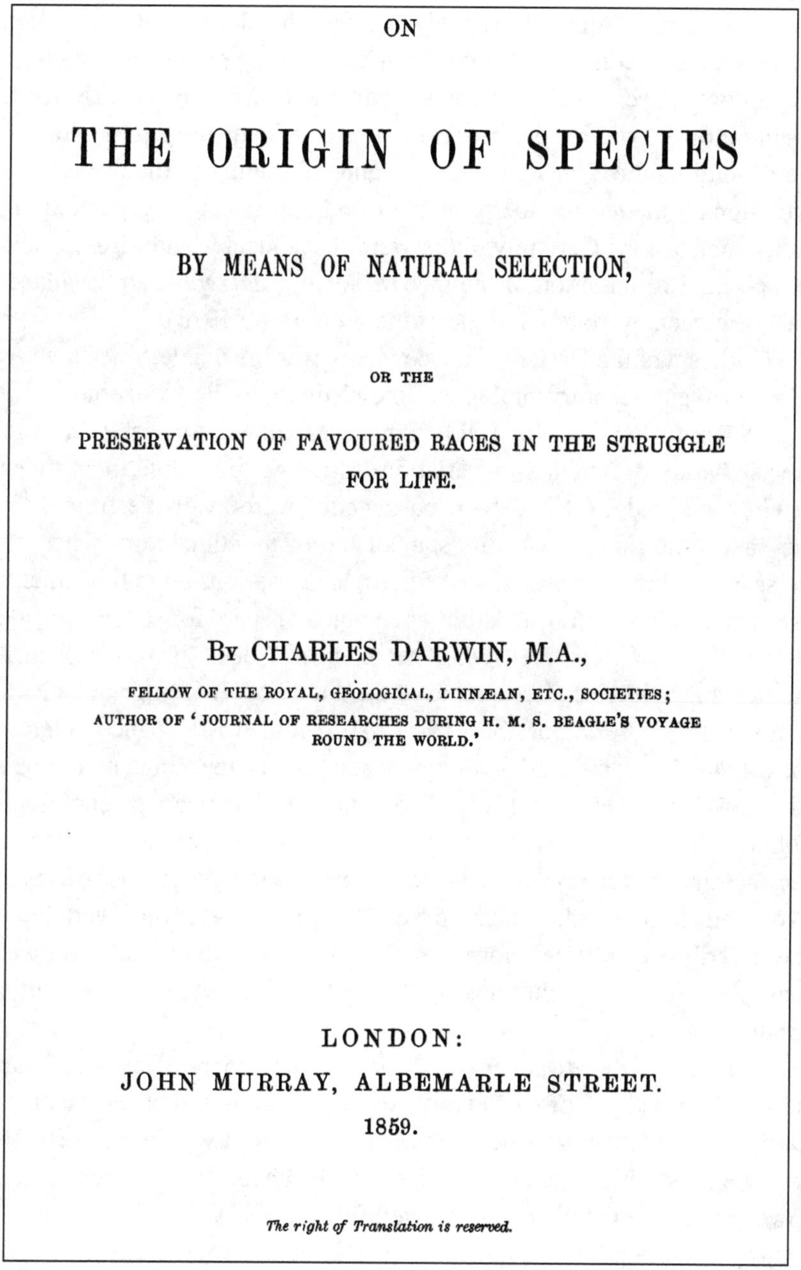

ON

THE ORIGIN OF SPECIES

BY MEANS OF NATURAL SELECTION,

OR THE

PRESERVATION OF FAVOURED RACES IN THE STRUGGLE FOR LIFE.

BY CHARLES DARWIN, M.A.,

FELLOW OF THE ROYAL, GEOLOGICAL, LINNÆAN, ETC., SOCIETIES;
AUTHOR OF 'JOURNAL OF RESEARCHES DURING H. M. S. BEAGLE'S VOYAGE ROUND THE WORLD.'

LONDON:
JOHN MURRAY, ALBEMARLE STREET.
1859.

The right of Translation is reserved.

Fig. 1. Title page of the first edition (1859) of Darwin's *The Origin of Species*. (London, 1859) tory of modern culture. Surprisingly, it seems that the announcement generated little immediate notice or response. It was the publication of *Origin* the following year that brought Darwin's and Wallace's theory to the attention of the wider scientific community and, equally significantly, to the general public. During the 1860s, and continuing to the present day, evolutionary theory and the principle of natural selection generated vigorous controversy. Certainly, there were inadequacies and discrepancies in the initial formulation of the theory. Some of Darwin's and Wallace's scientific peers were quick to point these out (Hull 1973).

Criticism of the Darwin-Wallace theory was inevitable, since much of mid–nineteenth-century biological knowledge relating to evolution was poor or confusing (Bowler 1990). Little was known, for instance, of the precise nature of heredity or variation. Although the revolutionary ideas of Gregor Mendel (1822–1884) on genetics were available from 1865, they had little impact until the start of the twentieth century. Thus, the cause and mode of transmission of heritable variations—the fundamental raw material for natural selection—remained obscure for several decades after the appearance of *Origin*. The science of ecology was still in its infancy and a fuller understanding of the nature of environmental adaptation was therefore impossible. The fossil record was still incomplete. It took time for convinced evolutionists, such as the English biologist Thomas Henry Huxley (1825–1895) and the American paleontologist Othniel C. Marsh (1831–1899), to trace plausible evolutionary trees of the ancestry of creatures such as the horse, so as to strengthen the case for evolution. Marsh's classification and description of extinct vertebrates (he described eighty new forms of dinosaurs, both giant and tiny) were major empirical contributions to the body of knowledge of evolution (Shor 1974, 134).

Claims for the discovery of fossilized human remains were—and remain to this day—often fiercely contested. Many scientists, including Darwin and Wallace's close friend and colleague Lyell, were perturbed for religious reasons about possible human historical coexistence with lower and now extinct creatures. Lyell felt that this would indicate a religiously unacceptable evolutionary prehistory for the human species. However, Lyell's important *Antiquity of Man* (1863) presented to the Victorian reader a broad array of evidence that indicated that man had probably evolved from lower animals over long periods of time. Lyell's evidence was drawn from sources such as skeletal remains (for example, the skull of the Neanderthal man discovered in Germany in 1857) and the gradual divergence of the European languages from a common ancestral root language. Despite his own ambivalence on the question of human evolution, Lyell's geological and biological writings were ultimately of immense support to Darwin, Wallace, and their advocates (L. Wilson 1972). Finally, the timescale of the earth's history as estimated by many geologists and physicists seemed grossly inadequate for the time needed for *gradual* evolution to have occurred (Burchfield 1975). These and other objections, however, have been rendered substantially ineffectual by subsequent scientific investigation. The theory of evolution by natural selection stands as one of the cornerstones of modern biology (Mayr 1991). During the Victorian era, natural selection was a highly influential ingredient in the complex debates both on the scientific status of evolutionary theory and the potential application of that theory to broader cultural and sociopolitical concerns. Understanding the genesis and details of Darwin's and Wallace's theory permits us to appreciate why evolution was destined to become a central focus for debating a wide range of issues in the Victorian period. Precisely because the very language and metaphors of nineteenth-century evolutionism were inextricably tied to the language and metaphors of social, political, economic, and religious thought, it appeared logical—indeed, mandatory—to the leading thinkers of that era to attempt to transpose evolution from the biological to the sociocultural realm (R. Young 1985).

Theories of Societal Evolution

The goal of employing the scientific method to elucidate and, hopefully, solve the problems of society dates back to the period of the Scientific Revolution. During the eighteenth century, efforts to create a "science of society," usually on mathematical and physical models, were common. These attempts reveal the complex methodological and ideological issues posed by the interactions between various concepts of natural science and social science (Cohen 1994). Victorian biologists and their wide audiences, therefore, were pursuing a well-established tradition when they sought to deduce sociopolitical guidelines from evolutionary theory (Bowler 1993; Jones 1980; Moore 1989). As historian of science Theodore Porter has noted, however, "the urge to elevate politics above *mere* politics by achieving a consensus of [scientific] experts" usually fails to "force a consensus, particularly when practical applications are at issue. In part, this is because conflicting political or social visions are often only masked by the ostensibly neutral language of science." Equally pertinent is the fact that the natural sciences, particularly evolutionary biology "present neither a unified nor any single readily-applicable model for social science. Science envy, far from elevating political and social thought above politics, has provided instead a pervasive idiom of debate" (Porter 1990, 1024).

Evolutionary biology conjured up variant, often conflicting readings within the British scientific community. Historians of science have now successfully challenged the view that Darwinism was the sole, or even dominant, evolutionary hypothesis in the Victorian era. In the later decades of the nineteenth century, many biologists became outspoken opponents of the theory of natural selection, erecting a variety of alternative mechanisms for evolution. Most influential were the neo-Lamarckian interpretations of the concept of the inheritance of acquired characteristics. These various neo-Lamarckian hypotheses were diverse, but all drew upon the assumption that characteristics, or traits, acquired by the adult organism can somehow be transmitted to the offspring and can ultimately be incorporated into the species' hereditary constitution. Lamarck's best-known work was the *Philosophie zoologique* of 1809. One of Lamarck's classic examples of such postulated evolutionary change is the neck of the giraffe, which grew as a result of the new habit of feeding from trees. The popularity of Lamarckism arose largely from the fact that it was a natural extension of a pregenetical way of looking at reproduction and development, in which growth and inheritance were seen as aspects of a single integrated biological phenomenon. The inheritance of acquired characteristics was, however, not the sole mechanism of evolutionary development in Lamarck's theory. He also postulated an automatic progressive trend that would produce what he believed was a linear sequence of animal and plant forms, from the simplest to the most complex, that is to say, most highly evolved species. In the absence of any well established theory of genetics, many nineteenth-century biologists continued to be drawn to aspects of Larmarckism. Darwin, himself, never ruled out the possibility that the inheritance of acquired characteristics might act as a supplementary mechanism alongside natural selection in some instances in the evolutionary process. Even some prominent Darwinists, such as Herbert Spencer and Ernst Haeckel, placed great emphasis on the inheritance of acquired characteristics (Keller and Lloyd 1992, 188–190). In addition to the rise of neo-Larmarckism, other mechanisms were proposed as being far more important than natural selection. These alternative hypotheses included nonadaptationist (orthogenetic) and discontinuous (nongradual) models (Bowler 1983). Political conclusions drawn from biology relied, however, mainly upon analogies to the two main mechanisms—Darwinian and Lamarckian—proposed for evolutionary change. The idea of progressionism was central to Lamarckism; since traits acquired by the purposeful behavior of animals were inherently adaptive, evolution would thus be guided along beneficial lines, as organisms gradually became fitter as they responded to changing environmental demands (R. Burkhardt 1981, 132, 206). In contrast, the theory of evolution by natural selection presents a more complex epistemological relation to progressionism (Nitecki 1988).

An ironic result of Darwinism's success, however, was that Lamarckism seemed more plausible after *Origin of Species* had given evolutionism greater credibility than it possessed in the first half of the century. Before the advent of the German zoologist August Weismann's (1834–1914) theory of the germ plasm (the concept that the transmission of heritable traits cannot be affected by the environment) in the 1880s, many biologists probably found it difficult to distinguish between natural selection and Lamarckism. Thus, there were a variety of evolutionary theories upon which biologists and other interested parties could erect rival "scientifically sanctioned" political systems. Despite the social, professional, theological, philosophical, gender, and empirical constraints that conditioned the shaping of biological theories, there was still a wide range of maneuvering for biologists to draw explicit cultural conclusions from their science (Bowler 1993, 16, 61, 89, 92).

When the Victorian debates are widened to include the German, French, American, and other scientific communities, the meaning of evolution becomes more complex still (Numbers and Stenhouse 1999; Glick 1974). An additional factor complicating the analysis of the interaction between biological and political thought is the ambiguity surrounding the crucial and value-laden term *progress* (R. Richards 1992; Desmond 1982). Particularly in the English-speaking world, progressive movements, progressive thinkers, and progressive political parties all benefited from the wide, if confusing, scope afforded by the concept of evolutionary progress. Of all terms in the evolutionists' lexicon, *progress* is one of the most problematic because of the multiplicity of meanings that have been attached to it. Darwin's and Wallace's changing conceptions of progress are reflective of their and their contemporaries' attempts to provide provisional definitions of the relation between so-called higher and lower categories of animals and plants. Indeed, the very adjectives *higher* and *lower* have been rejected by some recent biologists, such as Richard Dawkins, as so "mischievous" as to merit deletion from the discourse of evolutionary biology. In the Victorian period, mischievous as the term *progress* may have been because of its ambiguity and uncertainty in the hands of different thinkers, the concept itself was at the very heart—semantically and culturally—of Victorians' conception of evolutionism in both taxonomic and ideological dimensions (Keller and Lloyd 1992, 6, 263–272). The debates over the definition of progress among the scientific community contributed to the diversity of sociopolitical theories proposed to describe the course of human and social evolution. Moreover, the current scholarly reassessment of the so-called demarcation battles in attempting to define what constituted science as opposed to pseudoscience in the Victorian period (as well as in the present day) puts into question deeply ingrained notions of what constitutes orthodoxy versus heterodoxy in the life sciences in particular (Winter 1997, 24–50).

Science and Pseudoscience

The science/pseudoscience demarcation debates are particularly relevant to evolutionism and those fields, such as phrenology, mesmerism, and psychical research, which were associated in both the scientific community and the broader public with the potent issues of assessing human nature. These fields of theoretical and experimental investigation, which were often linked to evolutionism in the Victorian era, have been a problematic area of inquiry for historians. First, the claims and experimental procedures of what might appear at first glance to be marginal subjects were, in fact, embraced by a significant number of Victorians (scientists and laypersons) just as equally large numbers rejected them. Second, mesmerism, phrenology, psychical research, and spiritualism as fields of serious historiographic analysis have both attracted and repelled scholars throughout the twentieth century. The reasons for their mixed reception in the Victorian period are complex but comprehensible. The reasons for the generally marginalized status they have been accorded by both cultural historians and historians of science for much of the twentieth century are clearer but more disturbing. Mesmerism, phrenology, and what may be termed the other occult sciences naturally seemed less than completely reputable to early and mid–twentieth-century historians who were trained to emphasize the triumph of positivism. In the 1980s, however, a newer breed of scholars began to reconsider the significance of mesmerism, spiritualism, and related movements in the Victorian era (Oppenheim 1985).

The widespread appeal of spiritualist and related beliefs during the Victorian period is now established incontestably. Most crucially, the intellectual relevance and pervasive cultural importance of spiritualism, mesmerism, and phrenology in the Victorian period can no longer be construed as marginalized phenomena. They were simply too deeply ingrained in the minds and hearts of too many people from diverse social, economic, political, and theological backgrounds to be anything other than profound, if highly controversial, characteristics of late–nineteenth-century culture. The first revisionist studies of the 1980s emphasized the intense sense of religious doubt and questing that drew diverse individuals and groups of individuals to the teachings of spiritualism. These studies demonstrated that spiritualism and related psychic avenues into the realm of the nonmaterial universe were powerful elements in the worldviews of many thousands of men and women, including some of Britain's foremost intellectuals and scientists, such as Henry Sidgwick, Edmund Gurney, and Frederic W. H. Myers. Spiritualism, mesmerism, and psychic research afforded vehicles for mediating between the often competing claims of traditional religions and modern science (Mandler 1997, 77–84). More recent studies in the past decade have refined the new historiography by broadening the explanatory scope of spiritualism's and mesmerism's appeal to address issues of gender and sexual politics, rhetorical strategies, constructions of subjectivity and the "self," and the epistemological claims of science (Owen 1990; Barrow 1986; Winter 1998; Noakes 1998).

The spiritualist concerns of leading evolutionists such as Wallace were anything but singular. These individuals were part of a broad sociocultural as well as intellectual community for whom investigations into the spiritual realm were epistemologically significant, politically influential, and emotionally rewarding. The cultural context of evolutionism thus becomes an even richer subject. The enduring fascination with the marvelous and the mysterious, a constant in human history both before, during, and after the scientific revolution(s) of the seventeenth and eighteenth centuries, had lost none of its power during the nineteenth century (Daston and Park 1998). One may argue further that as positivism and kindred doctrines gained official ascendancy during the course of the Victorian era, fascination with the marvelous and the occult increased as counterpoints, especially among the general public. Parallels with the late twentieth and very early twentieth centuries are not hard to draw.

Evolutionism for the People: Victorian Science Popularization

The question of just how widely circulated evolutionary ideas were among the Victorian public is critical to understanding the full impact of evolutionism upon Victorian culture. In our contemporary world, particularly in the West, the ordinary person has developed an almost insatiable interest in the discoveries and implications of modern science. Evolution and its related fields, notably genetic and reproductive technologies, are front-page news. Cloning, genetically modified foods, and designer babies are merely the tip of the iceberg of popular interest in evolutionary science. Bookstores and magazine racks are lined with popular accounts of evolutionary topics. Stephen Jay Gould, Edward O. Wilson, and David Suzuki are but three of many biologists who have made a specific effort to translate the latest findings of evolutionary science into terms that fuel public interest and debate. Science documentaries on television command wide and enthusiastic audiences. As the well known Victorianist scholar Bernard Lightman has noted, "it is not possible to overestimate the importance of current popularizations of science, in all their varied forms, for our understanding of the relationship between contemporary science and culture. [But] can the same be said for the Victorian period, or is the popularization of science a phenomenon of significance only in the twentieth century?" As Lightman and a number of other recent scholars have demonstrated, the answer is clear: Science popularization had become a thriving industry in the Victorian era (Lightman 1997b, 187).

Large and small public lectures and scientific demonstrations, textbooks, atlases, dozens of popular magazines and pamphlets, as well as science fiction provided the Victorian public with ample opportunity to keep abreast of scientific discoveries and debates. Among the most controversial of these was the revolution in biology associated with Darwin and Wallace. Evolutionism was a potent ingredient in the already complex brew of nineteenth-century cultural and scientific changes. The insights and speculations concerning evolution showed nature to be not static but mutable in strange and unforeseen ways. Evolutionism, along with other major nineteenth-century scientific and technological developments, forced constant reassessments of the way in which people might understand life or locate themselves in the great scheme of things. These reassessments of self, society, and nature affected both scientists and members of a science-hungry and science-fearful public. Social Darwinism, with its various applications of ideas of species survival to human social and economic survival, was one of the most obvious and unsettling results of evolutionary thought. But the new manner in which Victorian scientists and intellectuals used the findings of evolution was far from simply objective statements of fact. Evolutionism, perhaps more so than any other nineteenth-century science, was a rich resource for data, speculation, and theories by which specific Victorians could put forward biased ideas or particular ideological visions of the so-called natural order. In fact, there was nothing natural about the conceptions of the natural order presented by leading Victorian evolutionists like Huxley, Darwin, and Wallace. Each of these central figures brought a unique take on the manner in which the natural and the social order existed in dynamic tension. If the leading evolutionists disagreed among themselves about the precise meaning(s) of their science for broader cultural issues, the popularizers of evolution presented an even greater smorgasbord for the public.

The numerous books, pamphlets, and illustrations that lined the bookstalls in cities on both sides of the Atlantic, and in the British colonial outposts worldwide, offered the interested reader for a few cents a bewildering variety of evolutionary ideas and interpretations. Similarly, museums—including the biological and medical freak shows that displayed fascinating and often disturbing and confounding monstrosities—served as conduits for passing scientific information to the public. Scholars are just beginning to mine this rich harvest of popular accounts of evolutionism, which likely affected ordinary Victorians more than the writings of the leading intellectuals on the same subjects. The Victorian natural history craze owed much to the popularizers who mediated between the professional scientific community and the general public. In some cases, the popularizers were themselves distinguished evolutionary scientists, notably Huxley and Wallace. More often, the popularizers were amateurs, often female, who collected butterflies, marine animals, and many species of plants to illustrate the cultural and ethical dimensions of evolutionism. They stood positioned between the secular implications of scientific naturalism and the theological underpinnings of Victorian culture. In a real sense, the public learned much of what they came to understand about evolutionism through the efforts of the science popularizers.

In the Victorian era, as in our own, it was often the popularizers who guided the general public through the complex web of theories, discoveries, interpretations, and implications (real or imagined) of evolutionism. The science popularizers of the nineteenth century not only educated an eager public, they romanced and controlled the public's perception of evolution because they knew how to read public concerns about what evolutionism meant in a practical sense. In disseminating science, the popularizers were not merely transmitting information. Their books, pamphlets, lectures, and exhibitions often assisted in the transformation or reinforcement of cultural attitudes of many Victorians. Well known author Eliza Brightwen's sentimental and proprietary attitudes toward animals, for example, reflected and encouraged the urge to domesticate and thus control species other than humans. In her writings, Brightwen's unexamined assumption that the educated classes had a right to control inferior species was apparent in the class and especially racial biases in her accounts of human evolution (Gates 1997, 179–186).

Gender Issues in Science Popularization

Gender issues also figured prominently in science popularization. While many science books and essays for general readers included women in their intended audience, by the middle decades of the nineteenth century a new genre expanded. Works targeted specifically toward introducing and disseminating science for girls, women, and "ladies" appeared with increasing frequency. Textbooks for informal and formal learning and coffee-table books adapted science to the perceived needs and abilities of women of different ages, social classes, and levels of expertise. The relationship between women and Victorian science is a subject of active interest among contemporary historians and sociologists of science. Recent scholarship has demonstrated that women were important fixtures on the cultural map of Victorian biology, particularly in the domain of botany, as audience, readers, writers, cultivators of science, investigators, and helpmates to their husbands and male relatives and friends (Shteir 1997, 236). Darwin's wife, Emma Wedgwood, devoted much of their long marriage to supporting his scientific work. In a very real sense, his priorities and prodigious research and writing endeavors had become a main priority in her own life (Desmond and Moore 1991, 644).

Science writing for and by women was part of the broader process by which the complex fabric of Victorian science culture was constructed. More critically, the question of what types of science women could write about and learn about had significant ideological overtones. There is a powerful gendered component to both popularizing science and professionalizing science in the Victorian era. Evolutionary biology was particularly important in this respect. Most nineteenth-century studies on human evolution were permeated by notions of "natural," that is, biological, as well as psychosocial differences between men and women. The image of the genteel woman, with her natural affinities for delicacy, nurturing, and ethical and religious duties was potent. In a typical work, *The Young Lady's Book: A Manual of Elegant Recreations, Arts, Sciences, and Accomplishments*, published in 1859, the same year as *Origin of Species*, ingrained gendered assumptions shine through. Chapters on topics such as embroidery, female deportment, and photography are interspersed with chapters on sciences such as geology, ornithology, and botany. Women were being encouraged to study science, but with a different slant than their male counterparts. The chapters on botany teach about plant classification but encourage female knowledge of plants because "there is something peculiarly adapted to feminine tenderness in the care of flowers." The book's author also recommends certain types of gardening as suitable for women, but carefully declares that it is inappropriate for females to "dig up the earth, study the modes of manuring it, or prepare compost." *The Young Lady's Book* does not exclude women from knowledge of nature. But it definitely presents women's relationship to science in terms of naturalized notions about gender and class in accordance with Victorian constructions and stereotypes of womanhood.

By the 1860s, however, certain shifts in attitude were becoming noticeable. With the rise of early feminism, the advocacy of certain limited political and social reforms, and the increased demands for higher education in science (and other subjects) for women, cracks in sexual stereotypes appeared. Lydia Becker, an activist in suffrage campaigns as well as a keen botanist and botanical writer, objected to female-specific books or pedagogies and linked science writing for women to larger debates about equity and women's education. She argued that the "mind has no sex," and that gendered features ("the conventional masculine type of mind") appear in both male and female bodies. Becker declared that biology provided "no necessary, or even presumptive connexion between the sex of a human being, and the type of intellect and character he possesses." In her continuing crusade for equity in education and professional opportunities, Becker posed a question in her essay "On the Study of Science by Women" (1869) that still resonates today: "Why are there fewer scientific women than scientific men?" Becker was part of a growing movement that sought to divest science, and especially evolutionary biology, from gender stereotypes and the notion of biologically innate differences in male and female character and intellect (Shteir 1997, 236–237, 249–252). Science popularization, particularly that relating to evolutionary and medical topics, both reflected and challenged key Victorian assumptions about uniquely feminine and masculine traits.

Science and Satire

Evolutionary theory clearly had numerous pathways by which its controversial ideas could make their mark on popular culture in the Victorian era. The two most widespread responses were either to embrace or to reject the theory and its many implications for human society. Another response was satire. In a lively piece in *Punch,* the most influential satirical periodical in Britain, evolutionism was satirized in September 1859, just two months before the publication of *Origin.* Evolution was burlesqued in cartoons of duck heads superimposed on fashionably dressed human female forms promenading in London's fashionable Hyde Park. Why should science have been the subject of satire? One part of the answer is that science *was* making significant inroads into traditional cultural domains such as religion and politics. As such, science was a powerful force for change. Victorians were confronted with both the promise and the perils of the increasingly pervasive naturalistic picture of the world displayed by scientists and their supporters. The new intellectual and institutional power exercised by the emerging professional scientific community generated friction and an extraordinary range of argument as Victorian society sought to accommodate science into traditional culture.

Journals such as *Punch* were both amusing and deadly serious. Satire and irony were effective tools with which writers and artists could call attention to conflicting cultural views of human nature and human society. Victorian society was in transition, and people were perplexed by the rapid changes taking place around them. Satirists could raise issues of immense importance by their use of striking, often shocking, imagery. Precisely because satirical distortion was able to caricature and juxtapose symbols of traditional worldviews with those of so-called progressive scientific worldviews, witty books and articles and devastatingly on-target cartoons recorded the conflicts of Victorian culture without the responsibility of resolving them.

Satire also made possible the widespread participation of Victorians at different social and economic levels in the science-generated intellectual traffic of the day. The bits and pieces of satire remain as records of cultural formation that help us understand science-related cultural conflicts as Victorians saw them. *Punch* often played with the progressive ideology of evolutionary biology, and science

Fig. 2. "The Lion of the Season" (Alarmed Flunky. "MR. G-G-G-O-O-O-RILLA!"). A cartoon from *Punch* two years after the publication of Darwin's *Origin*. It is a typical example of the social and intellectual uncertainties felt by Victorians when faced with the prospect of an evolutionary kinship between humans and other primates. (From *Punch* 40 [1861]: 213)

more generally, by contrasting the grand ambition with often meager practical results. It also sometimes lampooned the idea—dear to the emerging professional scientific community—that science is a selfless pursuit of knowledge by associating it with vanity and self-aggrandizement. Lyell, Huxley, and Darwin were among the evolutionists most frequently satirized. *Punch* editors, ever fascinated with the British obsession with social standing and hierarchy, had a field day with evolutionary themes. The 1860s and 1870s saw a virtual explosion of gorilla and other simian imagery in cartoons. Evolutionism added to the world of social hierarchy the powerful and unsettling hierarchy of biological ancestry. As new ideas were generated concerning human origins, the question of the precise relationship between humans and the higher primates (notably gorillas) became especially controversial. In one of *Punch*'s more famous and sensational cartoons, an alarmed looking servant at the door of a fashionable London evening party announces a surprising new guest, a "Mr. G-G-G-O-O-O-rilla," dressed in evening coat and tails. This cartoon, like many other pieces of Victorian satire, reflected deeply serious perplexities concerning the relationship of evolutionary science to racism, slavery, cultural relativism, and human identity itself (Paradis 1997, 143, 146–148, 155–158).

Victorian science popularization, therefore, provides a valuable insight into the conflicting images of science that marked that era. Questions such as who should participate in the making of science were still unresolved. How were the competing claims of the new professional scientists and those of clergymen, women, artisans, and nonprofessionals in general to be addressed and reconciled? Related questions of what kind of information should be transmitted to the general public concerning science also sparked vigorous controversy. What kinds of "stories" should be told about nature to the Victorian public? For most professional scientists, the answer was clear. Science should be presented in terms of a natural world described by the operation of demonstrable scientific laws and causes, particularly the theory of evolution, avoiding all reference to supernatural causes. Scientific naturalists like Huxley, John Tyndall, and Herbert Spencer all wrote popular accounts from this perspective. In the latter decades of the nineteenth century, the wider implications of evolutionary theory were addressed by them and other professionals. Their efforts can be seen as an attempt to control the public's understanding of modern science. In contrast, many, but by no means all, science popularizers sought to flesh out the portrait of nature of the scientific naturalists by emphasizing aesthetic, teleological, moral, and divine qualities of the natural world. At stake were competing interpretations of the cosmic significance of science. Books, magazines, children's literature, newspapers, popular lectures, museums, fairs, and exhibitions all served as vehicles of science popularization. All were crucial in defining, if not resolving, the tremendous cultural controversies engendered by evolutionism (Lightman 1997b, 204–207).

Darwinism and Lamarckism were the two dominant biological paradigms in the Victorian era. There were additional models, such as mechanisms of cooperative behavior among and between different species. All provided fertile ground for theorists of social evolutionism. Social evolutionists emphasize directional change in societies with time. The connection between directional change and the notoriously varied concepts of progress is complicated by empirical and logical ambiguities. Despite such philosophical and epistemological quandaries, social evolutionists of whatever stripe sought to justify their theoretical models as congruent with some version of natural, that is, "scientific," evolution. It was the profound political, technological, industrial, and urban transformations of Europe and North America in the late nineteenth century that generated the greater urgency, authority, and popularity of evolutionary accounts of societal change in Victorian culture (R. Smith 1981, 133–135). The stage was set for a proliferation of sociopolitical interpretations of evolutionary theory.

2

SOCIAL DARWINISM(S)

Varieties of Social Darwinism

Numerous thinkers in the Victorian period constructed evolutionary models of society. They ranged from the starkly competitive ideas of the American William Graham Sumner (1840–1910) to the collectivist views of the German communist Karl Marx (1818–1883). This wide diversity of political theories predicated upon biological hypotheses and metaphors indicates how such seemingly unambiguous terms as *Social Darwinism* in reality covered a vast and often contradictory set of interpretations and goals. There is, consequently, a large and controversial literature on the interaction between nineteenth-century evolutionary biology and sociopolitical theory (Bowler 1993; Pittenger 1993). It is of no little interest that the two discoverers of natural selection held contrary opinions concerning the application of that theory to the political realm. Wallace became an ardent crusader for land reform and socialism. Darwin, in contrast, subscribed to a more conventional usage of the concepts of struggle and competition in the political arena. Even here scholars are divided. Some see Darwin's own Social Darwinism as extolling unfettered competition and individualism. Others interpret his (admittedly ambivalent) pronouncements as reflecting an aversion to sociopolitical competition based on naked force. It is reported that Darwin was surprised to read in a newspaper that his theory justified Napoleon and every cheating tradesman (Bowler 1989, 285–286). Two of the most influential British theorists who applied evolutionary biology to the realm of politics were Herbert Spencer (1820–1903) and Francis Galton (1822–1911). The popularity of their writings testifies to the eagerness with which Victorians responded to sociopolitical concepts that seemed to bear the authoritative imprint of scientific foundation.

Spencer was one of the grandest systematizers of evolutionary thought. From his first book, *Social Statics* (1851), onward, Spencer elaborated an enormous lifelong project that he termed the "system of synthetic philosophy." Spencer's goal was to incorporate the entire realm of human knowledge and experience, from biology to religion, from psychology to sociology, into the framework of evolutionism. He had become a Lamarckian through his reading of Lyell in 1840. Although Spencer applied Malthusian principles of rapid human population growth and pressures to animal populations, deduced a struggle for survival, and coined the phrase "survival of the fittest"—thus incorporating natural selection partially into his grand system—his overall perspective remained Lamarckian. He continued throughout his career to maintain that the inheritance of acquired characteristics was the major mechanism of evolutionary change (Haines 1991). That Spencer explicitly defended Lamarckism against Weismann's critique of use-inheritance suggests one powerful reason why he refused to abandon his Lamarckian perspective: He would have undermined what he wanted most to maintain, namely, the fundamental identity of biological evolution and of psychic and social evolution (H. Spencer 1887).

Spencer envisioned higher forms of organic being emerging from earlier ones by a gradual process of adaptation to the environment. The mental development of the human species, he argued, had been from egotism to altruism. Correspondingly, society developed from a militant phase, in which rigid coercion was needed to hold individuals together, to an industrial phase. In this latter stage, Spencer suggested, altruism and a harmonious individualism permitted the decline of external state control and the emergence of a fully evolved and integrated social order, in complete and peaceful adaptation to its environment (Peel 1975, 569–572).

Needless to say, Spencer's vast evolutionary synthesis lent itself to the most diverse readings. He has been interpreted as providing a biological rationale for society as a ruthless struggle for existence. Spencer was seen by some to advocate relentless, individual competition as the engine for social progress in accordance with nature's laws, which put all alike under trial (Jones 1980, 56). It has also been persuasively argued that it was Spencer's use of the analogy of society as an organism, not his championing of competition, that best expressed his political views. This take on Spencer created a frequent opening for socialists to claim him as an intellectual comrade (Pittenger 1993, 20–22). Most scholars agree, however, that Spencer's synthetic philosophy was permeated not by materialism but by the notion that the ultimate goal of social development was a moral one (Bowler 1993, 65–69; R. Richards 1987, 287, 303–309). Spencer is significant for his efforts to provide a synthesis of much of the accepted physical and biological science of his day, coupled with the nascent social sciences, in the integrating framework of an evolutionary secularization of ethics. His unified vision contributed greatly to the acceptance of evolutionism as a major cultural force in Victorian society (Peel 1975, 570).

Francis Galton

Francis Galton, Darwin's cousin, adopted a more draconian approach toward the relationship between biology and politics than Spencer. More precisely, Galton simply subsumed politics under biology. He was fascinated by what he perceived to be the biological transmission of talent—scholarly, artistic, and athletic. In the 1865 essay "Hereditary Talent and Character," Galton assembled some of the first persuasive statistical evidence for the presumed inheritance of physical and mental traits in humans. In *Hereditary Genius* (1869), he further suggested that the races of mankind could be ranked according to the frequency with which each race produced individuals of high "natural ability." Galton defined this as intellectual capacity, eagerness for work, and power of doing superior work. Races that did not produce such individuals, he declared, would be swept away by their increasing contact with superior (read "advanced Western scientific") races as global industrialization proceeded (Mazumdar 1992, 39). Galton coined the term *eugenics* in 1883. His arguments for societal programs to foster talent, health, and other "fit" traits (positive eugenics) and to suppress "feeblemindedness" and other "unfit" traits (negative eugenics) became extremely influential in the closing years of the nineteenth century and the early decades of the twentieth century (Paul 1995, 3–5, 30–36). It was his brilliant disciple Karl Pearson who developed Galton's insights into the science of *biometry*. Pearson's goal was to apply precise and sophisticated statistical correlation techniques to biological phenomena, including human social evolution (Searle 1976, 7). Although both Galton and Pearson were skeptical—at times hostile—toward the implications of the rediscovery of Mendelian genetics in 1900, biometry became an important tool in the development of statistical methods for modern biology (Kevles 1985, 16–17, 35–39).

As is well known, Gregor Mendel's pioneering paper on heredity published in an obscure Czech natural history journal in 1865 lay relatively unnoticed for thirty years. Mendel, in his famous experiments with the garden pea, demonstrated that specific traits (such as size, color of the flower, whether the seeds are wrinkled or not) were inherited according to precise mathematical ratios. Moreover, each of these traits was transmitted from one generation to the next by a single particle or unit, what are now called genes. Finally, each gene exists in two forms ("alleles"), dominant or recessive. Each organism possesses two genes for a given trait. Two dominant genes—or a dominant and a recessive gene—will be expressed as the dominant trait in the organism. In those instances when an offspring inherits two recessive genes, then—and only then—will the recessive trait appear as the organism's observed characteristic. Mendel's brilliant twin-unit theory of heredity, coupled with the dominant-recessive relationship, what would come to be known as "Mendelism" or genetics, marked a decisive break with past explanations of how traits were passed from one generation of a given species to the next. Ironically, however, the crucial significance of genetics in the understanding of evolutionary change (such as mutations) was not immediately understood or endorsed by many biologists in the decades immediately following 1900. Indeed, it was only in the 1930s that what is now known as the "modern evolutionary synthesis" emerged. After 1930, biologists finally had a comprehensive theory which firmly integrated Mendelian and population genetics into the theory of natural selection (Bowler 1989, 270–281, 307–332). But in the late Victorian period, the situation with respect to the mechanism of inheritance was still in flux, and there was ample room for speculations about the mechanism of evolutionary change that could easily tie biology to overt sociopolitical agendas.

For Galton, therefore, eugenics was preeminently a scientific repudiation of conservative, aristocratic privilege. Politically, he shared the middle-class outlook of much of the liberal intelligentsia (Jones 1980, 35–36). As with Spencer's system, however, a wide variety of political strategies could be drawn from Galton's concept of eugenics and his emphasis on state intervention in human breeding. Socialist intellectuals, from Pearson to George Bernard Shaw, fashioned systems of "reform eugenics." Agitated conservatives, anxious to counter the emergence of social welfare politics in the years preceding World War I, saw in eugenics a scientific alternative both to the prewar Liberal government's programs and to socialism (Searle 1976, 112–115). Galton also impressed those who began to take the perceived threat of racial degeneration seriously in the 1890s, and who were receptive to hereditarian theories that appeared to justify imperialism and racism (Bowler 1993, 78, 90; Paul 1995, 50–84). The potency of eugenics as a political force is a significant feature of twentieth-century history (Kevles 1985).

As part of the wider movement of scientific naturalism, Galton's eugenics was "a celebration of the work of the professional elite [that] was also a bold attempt to colonise intellectual territory previously occupied by science's rivals" (MacKenzie 1981, 51). Convinced of the obligatory ideological function of biological evolutionism, Galton committed five hundred pounds a year to London's University College in 1904 for a research fellowship in "national eugenics." He defined this as "the study of agencies under social control that may improve or impair the racial qualities of future generations either physically or mentally" (Galton 1909, 81). In *Natural Inheritance* (1889), Galton declared that the statistics of heredity and their eugenic imperatives "are the only tools by which an opening can be cut through the formidable thicket of difficulties that bars the path of those who pursue the Science of man" (Galton 1889, 62–63). It was, however, an evolutionary science constructed upon an overt political ideological base.

Germany

Evolutionary theory became a common resource for political thinking not merely in Britain but throughout Europe. In Germany, for example, Ernst Haeckel (1834–1919) was converted to evolutionism upon reading the German translation of Darwin's *Origin*. For Haeckel, an eminent zoologist and embryologist in his own right, Darwin's theory provided the spark for constructing a broad philosophy integrating science with the study of human culture. Haeckel erected a synthesis of human knowledge within the framework of what he conceived to be the essence of the philosophical standpoint known as monism—what he termed "the unity of nature and the unity of science." Monism stands in marked contrast to philosophical dualism, or those systems that emphasize the fundamental separation, or distinctness, between mind and matter. Haeckel strove, as did many Victorians, for a scientifically based worldview that would incorporate philosophical, political, and religious elements. Although he adhered throughout his career to key Lamarckian hypotheses, Haeckel was regarded as the foremost advocate and popularizer of Darwinian evolution in Germany during the later decades of the nineteenth century.

His *General Morphology of Organisms* (1866) was predicated on the conviction that there are no absolute differences between organic and inorganic substances, only relative ones. Haeckel contended that the material basis of true life phenomena, nourishment and reproduction, lay in the very intricate chemical composition of the compounds of carbon and in the resultant unique physical properties characterizing living organisms. Haeckel portrayed the lowest creatures as mere protoplasm without nuclei. From them, he traced the evolution of one-celled forms with nuclei onward to the three biological taxonomic kingdoms: animal, vegetable, and the neutral, borderline "protista." Haeckel's obsession with ideal symmetries led him to outline numerous genealogical trees, sometimes supplying imaginative missing links or branches. His reconstruction of the human ancestral tree aimed at demonstrating humanity's evolution from the lower animals. Although Haeckel's leaps from experimental science to grander speculations were deplored by many of his scientific colleagues, his writings of science popularization attracted wide audiences. Haeckel's particular exposition of evolutionism is considered to have been one stimulus to the birth, and rise, of fascist ideology in Germany, Italy, and France (Gasman 1998).

Kropotkin's *Mutual Aid*

In Russia, Peter Kropotkin (1842–1921) developed a version of evolutionary theory that was the antithesis of Haeckel's, and indeed of most variants of Social Darwinism. Author of a number of highly respected scientific works on geography and geomorphology, Kropotkin was also deeply involved in politics. In the 1870s and after, he became one of the leading architects of anarchist ideology. Kropotkin's anarchism was a very libertarian one, similar to that of the radical French political theorist Pierre-Joseph Proudhon (1809–1865). Kropotkin's aim was to put anarchism on a scientific basis. His goal was a rational society in which people lived in harmony because of the existence of an attitude he termed mutual aid (altruistic cooperation). Much of Kropotkin's philosophical framework is contained in the famous *Mutual Aid* (1902), in which he synthesized his views on biology and society. Like many other Victorians, Kropotkin attempted to incorporate both Darwinian and Lamarckian elements into his sociopolitical evolutionism. He adamantly opposed all theories of societal evolution that stressed competition and conflict as prerequisites for survival and progress. Instead, Kropotkin claimed that cooperation, or altruism, was the chief factor in the evolution and success of a species. This idea was based on his empirical studies in Siberia during the 1860s, where he observed that sociability was the dominant feature at every level of the animal world. Especially among humans, Kropotkin believed that cooperation was the rule rather than the exception.

In human societies, Kropotkin argued that cooperation had evolved from the practices of primitive tribe to peasant village to medieval commune and, finally, to modern social/trade association (for example trade unions). He believed that even after the rise of bureaucratic government, people still maintained some sort of localized cooperation among themselves. Kropotkin interpreted the trend of modern history, therefore, as evolving toward a decentralized, nonpartisan, and cooperative society. In such an environment, humans could develop their creative faculties without interference from rulers, religious leaders, or soldiers. The ultimate goal of Kropotkin's theories was the creation of anarchist communism, a system in which society functioned on adherence to the principle of mutual aid. The ills of private property and economic inequality would be eliminated through the free distribution of goods and services. Kropotkin envisioned a utopian society where people could do both industrial and agricultural work. Society would be organized into communes, or cooperative communities, whose members would lead a comfortable life, working until forty or fifty years of age. Kropotkin's evolutionary scenario had all persons benefiting from a social organization based upon a "fair" division of labor.

To ensure the functioning of such a harmonious society, Kropotkin stressed the need for an integral education, in which both mental and manual skills would be cultivated. All children would be taught science and mathematics in school, but they would also be taught in the outdoors, by doing and observing firsthand various agricultural and other societal tasks. Kropotkin's brand of social evolutionism—with its central emphasis on altruism, not competition, as the driving force of evolution—was both radical and utopian. It bears certain affinities to Wallace's and the American philosopher John Fiske's (1842–1901) concepts of progressive evolution in human society, which also accorded a key role to cooperation as both a biologically and socially efficacious mechanism of development. Although he attracted devoted disciples, Kropotkin himself was persecuted for his beliefs, which were deemed dangerous in the climate of late Victorian capitalism and imperialism. He was arrested in Russia in the 1870s and again in France in the 1880s, and imprisoned for disseminating anarchist propaganda. Kropotkin found sanctuary in England in the 1890s. He lived in London until his return to Russia in 1917 after the communist revolution in that country. Rediscovered in the late 1960s with the creation of the counterculture and the New Left, Kropotkin's legacy endures among European and North American anarchist and liberal reform movements of the present day (Morland 1997).

Evolutionism, Professionalization, and the "Authority" of Science

Among the most important developments of the late eighteenth and nineteenth centuries was the institutional transformation of scientific activity. Especially in Europe and the United States, science changed, slowly but decisively, from a generally part-time pursuit of individual private researchers to a regular vocational enterprise of full-time research and development by professional scientists (Morrell 1990, 982). Of course, the career scientist had existed previously in history. We have only to recall the names of Galileo, Newton, Robert Boyle, and Antoine Lavoisier to recognize individuals whose lives were devoted to the pursuit of scientific knowledge. But even these exemplars of science either had to or chose to pursue other professional activities in addition to science. By the start of the nineteenth century, as science and industry joined forces to profoundly accelerate the transformation of European social and economic life, the number of individuals who practiced science as a full-time career began to increase dramatically. This development—*professionalization*—had major implications for the magnitude and direction of scientific activity. Not only did the rate of scientific discovery quicken—a prelude to the rapid growth of science in the twentieth century—but the very character of the social institutions of science underwent fundamental alteration. Science, by the mid-nineteenth century, had become a cultural as well as theoretical force of great impressiveness. It vied with other venerable institutions—notably religion, the legal profession, and traditional education—for the position of the definitive authority in many domains of public concern. Evolutionary biology, because of its intimate relation to human affairs, was one of the most obvious sciences to gain pride of place in Victorian intellectual and cultural life.

The full scope of the professionalization of science during the nineteenth century can be encapsulated in a brief look at the British Association for the Advancement of Science (BAAS), founded in 1831. The traditionally preeminent symbol of British scientific achievement, the Royal Society of London, had been officially chartered in 1662. It was a wholly private creation; royalty gave patronage, but nothing more. The Fellows of the Royal Society enjoyed neither privilege nor pension. They were granted no buildings or funds. Therefore, the Society remained, through the nineteenth century, impoverished and inadequately housed. It had never possessed laboratories or other than honorary means of promoting research. Furthermore, for over a century and a half the qualifications for a Fellowship included wealth and influence as well as scientific merit. Without such support from the politically powerful, the Society would have collapsed. During the eighteenth century, the Royal Society came to be seen more as a place for the privileged in search of notoriety than as a forum for scientists. By 1830, not one of the sixty-three aristocrats who held Fellowships contributed a significant scientific paper.

Criticism of the Royal Society increased, and when the BAAS was organized in 1831, it quickly moved to redress the defects for which the Royal Society had been attacked. From its inception, membership in the BAAS was restricted to practicing scientists, whether so-called amateurs of science or the newer breed of professional scientists. Meetings were held annually and in different cities throughout Britain. The BAAS served to disseminate the latest results of the different scientific disciplines, and was typical of the newer nineteenth-century organizations devoted to spreading scientific knowledge. It also provided a forum for the discussion of issues relating to the interaction of science and society. Such discussions frequently dealt with the ways in which scientific education and research should be organized and funded (Morrell 1990). By the middle of the nineteenth century, similar bodies existed in France, Hungary, Germany, Italy, Scandinavia, and North America. Governments became increasingly involved in science by supporting research and influencing what research was undertaken. In the United States, the Smithsonian Institution was founded in 1846 as a government center for research and publication in science. The establishment and development of the Smithsonian has been crucial to the evolution of federally supported science in America (Seitz 1992, 45–46). There was also a growing interaction between scientific research and technological/industrial advances.

This is not to imply that the newer professional institutions, like the BAAS, totally transformed the social basis of science. There is much evidence to show that the BAAS was still controlled by a gentlemanly group of professionals who were paid to do what they had wanted to do as amateurs. The BAAS and other leading British organizations, such as the Geological Society, were still very much collections of powerful, influential and well-connected scientists (Morrell 1990, 987–988). The crucial point remains, however, that members were now all becoming professional scientists.

In one important respect, however, the BAAS was socially innovative. From its foundation, the BAAS had to confront the "woman question." At first, females were permitted to attend the social functions but not the scientific meetings of the association. Finally, after much agitation and demonstration, in 1839 women were admitted to all the sessions (though confined to separate galleries or railed-off areas). Although the first official woman member was not admitted until 1853, from 1840 onward women were active participants in the BAAS. The Zoological Society of London (1827) and the Royal Entomological Society (1833) both admitted women on the same terms as men from the outset. This gender egalitarianism was not, however, characteristic of all scientific societies. The Linnean Society of London, the Royal Microscopical Society, the Geological Society, and the Royal Society itself did not admit women until the twentieth century. Moreover, women were barred, except in a few special cases, from the most important centers of nineteenth-century science, the universities. Similarly, the prestigious Ecole Polytechnique, founded in Paris in 1794, did not admit women until 1972 (Alic 1986, 178–181).

Why Was Professionalization So Crucial?

This transformation of scientific activity provides the backdrop for many aspects of the interaction between evolutionism and Victorian culture. The emergence of professionalized science created both opportunities and challenges for Victorian evolutionists. The overall process of nineteenth-century science professionalization has, moreover, a broader historical significance. First, the professionalization of science and the increased links between science and technology radically transformed the international balance of geopolitical power. At the start of the nineteenth century, Great Britain was the undisputed world leader in technology and industry and a giant of scientific achievements. The Crystal Palace Exhibition of 1851 attested to this fact. By the close of the century, however, several European nations (notably, Germany, France, and Belgium), the United States, and, to a more modest degree, Japan were rivaling Great Britain in scientific accomplishments and equaling or surpassing her technologically. Thus, an understanding of how particular cultural and sociopolitical factors influenced the advance (or retardation) of science and technology can help us understand much of the course of modern history.

A second (and related) reason for looking at the professionalization of science during the nineteenth century is that this period witnessed a marked growth of governmental involvement in the funding and organization of scientific research and education. With increased government funding came political, social, economic, or military pressures that influenced whether or not national—and often local—governments would, or could, support specific scientific and technological projects. Today, governments are increasingly involved in the funding and utilization of scientific and technological research and education. The more we learn about the reasons for which nineteenth-century governments became thus involved, the more we may understand our contemporary controversies surrounding science and technology. We come to understand how systematic arguments emerged (and have evolved over time) to justify such public expenditures; we also gain useful insights into the public pressures, or apathy, toward those arguments. The cultural debates about science and technology that first emerged clearly during the nineteenth century foreshadowed the intense debates of today. Such debates were intense in Victorian Britain and constitute the background against which controversies surrounding evolutionism are best understood (Ben-David 1991).

The specific aspects of science professionalization that are particularly relevant for assessing the Victorian debates include (1) changes in the cost patterns of scientific research and administration; (2) changes in the social backgrounds and occupational status and expectations of scientists (for example, the rise of a new technoscientific elite); (3) trends in the development of scientific education and advanced degree granting. Well into the nineteenth century, the older European universities such as Oxford, Cambridge, and the Sorbonne remained relatively indifferent, if not resistant, to the needs of a rapidly expanding scientific community. These traditional universities eventually did incorporate more science into their curricula, albeit gradually. However, a crucial nineteenth-century development provided the new educational institutions necessary for the further development of modern science. Universities independent of the medieval giants were founded that could welcome and encourage science. Notable here are the various colleges (such as University College and King's College) of the new University of London (1836), the German universities of Bonn and Berlin, and North American universities such as Johns Hopkins (1876) in Maryland and McGill in Canada (1821); (4) the growth of scientific journals and sophisticated communications networks (crucial for the development of modern science); (5) specialization of science. During the nineteenth century, the professionalization of science demanded a more rigorous specialization to ensure advances (Rudwick 1985). As the methods and theories of each field developed, it became necessary to define the particular nature of each discipline and to train its practitioners differently. Biology witnessed the emergence of specialized fields such as physiology, ecology, and paleontology, as well as evolutionary biology;

(6) new patterns in governmental support of science and technology under competing international economic and military pressures; and (7) a comparison of changing national attitudes toward science and technology within the broader context of cultural, sociopolitical, and economic developments. When, for instance, in the 1870s and 1880s the first large-scale industrial research laboratories were organized by synthetic dye manufacturers in Germany, revolutionary possibilities for the future societal role of science and technology were revealed (Basalla 1988).

All these steps through which science became increasingly professionalized by the mid- and late-Victorian period ensured that evolutionary biology would be clothed with a powerful mantle of authority. To those in the corridors of political and economic power, as well as to the general public, science became part of a shared cultural vocabulary. Somewhat ironically, however, as Europe and North America fostered the professionalization of science—and witnessed the social transformation of the scientist's role—their general populace became more divorced from any actual involvement in science itself. Because of specialization, the increasing social impact of science was accompanied by a decreased ability on the part of the general public to comprehend the scientific or technical basis of that impact. Consequently, while science grew more impressive, even highly educated laypeople were less able to participate directly in theoretical or institutional developments. A failure to achieve one of the Enlightenment's greatest goals, namely, a scientifically literate public, seemed to many an inevitable cost of the professionalization of science during the nineteenth century. The effective popularization of science was a challenging problem that certainly attracted much attention (see chapter 1) but met with only modest success. This situation persists up to our own times. The problem of educating the public about science and its societal and environmental consequences is emerging as a crucial area of research and action. In the Victorian era, the public's relationship to evolutionism was paradoxical: the concepts, vocabulary, visual images, and metaphors of evolutionary biology were bandied about with great frequency, but their precise meaning was often obscure or subject to conflicting messages.

The Ideology of Scientific Neutrality

Debates about the nature and implications of scientific knowledge, such as those involving the work of Spencer and Galton, were also important features of the early Victorian period. William Whewell, John Herschel, and David Brewster, among others, raised fundamental questions concerning the epistemological and cultural status of science. However, science in this earlier period did not enjoy the cultural and institutional security it acquired after mid-century. Accordingly, it is only after the 1860s that there emerged "a scientific culture, rather than science in culture" (Yeo 1993, 32). Numerous works appeared in the 1870s and after, attesting to the ascendancy and autonomy of science, what John Stuart Mill (1806–1873) termed the "general property of the age." Books, articles, lectures, and speeches made claims for the broad scope of scientific philosophy that would not have received a sympathetic hearing in the 1830s, nor even the 1850s (Butts 1993, 313–317). Evolutionism was a key ingredient in this new and potent cultural vision of a world permeated by science. However, since evolutionary biology was itself an ongoing, contentious discourse, it imposed no single set of conclusions on social and political thinkers and activists. True, evolutionism offered powerful new avenues for studying both nature and human society. Evolutionists certainly posed provocative questions concerning humanity. But the answers to those questions were often controversial, if not ambivalent (Pittenger 1993, 8). To those who were seeking to solidify science's status as a definitive source of convincing and accurate information about the natural and social worlds, a major challenge had to be confronted. How could science be shown to be superior to traditional institutions of cultural authority, such as religion, time-honored customs, or the law? To certain segments of the scientific community, most notably the *scientific naturalists*, the solution seemed clear. If science could be presented as an objective and ideologically neutral enterprise, it would appear as a truly unbiased approach to knowledge. Scientific pronouncements could then be taken as definitive in cultural as well as technical matters. Led by Thomas Huxley and his allies, the scientific naturalists embarked on a vigorous campaign to persuade all levels of the public that science was ideologically pure. If science was accepted as an objective quest, then its conclusions would carry an almost invincible authority. It was a bold and brilliant strategy. To the extent that it succeeded, the various theories of social evolutionism would gain greater prestige and a commanding voice in the minds and hearts of Victorians.

The representatives for scientific naturalism constituted one of the most vocal and visible groups on the Victorian intellectual landscape. With a combination of research achievements, polemic wit, and literary eloquence, this influential coterie—including Huxley, John Tyndall, Leslie Stephen, and John Morley, as well as Spencer and Galton—helped to create a largely secular climate of opinion in which the theories and metaphors of modern science penetrated the institutions of education, industry, and government. They preached a gospel of social and material progress allied to the advance of science and technology to enthusiastic audiences ranging from skilled mechanics to members of the aristocracy. From the 1860s onward, the Victorian scientific world was essentially transformed into a modern professional community (Turner 1993, 131–132, 179). By the 1870s, in terms of editorships, professorships, and offices in the major scientific societies, such figures as Huxley, Tyndall, Joseph Dalton Hooker, John Lubbock, Galton, and Lyon Playfair emerged as spokesmen for the new scientific elite. A key factor in the rising cultural status of these professional scientists was their insistence upon—and a growing popular acquiescence in—the authority of a thoroughly naturalistic approach to science. Such an approach would presumably free contemporary science of metaphysical and theological residues and exclude the kinds of troublesome questions, as well as answers, that characterized traditional approaches to natural knowledge. This posture of objective neutrality was seductive—although important evolutionists such as Wallace and St. George Mivart would have none of it—and attempted to discredit the wider cultural influence of organized religion (Desmond 1998, 373–375; Turner 1993, 180–182).

The interaction between biology and politics during this era, therefore, assumes greater significance when it is recognized as part of the broader process of the fundamental redistribution of cultural authority in Victorian society. Even those biologists who fought fiercely among themselves—

such as Huxley and Richard Owen in their celebrated "hippocampus minor" controversy on the significance of the comparative cerebral anatomy of humans, apes, and monkeys—joined forces against what they perceived to be a common enemy: the influence of theologians and members of other groups, such as the judiciary, whose authority was traditionally recognized as extending over questions of natural history. Owen wrote regularly for nontechnical periodicals, such as Dickens's *Household Words* and *Blackwood's Edinburgh Magazine*, to enhance the reputation of the emerging scientific profession in major public controversies. In these "gladiatorial shows," including debates over the great sea serpent, the longevity of man, and vivisection, Owen emerged as an expert who exposed traditional ignorance and demonstrated the superiority, as well as the usefulness, of scientific knowledge (Rupke 1994, 287–352). Similarly, the often acrimonious disputes between the rival (anti-Darwinian) Anthropological Society and the (pro-Darwinian) Ethnological Society during the 1860s must be seen in the context of their mutual aim of establishing the paradigm of the "scientific study of man." Huxley finally succeeded in amalgamating the two groups in the newly formed Anthropological Institute of Great Britain and Ireland in 1871 (E. Richards 1989a, 253–284).

The central issue, then, is the degree to which later Victorians accorded scientific authority to evolutionary naturalists in questions concerning human society and its politics. Precisely because biologists were becoming part of a more clearly defined professional scientific community, the cultural context of professional pronouncements is essential to understanding the political uses of evolutionary biology. The issue is further complicated by the fact that the biological and physical sciences were not the only fields aspiring to the status of professions. There was a widespread movement toward professionalization of many intellectual disciplines in the late Victorian period. There were rival voices claiming the status of experts in the field of the emerging moral and political sciences. The National Association for the Promotion of Social Science, founded in 1857, was one such Victorian institution that sought to promote the scientific investigation of fields ranging from legal reform to penal policy, education, public health, and social economy. Both Prime Minister William Gladstone and John Stuart Mill addressed some of the association's meetings. Such new professionals would, in Henry Sidgwick's phrasing, then be "able, to a certain extent, to pour the stream of pure science into the somewhat muddy channel of current [public] opinion" on a wide range of issues (Collini 1991, 200). That these new professional experts would also speak authoritatively to the public from the lofty plateau of academic neutrality underscores the importance of Huxley's—"Darwin's bulldog"—campaign to establish the objective neutrality of the emerging biology profession. The scientific naturalists wanted to claim a privileged voice among the competing groups of experts vying for the prize of preeminent knowledge source in both elite and popular Victorian culture (Collini 1991, 199–200, 204–205, 210–211, 224).

Huxley Leads the Charge

Thomas Huxley was one of the first to see the polemical advantage of adopting agnosticism as a public philosophy. Agnosticism is the philosophical doctrine that absolute truth or certainty is unattainable, and that only perceptual phenomena can be objects of exact knowledge. In theology, agnostics do not deny God's existence but deny the possibility of our knowing anything about Him. Huxley, who coined the term, adroitly used it as a tool to rise above what he regarded as the sterile debates between materialist and spiritualist worldviews (Paradis 1997, 166–167). Huxley did not simply want to disengage himself, and the scientific community and general public, from endless controversies about the relations between science and religion. He wanted to promote the superiority of a "cleansed" scientific naturalism. Darwin, too, soon appropriated the term, in part to extricate himself from the religious controversies that had long hounded his theory (Lightman 1987; Desmond and Moore 1991, 636, 657–658). However, Huxley's cloak of agnosticism, while undoubtedly effective, was not palatable to all evolutionists. For some, their science was inextricably and explicitly linked to ideological and cultural determinants. These contrasting views concerning the legitimate political, religious, and cultural roles of evolutionary biology will be explored in subsequent chapters.

Huxley's own career epitomizes the professionalization of Victorian science. He is one of the most central of all the evolutionary naturalists to the critical analysis of the relationship of biology to politics (Desmond 1998). Huxley's famous Romanes Lecture of 1893, "Evolution and Ethics," is generally taken as a humanistic distancing of the authority of evolutionary science from sociopolitical and ethical policy disputes. In contrast to most of his contemporaries and to his own earlier convictions Huxley is deemed to have discredited the pervasive utilization of biological analogy by rejecting nature itself as a moral or sociopolitical norm. However, it has been persuasively argued that Huxley's celebrated essay, rather than limiting and depoliticizing the authority of evolutionary science, subtly invoked it to support his own political views. In particular, Huxley deployed a version of competitive biological selection to sanction the centralized and paternalistic sociopolitical agenda of the Liberal party. The Liberals proposed to solve the problems of the decline in English prosperity that had occurred from the mid-1870s to the mid-1890s (Helfand 1977, 159–177). Huxley himself described the Romanes Lecture as an "egg-dance": He was well aware of the extremely controversial and tricky nature of the subject matter. In fact, the lecture was a rhetorical tour de force. Huxley appeared to deny that any ethical message could, or should, be drawn from evolutionary biology, in order to deny scientific legitimacy to the biological analogies used by both Spencerian individualists and the radical land nationalists and socialists. But he then subtly appropriated evolutionism for his own Liberal political agenda. In dismissing as unrealistic the socialist theories of Henry George and Alfred Russel Wallace, for example, because they challenged Malthusianism, Huxley was compelled to readmit a modified theory of biological struggle to political discourse. In a letter to Romanes, Huxley affirmed that "though there is no [direct] allusion to politics in my lecture," he would "never have taken the pains I have bestowed on these 36 pages" if his audience had failed to apply his ideas to draw appropriate "liberal" political conclusions (L. Huxley 1901, vol. 2, 375). Further, in 1887, Huxley had written an influential letter to the *Times* urging Britain to strengthen its economic and military position by relying more substantially upon the political expertise and vision of its professional scientific and technical elite (Turner 1993, 206).

Huxley's Romanes Lecture, therefore, was paradoxical but powerful. He and other advocates of the rising cultural authority of professional and naturalistic science had striven to *publicly*, but not privately, divorce science from ideology. This strategy worked at one important level. The scientific naturalists' efforts to ensure more adequate government and public recognition and support of science as a self-governing entity were largely successful. In his presidential address to the BAAS in 1874, John Tyndall announced that all "theories, schemes and systems, which . . . reach into the domain of science must, insofar as *they do this,* submit to the control of science, and relinquish all thought of controlling it. Acting otherwise proved disastrous in the past, and it is simply fatuous to-day" (Tyndall 1874, 61). Similarly, Huxley emphasized the presumed neutral and apolitical character of evolutionary biology in his influential efforts to promote acceptance of the view that science, including the human sciences, should be regarded as definitive in its conclusions (Fichman 1984, 471–485). Although Tyndall's and Huxley's views were controversial, their strategy to foster the concept of an ideologically pure, and hence objective, scientific naturalism had become a crucial tool of the Victorian lobbyists for the "cult of science."

Evolutionism and the "Cult of Science"

The cult of science had many interpretations in the Victorian era. The very term *cult* testifies to the historical reality that science has rarely if ever been a value-free enterprise. Certain socialist thinkers, notably Pearson, Wallace, Annie Besant, and Edward Aveling (anatomist and prominent popularizer of Darwinism), for example, explicitly used evolutionary theory to substantiate their collectivist visions of a scientifically grounded social order (Pittenger 1993, 23). Historians are now examining such rival "scientific campaigns"—as those involving competing pronouncements of the true meaning of social evolutionism—in terms of their different sociopolitical, professional, religious, and class contexts. This necessitates a more critical examination of the ways in which scientists and their disciples use language—in writing texts, giving lectures, preparing research reports, in conversations and correspondence, and in popularizing their concepts. New insights into the functioning of scientific metaphors and analogies in specific cultural circumstances permit fields as diverse as phrenology (Cooter 1984) and evolutionary biology to be interpreted, in part, as modes of discourse whose success stemmed from their manifest adaptability as ways of making sense of a wide range of human social relations. Discussion of the varied attempts at demarcation of biology and politics—or the denial that such demarcation is possible—must, therefore, rely upon these tools of semantic, rhetorical, and symbolic analysis (Fichman 1997, 94–118). Huxley's Romanes Lecture, finally, must be regarded as one of the most persuasive, if ultimately disingenuous, Victorian attempts to cloak evolutionary biology in the garment of objective neutrality. Despite its paradoxes, however, Huxley's lecture did serve to promote an enhanced cultural authority for science (Paradis and Williams 1989).

Owing to the intricate and subtle agenda and propaganda of Huxley and his colleagues, the prestige, indeed respectability, of science increased notably during the late nineteenth century. Certain disciplines such as history, psychology, sociology, and economics sought to divest themselves of aspects of their philosophical past and acquire the aura of science. An influential school of scientific historians, especially in Germany, sought to place modern historiography on a scientific basis by establishing canons of objectivity for the study of documents and sources. Although the debate on the precise relationship between the social sciences and the natural sciences continues to our own day, the Enlightenment ideal of a scientific model for all explanation exercised a profound allure for thinkers as diverse as Jeremy Bentham and Karl Marx (Gordon 1991).

With the growth of professionalization and specialization came a shift—gradual at first, but more perceptible by the close of the nineteenth century—in the social origins of scientists and engineers (Desmond 1998). Science in the seventeenth and eighteenth centuries had depended primarily upon individuals who either possessed independent means or who were fortunate enough to secure some type of patronage. By contrast, the late-nineteenth-century scientist was increasingly of middle-class origin. These persons needed to be paid a living wage for doing scientific work. Thus, an expectation of a certain level of remuneration and a recognizable career structure became part of the defining characteristics of a professional scientist. Interestingly, in this restricted sense, neither Darwin nor Wallace were professional (although they were brilliant) scientists. Darwin was independently wealthy. Wallace had to struggle throughout his long career to earn a living variously as an author, lecturer, and marker-grader. The success of scientists in achieving professional status—with the concomitant sense of corporate identity, level of minimum certified competence, and acceptance of certain social obligations—was a noteworthy development in nineteenth-century history.

The professionalization of science has implications that society is still assessing. This initial professionalization was limited mainly to Europe and North America. Those countries—such as Islamic lands subjected to colonial rule by Western powers during the nineteenth and early twentieth centuries—which did not share in this science-based industrialization were deterred from making similar advances (al-Hassan and Hill 1986, 283). The great global question posed by the accelerating integration of science, technology, and industry was stark: which nation or nations would most effectively support and exploit the new profession and to what ends? That question first became highly significant in the nineteenth century. It is, perhaps, an even more crucial question today. One nation that had embraced the new scientific and industrial ethos of the late nineteenth century was the United States. The reception of evolutionary ideas there during the Victorian period tells much about the controversial impact of the biological revolution of the nineteenth century.

TRANSATLANTIC EVOLUTIONISM

Evolutionism Comes to America

As in Europe, the first half of the nineteenth century in the United States witnessed occasional publications and debates concerning the possibility of organic evolution. It was the appearance of Darwin's *Origin* in 1859 that transformed this trickle of interest in such questions to full-scale involvement on the part of the American scientific community as well as educated laypersons. As noted historian Ronald Numbers has summarized it:

> The publication of the *Origin* moved the topic of organic evolution out into the open and made it a subject of intense discussion. Scientific societies debated it; college students titillated one another with it; critics railed against it. By the early 1870s the paleontologist Edward Drinker Cope was declaring evolution to be an "ascertained fact," and within a few more years scientific opposition in North America had diminished to a whisper. Naturalists continued to argue about the adequacy of natural selection to account for evolution, but with few exceptions they, like Darwin, had forever turned their backs on the special creation of species. (Numbers 1998, 24).

This assessment is accurate but implies, significantly, that the American response to Darwinism was scarcely unified. As in Europe, American proponents of evolutionism still had ample grounds for questioning or rejecting Darwin's specific mechanism of natural selection. Lamarckian explanations were often invoked to supplement or displace natural selection as the main agent of evolutionary change.

American neo-Lamarckism began to develop in the 1870s. During the last decades of the nineteenth century, it became a more formidable alternative, for a variety of scientific and cultural reasons, to natural selection than it did in Britain. Ironically, some of the most influential proponents of neo-Lamarckism, stressing the direct influence of the environment and use-inheritance in effecting evolutionary change, were students of the Swiss-born Louis Agassiz (1807–1873). Harvard's prestigious professor of zoology, Agassiz was America's most formidable opponent not merely of Darwinism but of evolutionism in general (Bowler 1985, 641–681). But he encouraged open discussion of evolutionary hypotheses, and some of his students gradually came to veer from Agassiz's fervent creationist stance (Numbers 1998, 31–34).

Enter William James

An instructive example of the early impact of British evolutionary thought in the United States comes from the pen of the famous American philosopher and psychologist William James. One of James's earliest published writings was a review of Wallace's "The Origin of Human Races and the Antiquity of Man Deduced from the Theory of Natural Selection" (James 1865, 261–263). James regarded, as did many others, Wallace's 1864 essay as a brilliant resolution of the bitter anthropological controversy in England and North America between the *monogenists* and *polygenists*. The monogenists (or "unity" theorists) held that humans constitute a single species and that the various human races are merely local and temporary variations produced by different environmental conditions. The polygenists argued that the different human races are separate species, each of which had always been as distinct as it was at present. Wallace, by an ingenious application of the principle of natural selection, provided a compromise between the opposing groups of anthropologists. His 1864 essay demonstrated that although racial differences do in fact antedate the recent historical period of human evolution, the several races ultimately derive from a common ancestor. Early humans, Wallace postulated, would have been subject to the harsh selective pressures of the environment. As they spread to different regions of the globe in prehistory, the extremes of food, climate, disease, and predators would have produced, through the mechanism of natural selection acting upon physical variations, those marked racial differences in appearance that remain so striking. But, Wallace suggested, these early humans were also subject to natural selection of intellectual variations. And once the human brain had evolved to a sufficient degree of power and flexibility, developing man would respond to the selection pressure of the environment by social, cultural, and intellectual adaptations. Wallace concluded that the development of the complex human brain essentially suspended the action of natural selection upon bodily structures. The different physical racial characteristics would henceforth remain fixed, and further human evolution would proceed by the selection of mental and cultural variants (Wallace 1864).

In his review, James applauded Wallace's theory and asserted that "Natural Selection, then, in its action upon man, singles out for preservation those communities whose social qualities are the most complete, those whose intellectual superiority enables them to be most independent of the external world. The physical part of him is left immutable, and his mental and moral advance is secured." So reasonable and irrefutable did

Wallace's essay appear to James, that he declared it astonishing that Wallace's solution to the monogenist/polygenist debate was "made so late." James felt that Wallace's insight, like the theory of natural selection itself, underwent a gestation period prolonged by the obstinacy of less forward-looking thinkers. He concluded the review by asking why "may there not now be lying on the surface of things, and only waiting for an eye to see it, some principle as fertile as Natural Selection, or more so, to make up for its insufficiency (if insufficiency there be) in accounting for all organic change?" (James 1865, 263). James is here picking up on Wallace's own concluding remarks in the 1864 essay that intimated that his "great leading idea"—as Darwin described it in declaring it "quite new to him" (Marchant 1916, 127)—"neither requires us to depreciate the intellectual chasm which separates man from the apes, nor refuses full recognition of the striking resemblances to them which exist in other parts of his structure" (Wallace 1864). Although Wallace had not yet (publicly) suggested that natural selection alone could not account for the origin and development of mankind—or ensure a utopian future of socioeconomic and ethical harmony—a reader as astute and prepared as James could read a great deal more into the 1864 essay than the undeniably significant theoretical resolution of the anthropologists' racial controversy.

James's review of Wallace, along with another review James wrote on Huxley's *Lectures on the Elements of Comparative Anatomy,* confirms that he was an active participant in the evolutionary debates in America as early as the mid-1860s. James also saw evolutionary theory as crucial to the emerging field of experimental psychology. From the outset, he regarded the problem of human consciousness as residing at the epicenter of evolutionary speculation. This review also reveals the influence of Charles Sanders Peirce and Chauncey Wright upon James's intellectual development. Peirce, along with James, founded the influential philosophical school of *pragmatism*. There are suggestive links between evolutionary theory and the origins of American pragmatism. In 1903, Peirce recalled the "immense sensation" that the Darwin-Wallace theory of evolution created among his circle of friends. Peirce emphasized its impact on Wright, the mentor of several Harvard intellectuals, including himself and James. Peirce regarded evolutionary theory as providing a new paradigm for scientific thinking in the late nineteenth century, one that extended beyond biological disciplines. He saw in evolution ideas of development that compelled philosophers as well as scientists to recognize "that there is a mode of influence upon external facts which cannot be resolved into mere mechanical action." Because for Peirce "all matter is really mind," he emphasized the teleological implications of evolutionary theory in his own extensive writings on the subject (Parker 1998, 15–16, 200). Both Peirce and James refused to divorce logic and epistemology from metaphysics, cosmology, and ethical and social commitment (Peirce 1997, 36, 65–75, 205–220). They embraced a holistic approach that entailed that scientific statements could not be left in their immediacy as fact but must rather be understood as manifestations of patterns of development underlying the evolution of the entire universe (Gavin 1992, 56–95, 99–109, 202, 209). Throughout their careers, James and Peirce wrote extensively on the methodological and metaphysical implications of evolutionary theory and the debates arising therefrom in England, France, and Germany, as well as in North America (B. Kuklick 1977; Taylor 1990, 7–33).

Peirce Contra the Scientific Naturalists

Like James—and like Wallace but unlike Darwin or Huxley—Peirce's scientific and philosophical writings came to be increasingly permeated by theistic reflections. In contrast to the more dogmatic scientific naturalists, Peirce did not regard science and religion, or reason and faith, as ontologically distinct fields of human thought and activity. Rather, he suggested that faith plays an important role in the process of scientific inquiry, particularly in formulating hypotheses. Peirce saw a "sublime manifestation of the Deity" in human attempts to analyze and solve problems through the utilization of scientific method (Croce 1995, 16; Potter 1996, 169–194).

Peirce and Wallace shared a commitment to *scientific theism*. Many other Victorian evolutionists also believed that science and religion were compatible, indeed deeply linked, modes of understanding nature and humanity's place therein (see chapter 7). The Americans Asa Gray (that country's leading botanist) and Joseph LeConte (1823–1901) and the influential Scottish educator and philosopher Henry Drummond (1851–1897) are representatives of this important facet of transatlantic evolutionism. Theistic convictions were fundamental to Peirce's worldview. They influenced not only his conception of science but also his metaphysics, cosmology, and sociocultural ideas. In turn, Peirce's theistic ideas were transformed by his commitments and discoveries in these other fields. Moreover, Peirce believed, as did Wallace, that the scientific spirit was not antagonistic to but rather wholly compatible with a "community of faith." Despite his (and Wallace's) frequent condemnation of the traditional institutional aspects of historical religions—with their dogmatic and divisive creeds—Peirce deeply believed that some sort of "community of faith" was necessary for the development of a better society in the late Victorian period (Raposa 1989, 11–13).

Not surprisingly, Peirce considered Wallace to be one of the major scientific notables of the late Victorian period. He praised Wallace's lucid powers of argumentation on a bewildering range of issues. What others perceived as paradoxes or eccentricities in Wallace's life and work, Peirce regarded as the necessary attributes of a bold and original thinker. Declaring that Wallace's concept of natural selection (see chapter 5), as it began to emerge more explicitly in the 1880s, was "superior to Darwin's," Peirce considered Wallace's evolutionary teleology and cultural vision a notable achievement (Peirce 1906, 160–161).

Peirce's relationship to Darwin is complex. But there is no doubt that he regarded the evolutionism of Darwin—as well as that of Spencer, Huxley, and other prominent scientific naturalists—as critically important but limited. Fairly or not, he characterized the principle of natural selection in this phrase: "Every individual for himself, and the Devil take the hindmost" (Hartshorne and Weiss 1958–1965, vol. 6). One of Peirce's clearest expositions of evolutionary theism is the series of essays he wrote for the *Monist*, during the period from 1891 to 1893. "Evolutionary Love," the last of the *Monist* essays (and published in 1893), is the single most succinct statement of Peirce's scientific theism (Raposa 1989, 72–87). One of the principle objectives in writing that essay was to demonstrate that wholly naturalistic perspectives on evolutionary theory were inadequate. While accepting the scientific significance of the principle of natural selection, Peirce, like Wallace, held it to be incapable of accounting for the whole of the evolutionary process. For Peirce, strictly naturalistic evolutionary theories were "pseudo-evolutionism." He deemed them "scientifically unsatisfactory" because they gave "no possible hint of how the universe came about" and were "hostile to all hopes of personal relations to God" (Hartshorne and Weiss 1958–1965, vol. 6). In "Evolutionary Love," Peirce offers an alternative to such theories. His own evolutionary cosmology is predicated upon the central role played by both divine and human love, and social cooperation, in the course of evolution.

Peirce's scientific theism ultimately rests upon the criterion of teleology or purposefulness. A teleological perspective implies that there is a goal toward which a particular process is tending. Teleological views were central to much of Victorian evolutionary speculation. John Fiske, a highly successful American evolutionary popularizer, emphatically told his broad audience that the goal of evolutionary development was the perfection of human spiritual and material nature (Russett 1976, 48–55). In Peirce's conceptualization, the development of the entire cosmos represents nothing more than evolutionary teleology, worked out "in living realities" (Hartshorne and Weiss 1958–1965, vol. 5). Peirce's conviction that the scientific method demanded elements of faith, as well as the acknowledgment of both certainty and uncertainty in scientific inquiry, found strong resonance not only in William James but among an influential group of New England intellectuals. Peirce had been one of the leading members of the Metaphysical Club, a discussion group of Harvard students and graduates active in the 1860s and 1870s. In addition to Peirce and James, the Metaphysical Club's members and visitors included the future Supreme Court Justice Oliver Wendell Holmes, Fiske, and the liberal theologian and evolutionist Francis Ellingwood Abbot (Fisch 1964; Russett 1976, 29, 47, 59). Wallace would thus have found in the New England Cambridge setting an extremely hospitable environment for the evolutionism he expounded during his North American tour from 1886–1887.

Wallace and Transatlantic Evolutionism

Toward the end of 1885, Wallace received an invitation from the Lowell Institute in Boston, Massachusetts, to deliver its prestigious annual course of lectures in the late autumn and early winter of the following year. The eight Lowell lectures were highly successful. Wallace spent the next year traveling across North America, repeating the Lowell lectures with equal success in major American cities and in Toronto and Kingston in Canada. Wallace met many of the United States' most distinguished scientists as well as a host of leading political, social, and intellectual figures. In addition to speaking on scientific subjects and observing firsthand the flora and fauna of North America, Wallace spoke publicly on his political and theistic views. Wallace's North American sojourn thus situated him squarely within the broader context of late Victorian cultural controversies on both sides of the Atlantic. The tour constituted a high point of Wallace's public career both as statesman of evolutionary science and as ardent proponent of the controversial doctrines of increasingly socialist politics and of spiritualism (Wallace 1905, vol. II, 105–106, 129, 160, 200). As both his admirers and critics in North America would attest, Wallace's declarations on matters scientific and sociocultural always aroused heated debate. His North American tour afforded American and Canadian audiences a firsthand opportunity to witness one of the Victorian era's most brilliant and contentious public figures in action.

Wallace's first Lowell lecture, "The Darwinian Theory," was delivered before a crowded and attentive audience. Among the several newspaper notices the next morning, the *Boston Evening Transcript* captured the essence of Wallace's successful debut: "The first Darwinian, Wallace, did not leave a leg for anti-Darwinism to stand on when he got through his first Lowell lecture last evening. It was a masterpiece of condensed statement—as clear and simple as compact—a most beautiful specimen of scientific work." The account concluded, however, with the statement that "Dr. Wallace . . . then defined his own position upon the relation between man and the lower animals. Physically they are connected, but mentally there are powers which never could have been developed from lower animals" (*Boston Evening Transcript*, 2 November 1886, 4). Thus, from the outset of his North American tour, Wallace made it clear that his explication of Darwinism was orthodox in major respects but differed radically from Darwin on fundamental aspects of human evolution.

Fig. 3. Photograph of Alfred R. Wallace in 1878, at age fifty-five. (From his autobiography, *My Life* [London, 1905], vol. 2: facing p. 98)

Wallace stayed in Boston, with excursions to Poughkeepsie, New York, and Baltimore, Maryland, for the next two months. At Vassar College in Poughkeepsie, Wallace's invited lecture on evolutionary theory in 1886 was part of the process by which the new biology was introduced into the curriculum of that prestigious women's college. Evolutionism apparently encountered few objections, religious or otherwise, at Vassar. In the fall of 1889, Marcella O'Grady was appointed instructor in biology and by 1893 was named full professor, a status a woman could rarely attain except at a women's college. O'Grady's course on higher biology, introduced to the curriculum in 1894, was the first formal course at Vassar to deal explicitly with evolutionary theory (Wright 1997, 631–635).

During Wallace's sojourn in Boston, in addition to completing his Lowell lecture series, he had the opportunity of meeting most of Boston's (and Cambridge's) intellectual elite: William James, Asa Gray (who dined on several occasions with Wallace in order to introduce him to most of the important biology professors at Harvard), Oliver Wendell Holmes, the geologist James Dana, James Russell Lowell, and Alexander Agassiz, who gave Wallace an extensive personal tour of the Harvard Museum of Comparative Zoology. Although that museum was a bastion of antievolutionism under Agassiz's father, Louis, until the 1870s, its tone became somewhat more moderate when Alexander succeeded Louis as curator in 1874 (Winsor 1991). By the time of Wallace's trip to America, museums had ceased to be static organizations designed to preserve the status quo. Rather, there was an active transatlantic debate in the 1880s as to how best to ensure that natural history museums could best serve their scientific functions. The debates extended as far as the colonial outposts of the European empires (Sheets-Pyenson 1989). Wallace was deeply impressed with certain of the institutional aspects of the Harvard Museum and provided a detailed account of it upon his return to England. Since Darwin's death in 1882, Wallace was perhaps the most eminent of the original British evolutionists. His direct contact with North Americans was a further catalyst to the transatlantic exchange of views on the professional and public dimensions of Victorian evolutionism.

If the interpretation of Darwinism was exceedingly controversial in Britain and Europe, it was possibly even more so in North America. The American situation becomes more complex still when it is recalled that Gray, Darwin's preeminent advocate in the United States and the leading opponent of Louis Agassiz in the American Academy of Arts and Sciences during the heated debates of the 1860s, was a theistic evolutionist. The remarkable fact, however, is that both theistic evolutionists and antievolutionists studiously refused to call themselves creationists. That the label "creationist" languished in relative obscurity throughout the later decades of the nineteenth century in the United States takes on ironic dimensions when one notes that the same term erupted in the early decades of the twentieth century as a rallying cry for American antievolutionists. It was reborn yet again in the guise of *creation science* at the close of the twentieth century (see chapters 7, 8; Numbers 1998, 50–57).

Evolutionism in Crosscultural Contexts

The impact of evolutionary thought upon the world of the Victorians on both sides of the Atlantic was pervasive yet ambiguous (Moore 1989). The scientific contributions of Darwin and Wallace were used, transformed, and sometimes distorted in a bewildering variety of contexts. The interaction between biology and social thought was complex because scientific terminology and concepts were incorporated into existing sociopolitical debates that themselves were complex (Bowler 1993; R. Young 1985; Jones 1980). As we have seen, many of the British political thinkers who dealt with evolution held deeply opposing views. In the United States, the situation was equally divisive. Between the extremes of rugged and ruthless individualism championed by William Graham Sumner and the potential for cooperative and egalitarian collectivism advocated by Lester Ward, Americans looked to a spectrum of sociopolitical prophets not only to accommodate but to tame the tiger of evolutionary biology (Russett 1976, 83–123).

Underlying these attempts to clarify the cultural meaning of evolutionism was the vexing question of whether nature or nurture is the most important factor determining a person's character. The role of nature in determining behavior was central to the Victorian controversies concerning biology and gender (Russett 1989) and other areas of overriding public interest. Those who believe that nature, that is, heredity, is the most basic influence will most likely reject any strong claim that improved conditions and education can have a lasting, beneficial effect on character. If a person inherits "bad" traits, then nothing can help him or her and it is a waste of time even to try. This attitude led to conservative political arguments based on the need to limit the numbers and influence of the lower classes. This latter group presumably represented the chief reservoir of bad traits. Elimination of the unfit could take place by either natural or artificial selection. Darwinism and Mendelism both could seem to offer a scientific basis for eugenics because both theories stressed the power of heredity to determine the biological character of an individual (Paul 1995).

At the other end of the biological spectrum, the Lamarckians' emphasis on the inheritance of acquired traits provided a basis for believing that nurture could triumph over the limitations imposed by nature. Followers of Lamarckian ideas maintained that human beings can be improved by exposure to better environmental (including economic and educational) conditions. They further believed that such acquired improvements could be inherited and thus become fixed in succeeding generations. In the Soviet Union, Lamarckism enjoyed a major, if unfortunate, vogue under Stalin during the 1930s and 1940s. T. D. Lysenko (1898–1976), controversial Soviet biologist and agronomist, even succeeded in having his version of Larmarckian biology incorporated into official communist ideology of that period. Lysenko argued that Western, or capitalist, genetics, with its emphasis on the immunity of genes to most external influences, locked biology into patterns of heredity that could change only very slowly over time. According to Lysenko, Lamarckism, in contrast, provided communism with a science capable of quickly changing genetic characteristics by appropriate environmental manipulation (Graham 1993). The debate as to whether nature or nurture—or, most likely, some interaction between the two—is the dominant force influencing human development continues to the present (Rose, Lewontin, and Kamin 1990).

The nature/nurture puzzle was never far from Wallace's thoughts during his North American tour. Like most new American cities he visited, Wallace described Chicago—which struck him as a London run riot—as an architectural as well as sociological hybrid. Slums coexisted side by side with handsome shops and palatial townhouses. In the intense heat of a Midwestern summer, screeching engines of various types and function poured "out dense volumes of the blackest smoke, and at this time of year the grass is dried up, and the trees all blacker than in London" (Wallace 1905, vol. II, 185–186). As his journal recorded, Wallace believed he had witnessed in the United States evolution run amok. He suggested that a "nation formed by emigrants from several of the most energetic and intellectual nations of the old world, for the most part driven from their homes by religious persecution or political oppression, including from the very first all ranks and conditions of life—farmers and mechanics, traders and manufacturers, students and teachers, rich and poor—the very circumstances which drove them to emigrate led to a natural selection of the *most* energetic, the *most* independent, in many respects the *best* of their several nations" (Wallace 1905, vol. II, 198). Such a population would necessarily, Wallace declared, develop both the virtues as well as the prejudices and vices of the parent stock in an exceptionally high degree. To him, America seemed an experiment of civilization in the making, one in which the cult of progress had obliterated any constraints, ethical as well as socioeconomic. The only hope, in Wallace's eyes, lay in the fact that it was an American, Edward Bellamy (author of the famous utopian novel *Looking Backward: 2000–1887* [1888]), who "first opened the eyes of great numbers of educated readers to the practicability, the simplicity, and the beauty of Socialism." (Wallace 1905, vol. II, 199). Bellamy's works, with their evolutionary themes, were widely read both in North America and Britain. In yet one more instance of the two-way flow of transatlantic evolutionism, Wallace returned from North America with the spirit of a crusader (Wallace 1905, vol. II, 191–199). The people he met, the lectures he delivered, and his firsthand exposure to industrializing America all contributed to Wallace's increasing commitment to programs of evolutionary societal reform in Britain during the last twenty-five years of his life.

Another attitude that was reinforced by some versions of evolutionism was the tendency to think of human differences in hierarchical terms. In principle, Darwinian evolution should have undermined the concept of a strictly linear ranking of species. Darwin and Wallace had emphasized the branching nature of evolution, using the metaphor of a broad tree of life. The new species that arose from ancestral species were not necessarily *higher* on the evolutionary stage but appeared *later*. It was easy, however, for many thinkers to equate automatically more recent species (or varieties) with a more highly evolved status. This hierarchical ranking was most often applied to the human species itself. The question of the scientific status of the races of mankind became a hotly debated subject in the nineteenth century, especially in the United States (see chapter 4). It seemed only too obvious to some writers that those groups of people that had more primitive levels of technology and industry than other groups were therefore lower in the evolutionary scale. Europeans almost invariably assumed that other races were inferior to themselves. Moreover, the degree of inferiority was best measured by the level of technological and social development. It was all too simple for some biologists and anthropologists to assume that these so-called lower races corresponded to earlier stages in the evolutionary process by which the highest form of humanity had gradually emerged from prehuman ancestors. Some of the major founders of nineteenth-century anthropology argued, explicitly or implicitly, that the races of mankind could be ranked scientifically along a linear scale from least to most highly evolved (Stocking 1987).

The legacy of such views, which tend to see human racial differences in hierarchical terms, has been an unfortunate one. Most biologists and anthropologists of the late twentieth century would regard any attempt to buttress racism with pseudoscientific arguments as invalid and unscientific. But the tendency to rank different races hierarchically had, and continues to have, an appeal in certain political circles. Eugenics and some types of Social Darwinism were based on a similar ranking of individuals within a single society or nation. Some groups within a given society were assumed to be "naturally" more able than others. This was deemed to be a fact, reflected in the natural division of society into higher and lower classes (Adams 1990). Gender issues also bore the imprint of hierarchical arguments. For all these reasons, there is no clearer example of the complex interaction between scientific theory and sociopolitical thought than evolutionism. The debates over the implications of evolutionary theories and speculation were especially bitter and consequential in those matters touching on the increased contact between different races, which was so notable a feature of the Victorian era. In America, *scientific racism* became a formidable, if dubious, rationale for the oppression of blacks by whites. Scientists and social scientists, intoxicated with what they interpreted as the "truths" of evolution, helped to create and justify an institutional racism in the United States during the period from 1859 to 1900. The evolutionary sanctioning of the Negro's alleged inferiority to the Caucasian race entered into popular as well as elite culture. The tragic repercussions of this particularly offensive version of social evolutionism are still felt today (Haller 1971). In the global arena, many Victorian evolutionists confronted traditional politics and international affairs head-on. The vital, and often violent, domain of imperialist thought and practice was permeated with evolutionist ideas and ideologies.

Evolutionism and Imperialism

The global expansion of European influence in the seventeenth and eighteenth centuries, made possible by a potent mixture of conquest, politics, and trade, is a dominant feature of world history. The role of "knowledge workers"—geologists, naturalists, cartographers—was instrumental in this process. These scientific adventurers shared their basic subject matter with imperial administrators and entrepreneurs, and a rich symbiosis developed between scientific exploration and political and economic expansion. The result was an intimate relationship between the intellectual and the more explicitly practical faces of the imperializing mission. This relationship was destined to have a profound impact on the manner in which the conquest of the globe and the conquest of peoples became inextricably linked by the start of the nineteenth century. Institutional factors were becoming key to imperialist achievements. The British Admiralty cooperated with the Royal Society in sponsoring the voyage of the *Endeavour*, which sailed for the South Seas in 1768. The *Endeavour*'s scientific mission was to observe the transit of Venus. But there were more than simply astronomical interests at work. According to Captain Cook's official instructions, because "the making Discoverys of Countries hitherto unknown and the Attaining a Knowledge of distant Parts . . . will redound greatly to the Honour of this Nation as a Maritime Power . . . and may tend greatly to the advancement of the Trade and Navigation thereof." The growing alliance of imperialism and science set the stage for the numerous voyages by which Victorian naturalists like Darwin, Hooker, Huxley, and Wallace were able to travel aboard government naval vessels (or on private ships through financial support from official sources) to distant lands and observe the exotic flora and fauna that were crucial to their articulations of evolutionary theories (Ritvo 1997, 334–335).

Many Victorian naturalists who became evolutionists benefited significantly from voyages of exploration at early phases of their careers. Not only Englishmen, but numerous Europeans were guided by the writings and vision of Alexander von Humboldt. A Prussian scholar and adventurer, Humboldt's own voyages of explorations to many regions of the globe in the first decades of the nineteenth century provided a model for aspiring naturalists. The Humboldtian ethos, which eloquently argued for the importance of expeditions that observed, measured, and collected everything they encountered, became a touchstone for nineteenth-century scientific advances in fields as diverse as climatology, zoology, botany, and physical and human geography. Humboldt provided a standard and a center of gravity for the proliferation of scientific exploring expeditions beginning in the middle third of the century (Pyenson 1999, 258–260).

Humboldt's *Personal Narrative* of his journey through the Brazilian rain forest (made from 1799 to 1804) inspired many readers to ponder some of the most important philosophic, artistic, and scientific questions of the age. Darwin, who devoured Humboldt's evocative account while a student at Cambridge, developed a "burning zeal to add even the most humble contribution to this noble structure of natural science." A beautiful edition of Humboldt's *Narrative*, given by Darwin's professor and friend J. S. Henslow as a gift to the young naturalist as he set sail from Britain in 1831, was Darwin's constant companion on the *Beagle* voyage (Browne 1995, 133, 176). Similarly, Wallace considered von Humboldt's *Personal Narrative* "the first book that gave me a desire to visit the tropics" (Wallace 1905, vol. I, 232).

For Darwin, Wallace, Huxley, Hooker, and numerous others, training in the field served as a rite of passage en route to becoming scientists of eminence. While these figures came from different social backgrounds and possessed varying agendas, all were united by the social worlds and institutions that shaped the development of nineteenth-century natural history research. Specifically, it was the twin mantras of Victorian Britain, colonialism and industrialization, which established the context for that network of relationships, opportunities, and obligations within which the naturalist-scientists lived and functioned.

The relationship between natural history and the Royal Navy was crucial in the paths to evolutionism for Darwin, Hooker, Huxley, and many others. Wallace's field explorations were situated in that same nexus of relationships that marked mid-Victorian voyages of scientific exploration. Wallace's activities in the second of his two major voyages of discovery were inseparable from the British colonial network in the East Indian regions and in the lucrative trade in naturalists' specimens when shipped back to London (Camerini 1997, 354–377; Camerini 1996, 44–65; Allen 1994; Browne 1992, 453–475; Jardine, Secord, and Spary 1996; Rehbock 1983; Rupke 1994). Darwin's voyage to South America and Wallace's travels to the Amazon and to the Malay Archipelago were in the deepest sense journeys of both scientific and personal exploration and development. For the young Darwin, the tropical forests and the islands of the Galapagos alike revealed physical resemblances between animals and plants that made it hard to ignore the possibility of common ancestry. During the voyage of the *Beagle*, the puzzles of nature that beset him in England as a student at Cambridge and Edinburgh universities came together in the daring concept of transmutation. Unspoiled nature seemed almost to permit Darwin to abandon himself to new speculations that everything, including mankind, was part of one ancestral chain (Browne 1995, 361).

Imperialism and the New Biogeography

One of the most important consequences of the culture of imperialism for the development of Victorian evolutionism was the impetus given to reopen traditional explanations in biogeography. The study of the geographical distribution of animals and plants was a familiar one at the time of Darwin's voyage to South America and Wallace's journeys to the tropics. Explanations of distributional data were, however, as discussed in chapter 1, generally embedded within the framework of some version of the argument from design. Consequently, the detailed and precise observation and analysis of the boundaries separating closely allied species and varieties in the regions to which Darwin, Wallace, and their colleagues traveled—often to areas of the globe previously not studied firsthand by European naturalists—were crucial steps in the transition from creationist to evolutionary paradigms.

Both Darwin and Wallace were acutely aware that geographical boundaries had scientific and political significance. A long history predates their preoccupation with mapping biological regions. Prior to the advent of evolutionism, however, maps of floral and faunal regions usually depicted a static portrait of organic distribution, in keeping with the theory of special creationism. Animals and plants had been placed to occupy those particular environments on the earth's surface and beneath the seas for which they had been preadapted by the divine creator. Nineteenth-century naturalists' maps generally showed the same regions as earlier maps, but came to be interpreted quite differently. Victorian maps were increasingly seen as reflecting the outcome of the evolutionary transformation and history of organisms themselves, in addition to the past migrations of certain species. This persistent attention to mapping biogeographical regions and determining their precise boundaries, as far as observations permitted, was part of a broader process of visualizing biology that gathered force in the mid-nineteenth century. The shift in the way in which the distribution and characteristics of organisms (including humans) was perceived, triggered in part by the theory of evolution, is now the focus of increasing scholarly interest (B. Stafford 1984; Lynch and Woolgar 1990; Myers 1990; Camerini 1993, 700–727; Baigrie 1996; Rubino 1997, 118–130). Two main issues are raised by this recent scholarship. The first centers on the capacity for maps to look the same but radically change in meaning over time. The second involves the notion that the meaning of a map resides not only in the map but in relation to the written text of which it is a part and the larger historical context in which it appears. Maps were obviously an indispensable practical tool for traveling naturalists. The far more profound, and illuminating, question is, what was their value as tools of biogeographical thought (Camerini 1993, 702, 709)? The famous "Wallace's Line," which depicts the postulated boundary between that portion of the Malay Archipelago whose animal and plant species were of Indian evolutionary descent, from that portion whose indigenous flora and fauna were related to those of Australia, is but one of many examples of changes in the conceptualization and interpretation of maps.

Darwin's classic investigation of closely related species of finches in the Galapagos Islands is emblematic of the theoretical as well as practical link between imperialism and evolutionism. For the *Beagle*'s captain, Robert FitzRoy, one of the primary aims of the voyage was to chart more precisely the South American coast. For FitzRoy, the *Beagle*'s journey would enable the British Admiralty to make better-informed decisions on naval, military, and commercial operations along the relatively unexplored coastline south of Buenos Aires. Britain's main goal for the *Beagle* was to provide information that would be immediately useful in aiding British efforts to establish strong footholds in those areas so recently released from their commitment to trade only with Spain and Portugal. The untapped riches of South America could add great weight to the

British position in the nineteenth-century European balance of power (Browne 1995, 180–183). But Darwin had significant additional goals as the ship's young naturalist. One of Darwin's most recent biographers, Janet Browne, has described his experiences in South America as a turning point in his life: "[The] finches joined the fossils as extraordinary and intriguingly complex problems he yearned to solve. . . . Why should different finches inhabit identical islets? The Galapagos iguanas . . . similarly divided themselves among the islands, and the heavily built tortoises with their individual shells again came to mind. In a . . . Paleyian world where animals and plants were perfectly adapted to their surroundings, things like this did not happen" (Browne 1995, 360–361). Darwin's mind had been spinning from the very outset of the *Beagle* voyage. The first place he disembarked, the first foreign soil he stepped on as a natural history explorer, was in early 1832 at St. Jago, one of the charred volcanic islands of the Cape Verdes group, about 450 miles northwest off the coast of Africa in the Atlantic Ocean.

For Darwin, this desolate location was one of the most important stops of the entire *Beagle* voyage. Its strange geological and biological features prompted him to begin pulling together all his diverse natural history experiences in Britain and enter the world of investigative science. St. Jago, otherwise so insignificant an outpost on the globe, forever remained in Darwin's memory as the site of a philosophical and personal initiation into the world of evolutionary speculation. Almost immediately upon his return to England in October 1836, Darwin's *Beagle* specimens and his notes and journals from the voyage seemed to force him to articulate more fully his evolutionary premonitions. From early 1837 onward, Darwin began that intellectual and ideological journey that would culminate in the publication of *Origin* two decades later. The lessons of geographical distribution were translated into a theory of evolutionary relationships over time (Browne 1995, 183–186, 361–371).

Wallace's extensive observations of the geographical distribution of the human as well as animal and plant inhabitants of the Amazon and Malaysian regions had a similarly decisive effect. The role of geographic *and* cultural boundaries was crucial in Wallace's developing evolutionary views. As with Darwin, biological data mingled with the political, social, and economic aspects of exotic lands and peoples that Wallace encountered in his travels (Wallace 1853; 1891; 1905, vol. I, 264–288, 302–384). Imperialist realities were part of the natural world that Darwin and Wallace investigated so brilliantly.

Victorian Science and Imperialism: Achievements and Ambiguities

Throughout the nineteenth century, European nations used the ideology of imperialism to motivate and justify quite specific kinds of scientific work. In fields such as geology, botany, zoology, and anthropology, the exploration of new lands and the cataloguing of new plants, animals, and peoples were central components of Victorian natural history. By the end of the nineteenth century, the British Empire alone took in more than one-fifth of the world's land area and a quarter of its people. The empire included such vast regions as Africa, India, Canada, Australia, New Zealand, and substantial portions of Latin America, the Middle East, and east Asia. This enormous empire affected virtually every aspect of Victorian life and provided one of the major contexts for Victorian science. The standard picture is one of science reaching out into the empire and bringing its prizes back to the metropolitan center. The monumental museums of Britain, especially those in South Kensington in London, testify to this version of the adage "to the victor belong the spoils." But recent scholarship has revealed that imperial science was a two-way street: The scientific work done in colonial countries must also be examined on its own terms and in relation to the development by such countries of scientific traditions and institutions of their own. As Britain and other European powers moved to appropriate and subordinate large regions of the globe, their very success posed problems and questions. Exploration, conquest, and commerce during the Victorian era were creating, literally, a new world order. Data and specimens gathered in the empire often posed new puzzles and offered new insights into old ones. The pursuit of science in a global enterprise cast existing knowledge in a new light, and in so doing sometimes set in train fundamental shifts in scientific belief and practice. The profound, and often emotionally and intellectually harrowing, reorientation that biological thought underwent in the Victorian era was the product of many causes. A major one was surely the confrontation, propelled by the needs of a growing empire, with new facts about the distribution and affinities of living things around the world (Hunt 1997, 312–315).

Inseparable from this process were the crucial issues raised by the sheer contact of European society with other cultures. Victorian imperialism posed highly contentious questions as to the nature of the relationship between colonizing and colonized nations and peoples (R. Stafford 1989). Significantly, just as Darwin and Wallace came to differ on their interpretations of the scope of natural selection, they also began their evolutionary investigations with somewhat different approaches to the culture of imperialism. Darwin's comfortable and confident view of British imperial exploits was more typical than Wallace's critical attitude in this ideologically laden realm. Darwin brought "to his work unthinking assumptions of superiority that encouraged him to believe the natural commodities of unexplored countries were his to claim by intellectual and cultural right. The distinction between collecting, hunting, and plundering was never very clear in the activities of any naturalist" in the Victorian era. The declared scientific nature of the *Beagle*'s voyage "and the argument that geographical exploration contributed to the general advancement of learning, however sincerely maintained, was only the outer varnish of a far deeper and more widely shared conviction that European countries—Britain in particular—were fully justified in exploring and exploiting the rest of the world. The absolute right to curiosity expressed their absolute control. The ultimate aim was always national expansion" (Browne 1995, 232). Wallace, in contrast to Darwin and many of their contemporaries, was uncertain that imperial contacts necessarily implied a hierarchical relationship between colonizer and colonized. His journeys made Wallace less convinced of the generally accepted notion—by Europeans—of the superiority of European civilization (Wallace 1869b, 455–457).

During the Victorian era, the acquisition of territory and of further trading concessions worldwide continued the pattern of eighteenth-century

British foreign policy. Darwin got a vivid glimpse of the growing reality of the British Empire when the *Beagle*, nearing the end of its long voyage around the globe, crossed the vast Indian Ocean and, on 31 May 1836, landed at the Cape of Good Hope on the southern tip of Africa. On the next day, he visited Cape Town, which, as he noted, was a "great inn on the highway to the east." The town, which had long been under Dutch rule, was becoming anglicized, with peoples of all nationalities now speaking English. Cape Town appeared to Darwin as one of the numerous colonies seeded around the globe in which "little embryo Englands are hatching." The aboriginal inhabitants, including Hottentots, had either been forced to retreat further into the interior of the region or were pressed into service by their colonizers. The young Darwin even had an impeccably dressed and well-mannered Hottentot groom, in white gloves, to guide him.

Although Darwin had mixed feelings about enforced servitude, he shared with many of his contemporaries the view that British expansionism was ultimately beneficial in terms of human evolution. Patrick Matthew, an advocate of both free trade and evolutionism, wrote a work whose title alone told volumes: *Emigration Fields: North America, The Cape, Australia, and New Zealand, describing these Countries, and giving a Comparative View of the Advantages they present to British Settlers* (1839). Matthew, like many of his contemporaries, saw imperialism and emigration as twin solutions to the perceived Malthusian dilemma facing a Britain whose home population was rapidly growing. His *Emigration Fields* announced that "the whole of the unpeopled regions of the earth may now be said to be British ground." Matthew counseled British policy makers that by encouraging emigration they could reduce poverty at home, raise wages, and transform the Malthusian "curse" into "a blessing." These hordes of emigrants (about 400,000 were leaving the British Islands annually in the 1830s and 1840s) would, Matthew predicted, create new markets abroad, so that "our paupers would be transformed into rich customers." Since "change of place . . . seems to have a tendency to improve the species equally in animals as in plants," he continued, it could not "therefore be doubted that the increase of the British race . . . and their extension over the world, and even the vigour of the race itself, will be more promoted by this colonizing system." Matthew's vivid mixture of social and organic evolution, dominated by Malthusian competition and selection, mirrored aspects of Darwin's own musings on the broader implications of evolutionism (Desmond and Moore 1991, 184, 266).

Evolutionism and the "White Man's Burden"

Imperialist sentiment, promoted by strategic considerations and aided or justified by philanthropic motivations, reached its peak when Queen Victoria, at Prime Minister Benjamin Disraeli's suggestion, had herself crowned Empress of India in 1876. Disraeli, who had arranged the purchase of the Suez Canal for Britain in 1875, was Britain's pre-eminent Conservative party leader of the nineteenth century and the Liberal William Gladstone's great rival during Victoria's reign. Disraeli epitomized certain aspects of nineteenth-century imperialism. He championed a paternalistic connection between the landed and the "lower" classes domestically, and between imperialist and subject peoples internationally. Advocates of imperialist foreign policies often justified them by invoking such paternalistic and racist theories. Although these theories had long histories prior to the nineteenth century, evolutionism seemed to afford them scientific sanction in the minds of many Victorians. An ostensibly biologically buttressed imperialism could be easily seen as a manifestation of what Rudyard Kipling would refer to as "the white man's burden."

In Kipling's work, as in his life, the British Empire assumed a complex mythical or legendary function, which he passed on to his readers. The "white man's burden" was, as far as culturally patronizing imperialists of Kipling's type were concerned, a genuine burden. Kipling viewed imperialism, predicated on deeply held political, racial, and religious beliefs that sustained a feeling of innate British superiority, as being primarily a moral responsibility. Imperialism could also be profitable, but Britons had a moral responsibility to maintain, defend, and protect the empire from rival world powers and from the rebellious governed. Kipling hoped that the colonized peoples would recognize their inferiority and freely obey their superiors. The implication, of course, was that the empire existed not only for the economic or strategic benefit of Britain itself. Its existence also guaranteed that "primitive" peoples, incapable of self-government, could, with British guidance, eventually become civilized and Christianized. The truth of this doctrine was accepted naively by some, and hypocritically by others. It served, in any case, to legitimize Britain's acquisition of portions of central Africa and her attempt to subjugate, in league with other European powers, China and other Asian regions (Parsons 1999; Stocking 1987; Reingold and Rothenberg 1987; Karlin 1999). It was precisely such categorical assertions of racial and national superiority and inferiority, based on a widespread interpretation of evolutionary doctrines, that both informed and inflamed Victorian debates on the history, current status, and potential future prospects of human culture and society.

4 DEBATES ON HUMAN EVOLUTION

The subject of human origins could scarcely fail to provoke a controversy with profound cultural as well as strictly biological implications (Lorimer 1997, 212–235). The publication of *Origin*, and more pointedly of *Descent of Man* (1871), were crucial but not isolated documents in the broader Victorian quest for an answer to what Huxley in 1863 termed the question of "man's place in nature." If humans were part of the animal kingdom, then evolutionary theory was expected to provide guidance, and, hopefully, definitive answers to three major facets of our relationship to other animals: (1) classification of the human species, (2) antiquity of the human species, and (3) the origin and significance of the different human races. All three of these issues had been discussed and debated prior to the nineteenth century, but they had been subsumed under an overarching religious creationist framework. It was the theory of evolution that brought these issues to the epicenter of scientific debate. Indeed, the reception accorded *Origin*, particularly by those who condemned it, was in large measure dominated by what evolutionary theory was perceived to entail regarding humans. Aside from naturalists, professional or amateur, few people were preoccupied with evolutionary hypotheses pertaining to the origin and development of other animal and plant species. The bitter public debates in the Victorian period over the status and implications of Darwinism and other evolutionary paradigms were largely fueled by the participants' views on human evolution. This pattern would persist into the twentieth century, in such celebrated episodes as the 1925 Scopes trial in Tennessee (See chapter 7).

Classification and Antiquity of *Homo Sapiens*

With respect to the task of classifying humans biologically, two major possibilities existed. *Homo sapiens* could be considered simply as part of the hierarchy of the animal kingdom. Alternatively, humans could be deemed to constitute a separate unit of creation, wholly distinct from animals and plants. By the eighteenth century, the physical similarities between humans and the higher primates were clear to all anatomists, as well as to the general public. Carl Linnaeus classified humans as one genus in the order *Anthropomorpha*, which also included apes. Although Linneaus was willing to concede that man was morally and intellectually superior to the higher primates, he believed that physical characteristics should be the scientist's main criteria for biological classification. Other taxonomists, however, such as Johann Friedrich Blumenbach (1752–1840), were inclined to set man apart more radically. Blumenbach maintained that humans constituted a distinct order, *Bimana*, or two-handed animals, thus distinguishing them definitively from the order of *Quadrumana* to which apes belonged (Haller 1971, 6; Bowler 1989, 93). Aside from notable exceptions such as Georges Louis Leclerc, the comte de Buffon, and Jean Baptiste de Lamarck, most pre–nineteenth-century naturalists tended to regard humans as uniquely different from the rest of the animal kingdom, physically and (especially) morally and intellectually. In the absence of any compelling theoretical or evidentiary claims for evolution, such a distinct status for humans seemed valid on scientific as well as philosophical and religious grounds. Moreover, to place humans biologically in the naturalists' classification schemes raised troubling questions such as whether they alone possessed a unique and immortal soul.

By the nineteenth century, the development of powerful evolutionary hypotheses made it impossible to dismiss the view that humans arose in the course of development from remote, nonhuman anthropoid ancestors. Nonetheless, even so eminent a supporter of evolutionary theory as Charles Lyell would write Darwin that he could not, or did not want to, "go the whole orang." Darwin had little sympathy for the presumed loss of human dignity that troubled Lyell and others who found the concept of a completely naturalistic account of human evolution disturbing. Darwin teased his old friend by informing Lyell that "*Our* ancestor was an animal which breathed water, had a swim bladder, a great swimming tail, an imperfect skull, and undoubtedly was a hermaphrodite! Here is a pleasant genealogy for mankind." Darwin added, optimistically, that "mankind will probably progress to such a pitch" that nineteenth-century gents will probably be looked back on "as mere Barbarians" (Desmond and Moore 1991, 505). Lyell was neither amused nor convinced. He later wrote Darwin that he felt that human evolution could not be entirely explained by natural selection. Lyell here sided with Wallace, informing Darwin that a "Supreme Will and Power" had intervened in the development of the human species ([K. M.] Lyell 1881, vol. II, 442). Victorian evolutionism forced people to consider in a far more pointed and immediate manner than ever before precisely where humans fit in the great scheme of things.

The issue of the development of modern human beings from nonhuman and extinct prehuman (hominid) forms lay at the center of the broader Victorian controversy over evolutionism. Today, paleontologists and anthropologists agree that human beings, extant and extinct, comprise the zoological family *Hominidae.* The single living human species, *Homo sapiens*, is itself one of the nearly 200 species of the order primates. Among the past and present diversity of primates, hominids are now recognized as having the closest resemblances, and, by implication, evolutionary affinities, to the African great apes (pongids). In 1863, Huxley was prescient in noting in *The Evidence as to Man's Place in Nature* that "whatever system of organs be studied . . . the structural differences which separate Man from the Gorilla and the Chimpanzee are not so great as those which separate the Gorilla from the lower apes [monkeys]" (T. Huxley 1863). Huxley's speculations—and they could only be regarded as such given the limited anatomical, behavioral, and fossil evidence available in the nineteenth century—found as many advocates as critics.

A second major aspect of the Victorian debates on human evolution was the vexing question of the antiquity of man. When did humans originate? This issue had to be decided in terms of the discovery of human remains or human artifacts in the sequence of geological strata—the fossil record. As late as the first decades of the nineteenth century, no traces of human remains were found to be anything but recent in terms of geological timescales. Thus, insofar as one was to be guided by the empirical evidence available in the first half of the nineteenth century, no fossils were found to suggest connections with humans and earlier human-like creatures. Hypotheses such as those of Lamarck as to how man *might* have arisen from the anthropoid apes were justifiably regarded as mere speculation, without any plausible fossil evidence. Early nineteenth-century geology seemed to show conclusively that humans had come on the scene in relatively recent times, and certainly long after the existence of the latest prehistoric species then known. More crucially, what scant fossil evidence then existed appeared to demonstrate that humans had shown no development (or change) in their physical characteristics from the postulated dates of their first appearance on the planet to the present. Thus, in the early Victorian period it seemed natural that humans should be viewed as specially created and therefore not covered by any transmutation hypotheses. Human evolution ranked as an implausible scientific hypothesis and moreover contradicted powerful traditional religious and cultural precepts.

It was Lyell's *Antiquity of Man* (1863), published two years after

Origin, which shook these traditional assumptions. Lyell eloquently amassed the scientific evidence that was rapidly accumulating by mid-century to demonstrate two crucial facts. First, there was sufficient geological time for man to have evolved, gradually, from earlier prehuman ancestors. Second, the fossil remains being uncovered pointed to direct links between humans and their posited nonhuman ancestors. *The Antiquity of Man* ran through three editions in one year (C. Lyell 1863). In this masterly work, Lyell gave a general survey of the arguments for man's early appearance on the earth, derived from the discoveries of flint implements in post-Pliocene strata in the Somme valley and elsewhere. Lyell also discussed the deposits of the Glacial epoch, and in the same volume gave his first major public endorsement of Darwin's theory of the origin of species. The fossil record known to Victorian science was, however, only rudimentary. Darwin devoted an entire chapter in *Origin* to the paucity of evidentiary fossil support for evolution. In the famous chapter IX, "On the Imperfection of the Geological Record," Darwin cited such gaps as the absence of any clear intermediary, or transitional species, for any posited animal or plant evolutionary development. Although *Antiquity of Man* did not prove the case for human evolution—Lyell himself was tormented by the possibility of an apelike ancestry for humans—it certainly placed the debates on a far more potent scientific basis than ever before.

Human Fossils

Two major fossil finds in the Victorian era rendered the concept of the evolution of modern humans from nonhuman ancestors powerfully persuasive, if still not yet conclusive. The first was when workers unearthed portions of what appeared to be a human skeleton of great antiquity in a cave in the Neander Valley, near Dusseldorf, Germany, in 1856. Immediately, there was disagreement as to whether the bones represented an archaic and extinct human form or an abnormal, pathological specimen of a member of the modern human species. Quickly dubbed Neanderthal man, the fossil reconstruction of the specimen revealed that while it presented certain apelike features, such as thick brow ridges, its cranial capacity was as large as the skull of modern humans. Huxley, among others, opted for the verdict that Neanderthal man was human, and thus could not represent the elusive missing link between *Homo sapiens* and small-brained apelike animal ancestors (Desmond 1998, 299–301, 304–305, 313–315, 326, 333, 581). This view was shown to be correct. In 1886, two Neanderthal skeletons were discovered in a cave at Spy, Belgium, associated with Middle Paleolithic stone tools and an extinct subarctic fauna. The discovery of Neanderthal man certainly strengthened the case for the antiquity of humans, and, by implication, afforded the lengthened timescale in which evolution would have operated. But it was not until the last decade of the nineteenth century that a fossil find was made that rendered the case for human evolution from nonhuman ancestors virtually irresistible.

In 1891, the Dutch paleontologist Eugene Dubois discovered the skullcap and thighbone of a creature far more primitive than Neanderthal man, on the island of Java. Dubois originally classified his find as *Pithecanthropus erectus*, borrowing Haeckel's term for the elusive link between prehuman and human species, although it quickly became known more commonly as "Java Man." This specimen was characterized by a cranial capacity of approximately 900 cubic centimeters, substantially smaller than that of *Homo sapiens*. Java Man's skull was flat in profile, with little forehead, very thick skull bones, heavy brow-ridges, a large palate, and a massive jaw with no chin. The teeth were essentially human but showed some definite simian characteristics, such as large, partly overlapping canines. The thighbone showed that Java Man walked fully erect, like modern man, and attained a height of about five feet eight inches (1.72 meters). Here, then, were fossil remains of a creature that seemed, at last, to be a convincing candidate for the honor of representing a clear stage in the evolution of humans from nonhuman forbears. Further discoveries in the early twentieth century confirmed that *Pithecanthropus erectus* occupied Java during the middle Pleistocene epoch, about 1 million to 500,000 years ago (Clark 1964, 88–95). Java Man also lent support to those naturalists, including Darwin and Ernst Haeckel, who held that the ancestors of modern humans walked upright *before* they acquired a larger brain capacity.

By the close of the nineteenth century, then, impressive evidence had been produced for the claim that modern humans had indeed evolved from ancient nonhuman forms. To be sure, the evidence was still scanty. Such issues as whether human evolution represented a clear linear progression from nonhuman to human species or, rather, a more complex branching pattern in which certain ancestral species succumbed to other, more adept species and became extinct, were left unresolved. Related notions of parallel evolution of different but allied ancestral human species and the complex issue of the link between biological and cultural evolution would not be capable of decisive scrutiny until the twentieth century. Even today, such issues are still the matter of vigorous scientific and popular debate; that they were even more confused and contentious in the Victorian era was inevitable (Bowler 1989, 231–233, 322–325).

Piltdown Man

The first Neanderthal fossil discovery in 1856 gave a tremendous impetus to uncover other, more definitive, fossil remains of human ancestors. In the next half century finds were made in continental Europe and in Asia, but not in Britain. In 1912, however, fossil remains of what appeared to be an ancient Pleistocene hominid were found in the Piltdown quarries in Sussex, England, coincidentally, not far from Darwin's country home. Three years earlier, amateur English geologist Charles Dawson had obtained from a laborer digging in a gravel pit on a farm near Piltdown Common, Sussex, a small fragment of a human cranium's parietal bone. In 1911, on another visit to the site, Dawson picked up a larger piece belonging to the frontal region of the same skull. These fragments of the Piltdown skull promised to provide the elusive "missing link" in human evolution. Consequently, Dawson accelerated his searches at the site. After uncovering flint tools and the fossil remains of various prehistoric animals, he took the set of specimens to the paleontologist Arthur Smith Woodward of the British Museum in London. Soon Woodward was in Piltdown with Dawson, conducting a systematic excavation of the site. During the summer of 1912, they found more fragments of the Piltdown skull, additional prehistoric animal fossils mixed with human tools, and part of a jawbone with two intact molars. These pieces carried tremendous potential significance (Bowler 1989, 323).

Owing to its size and shape, the Piltdown cranium clearly came from a hominid. The flint tools reinforced this conclusion. The animal remains and the geology of the site suggested that the skull dated from the Pleistocene epoch, at some point midway between the supposed date of the so-called ape-man of Java and the emergence of modern humans. The jaw, however, appeared to come from a type of ape never known to have lived in Europe, and the teeth were worn down in a human fashion. Pieced together by Woodward, the picture emerged of a new species of extinct hominid that he called *Eoanthropus dawsoni*, or the "dawn man" of Piltdown. Dawson and Woodward unveiled their discovery on December 18, 1912, before a packed house of Britain's scientific elite at the Geological Society of London. "While the skull, indeed, is essentially human, and approaching the lower grade in certain characters of the brain," they emphasized that "the mandible appears to be almost precisely that of an ape, with nothing human except the molar teeth." Sir Arthur Keith, one of the world's leading experts on human antiquity and anatomy, attended the presentation by Dawson and Woodward. Keith generally concurred with their conclusions, as did the renowned neurologist Grafton Elliot Smith and the famed biologist Boyd Dawkins. Perhaps Dawkins best expressed the collective response of the British audience when he declared during the discussion period, "The evidence was clear that this discovery revealed a missing link between man and the higher apes" (Larson 1997, 11–13).

The initial reaction to the finds was mixed. On the whole, British paleontologists were enthusiastic. Many English scientists felt left out by discoveries on the continent. Neanderthal man had been found in Germany in 1856, and Cro-Magnon man in France in 1868. French, German, and American paleontologists, in contrast, tended to be skeptical. These skeptics objected that the jawbone and the skull were obviously from two different animals and that their discovery together was simply an accident of placement. However, the report in 1917 of the discovery of a second skull, Piltdown II, converted many of the skeptics. One accident of placement was plausible; two were not. National differences in paleontological communities were reconciled by what appeared to be the most significant fossil find supporting the hypothesis that humans had evolved from ape-like ancestors. For nearly forty years the Piltdown skulls were considered archaeological finds of the greatest significance: the fossil remains of the presumed missing link in human evolution. Then, in 1953, a group of scientists, lead by Kenneth Page Oakley, attempted to use the new method of fluorine testing to get a more exact date on the bones. What the test showed surprised them. The jaw was modern, and the skull only six hundred years old. Additional analysis soon confirmed the fluorine tests. The jaw was really that of an orangutan. It had been filed down and parts that might have suggested its simian origin were broken off. Both pieces had been treated to indicate great age. What had for four decades been a name associated with momentous advance in the scientific evidence for human evolution now became synonymous with the worst type of scientific fraud. The Piltdown hoax certainly did not destroy the overwhelming body of evidence that had been legitimately obtained and that demonstrated the path of human evolution. But it still serves as an eloquent reminder that when the scientific stakes are high, objectivity and verification may sometimes suffer (Clark 1964, 87).

Fraud in Science

The question arises as to how these faked fragments of bone fooled the best scientific minds of the time? Perhaps the desire to be part of a great discovery blinded those charged with authenticating it. National pride may have kept the researchers from noticing the scratch marks made by the filing of the jaw and teeth. Also, the sophisticated tests conducted by Oakely, which provided the first conclusive evidence of fraud, were not undertaken until 1953. Who perpetrated the hoax? Many historians point to Charles Dawson. Others, though, lay the blame at the feet of people as diverse as a young Jesuit priest, Teilhard de Chardin, who assisted in the dig, to the author Sir Arthur Conan Doyle, who lived in the area. Whatever the ultimate answer, the Piltdown saga exemplifies the lengths to which some individuals and groups may resort to buttress competing claims concerning the scientifically as well as culturally potent consequences of different views on human evolution (Blinderman 1986; Broad and Wade 1982; F. Spencer 1990).

Fossil evidence, thus, could be—and vehemently was—variously interpreted, particularly so with respect to the then scant prehuman remains. Some thinkers sought to capitalize on this uncertainty to argue the case for creationism. They postulated that the gaps in the fossil record and the apparent sudden appearance of new species in the geological record argued for continued, direct interventions of divine activity in the history of life. Even the growing number of proponents of evolutionism were themselves far from united on details. The celebrated German embryologist and anatomist Ernst Haeckel did not share the Darwinists' enthusiasm for natural selection as the main mechanism for generating the diversity of the biological world. He subscribed to a version of Lamarckism, maintaining that the environment acted directly on organisms, producing new races and species. Haeckel also stressed an inherently progressive interpretation of evolution, culminating in humans as the highest product of the evolutionary process. His seductive deployment of fossil data coupled with the impressive findings of the advancing sciences of embryology and comparative anatomy were powerful ingredients in the growing popular as well as scientific endorsement of an evolutionary lineage for the human species. Even if one accepted evolutionism, however, certain ambiguities remained. Questions as to whether evolution implied a linear, direct progression from earlier to later forms, or whether evolution involved parallel or branching lines of species change were argued throughout the Victorian period (Bowler 1989, 180–181, 201–204, 322–323; D. Young 1993). Haeckel also stated that "politics is applied biology," a quote later used by Nazi propagandists. Haeckel's Monist League, which popularized certain presumed justifications for racism, nationalism, and extreme Social Darwinism, has been cited as an influence on the genesis of certain components of Nazi thought (Gasman 1971).

HUXLEY VS. OWEN

The increased willingness to accept claims for human evolution was reinforced by discoveries regarding other species. Thomas Huxley was among the first to demonstrate the significance of fossil links between reptiles and birds as providing powerful support for the common ancestry of these two important classes of the animal kingdom. The American paleontologist Othniel Charles Marsh discovered the first pterodactyl (a flying reptile) found in the North America. He made extensive scientific explorations of the western United States and contributed greatly to knowledge of extinct North American vertebrates. Credited with the discovery of more than a thousand fossil vertebrates and the description of at least 500 more, Marsh published major works on toothed birds, gigantic horned mammals, and North American dinosaurs. But it was Marsh's studies of fossil horses that presented some of the most compelling evidence for the evolutionists' paradigm in the late nineteenth century. He and his associates uncovered a remarkable series of fossil remains that dramatically linked the modern horse through clear stages back to a small, multi-toed ancestor. This was given the name *Eohippus*, or the "dawn horse" from the Eocene epoch. The actual evolution of the modern horse has turned out to be more complicated than Marsh's reconstruction suggested. Certain fossils are now known to represent extinct side branches rather than direct stages in the evolution of the modern horse. But Marsh's discoveries were sufficiently striking to encourage the ardent Huxley to declare them as "demonstrative evidence of evolution" in 1888 (Bowler 1989, 204–205).

Twenty years earlier, Huxley had exploited the announcement of the find of the earliest-known fossil bird *Archaeopteryx*, which was discovered in Upper Jurassic deposits in Bavaria, to advance the evolutionists' cause. This bird, about the size of a magpie, resembled some reptiles and differed from more recent birds in many ways. Its reptilian features included jaws containing teeth set in sockets and a long tail made up of a series of free vertebrae. The most obvious avian characteristic of *Archaeopteryx* was the possession of feathers. It thus represented an intermediate stage between reptiles and modern birds. In a dramatic lecture at the Royal Institution on 7 February 1868, Huxley juxtaposed two striking images for his audience. One was that of a heavy *Archaeopteryx,* flapping clumsily overhead. Below, Huxley placed the image of a diminutive dinosaur, hopping like a bird, neck bobbing, and snatching prey with its small arms. His vivid depiction of the process by which tiny dinosaurs might have evolved gradually into ancient flightless birds and thence into modern avian species was brilliant strategy. It was hailed in the popular and scientific press as a stunning propaganda stroke for the proponents of emergent evolution. Richard Owen (1804–1892), one of England's most eminent paleontologists but an arch antievolutionist, was not impressed. Owen angrily termed Huxley's visual show a "sensational trick" (Desmond 1998, 359).

The ongoing disputes between Huxley and Owen expose, again, the complex mix of scientific, religious, cultural, and ideological factors in Victorian evolutionary debates. Owen synthesized diverse French anatomical work, especially from naturalists such as Georges Cuvier and Etienne Geoffroy Saint-Hilaire (1772–1844), with German transcendental anatomy to develop a system of biological classification based on the common structural plan of certain broad groupings of animals—what he termed *homology*. One classic example of homology was that organs as visibly and functionally different as a bat's wing, a seal's flipper, a cat's paw, and a human hand nonetheless display a common internal structure. They all have identical or very similar arrangements of bones and muscles. Owen concluded that there must exist a common structural plan for all vertebrates, which he termed the *archetype*. But his concept was explicitly framed so as to preclude any evolutionary connotation. For Owen, the archetype represented an idea in the divine mind, which also planned all its modifications in different species.

Among Owen's most contentious taxonomic hypotheses was that related to apes and humans. He described the anatomy of a newly discovered (1847) species of ape, the gorilla. Owen's antimaterialist and antievolutionist views led him to state that gorillas, and all other apes, lack certain parts of the brain that humans possess. Owen focused specifically on a structure known as the hippocampus minor. The uniqueness of human brains, Owen thought, showed that humans could not possibly have evolved from apes. Owen persisted in this view even when Huxley conclusively showed that Owen was mistaken—apes do have a hippocampus (Desmond 1982, 74–81). Victorian author Charles Kingsley satirized the dispute in his childrens' classic *The Water-Babies*:

> You may think that there are other more important differences between you and an ape, such as being able to speak, and make machines, and know right from wrong, and say your prayers, and other little matters of that kind; but that is a child's fancy, my dear. Nothing is to be depended on but the great hippopotamus test. If you have a hippopotamus major in your brain, you are no ape, though you had four hands, no feet, and were more apish than the apes of all aperies. But if a hippopotamus major is ever discovered in one single ape's brain, nothing will save your great- great- great- great- great- great- great- great- great- great- great- greater- greatest- grandmother from having been an ape too. (Kingsley 1885)

The wide-ranging Owen-Huxley confrontation was but one of numerous examples of the highly charged atmosphere surrounding evolutionism. At issue was the contest for cultural authority between rising professional and increasingly secular science and traditional centers of authority such as church and landed aristocracy (Turner 1993). *Essays and Reviews*, published in 1860, was yet another entry onto this Victorian battleground. A collection of seven essays on religion by liberal Anglicans, it covered such inflammatory topics as the "Higher Criticism" of the Bible, the irrationality of belief in miracles, and evolutionism and the cosmology and geology of Genesis. *Essays and Reviews* was highly controversial—the essayists were called "The Seven against Christ"—and exemplified the mounting challenge to biblical literalism by both liberal religious philosophers and the newer breed of scientists. In comparison to *Essays and Reviews*, *Origin* seemed almost tame, although both were powerful agents for change in Victorian thought. *Essays* sold 22,000 copies in just two years, as many as *Origin* sold in two decades (Desmond and Moore 1991, 500–501).

The Owen-Huxley debates on human ancestry were a notable part of the often agonizing public, as well as technical, struggle to define the moral and political order of a new society. The two eminent rivals knew that a dominant feature of the new society would turn on just how people reacted if human origins were shown to be not unique but, in evolutionary parlance, "bestial" (Desmond 1998, 295–297). For Owen and many others, the term *bestial* had the direst overtones. For Huxley and many others, bestial was a scientific metaphor, without sinister connotations. Evolutionary metaphors became pervasive in the Victorian period. Their impact is still being debated.

Despite his embarrassing defeat by Huxley in the hippocampus debate, Owen's overall anatomical and taxonomic work, and his role in founding the British Museum of Natural History, left a lasting legacy to scientists and laypersons alike. His concept of homology, when reinterpreted in evolutionary terms, remains an extremely important biological concept. And, as a public figure, lecturer, and expert, he helped biology grow in prestige and public understanding (Rupke 1994). The scientific and popular influence of such diverse, and competing, figures as Owen and Huxley demonstrates how fluid the evolutionary debates were in the Victorian era, and how high the scientific and cultural stakes were.

Evolutionary Images of Race

It is now generally agreed that all the modern human races are variants of one species, *Homo sapiens*. This view was certainly not unanimously held in the nineteenth century. Quite apart from the problem of the antiquity of *Homo sapiens* as a species, the origin and significance of the different races puzzled Victorians. Given the lack of agreement on the definition of the fundamental taxonomic terms themselves—species, races, varieties, subspecies—in both the Victorian and present eras, it is hardly surprising that racial theories were extremely contentious. Although evolutionism added crucial new dimensions to racial categorization, Victorian images of race invoked a complex web of traditional biological and cultural concepts. While no field of evolutionary speculation was exempt from ideology, racial discussions were particularly vulnerable to historical and ideological realities.

Racial stereotypes had a long history prior to the nineteenth century. Troubling ethical and political issues associated with these historical images of race could not be ignored. The increasing impact of colonialism and other forms of racial contact and conflict exacerbated the dilemma of studying races from an evolutionary perspective. The quest for an objective science of race involved, consciously or not, a selection from existing racial stereotypes. Questions abounded. Were "savages" encountered by European explorers and settlers inferior to "civilized" races? Or did they merely represent alternative cultural patterns of development? More pertinent was the question of whether the obvious characteristics of different races represented fixed, biologically determined products of nature. This inalterable image of race was embraced by the strident advocates of evolutionary racial hierarchies. Alternatively, did racial differences represent malleable cultural accretions that could be harmonized within a nonhierarchical framework? This was the answer welcomed by members of many humanitarian missionary and antislavery movements. The politics of race and science was as explosive an issue in the Victorian period as it is today (Lorimer 1997, 212–215; Numbers and Stenhouse 1999).

The conviction that the scientific study of humans—by means of anthropology, psychology, sociology, and related disciplines—depended largely on the successes of evolutionary biology is widely held. But this plausible view is open to critical scrutiny and serious qualification. Evolutionism, particularly in the later Victorian period and early twentieth century, played a powerful role in shaping the development of the social sciences. However, approaches to the study of human nature and society predate the nineteenth century. That humans shared with other species important physical similarities was no innovative discovery of the nineteenth century. Indeed, the striking physical resemblances between humans and the higher primates were recognized since antiquity. One of the reasons why Aristotle, the medieval Judeo-Christian religious philosophers, and the towering seventeenth-century French thinker René Descartes—to cite but three of the more influential examples—stressed the definition of humans primarily by the possession of an immortal soul or intellect was precisely their desire to isolate a nonphysical characteristic by which the uniqueness of humans could be affirmed. Moreover, it was the possession of a soul or rational faculty that made humans, alone among all other animal species, moral agents. By elevating the spiritual/rational aspect of humanity above its physical characteristics, man's position at (or near) the apex of the Great Chain of Being was both explained and justified. Consequently, in reports of European travelers in ancient and medieval times, it was generally cultural, rather than physical, differences of non-European races that struck the imagination of the first social scientists.

By the beginning of the eighteenth century, however, a crucial shift occurred. An emphasis upon the obvious physical differences that marked and separated different races was emerging. Students of the human condition began to focus, far more than previously, on physical characteristics such as the shape of the nose, lips, and cheekbones; the facial outline and form of the cranium; the texture of the hair; and the color of the skin. In 1735, this trend was sanctioned by the appearance of Linnaeus's magisterial *Systema Naturae*. Linnaeus's depiction of *Homo sapiens* placed great weight upon physical classifications. He went so far as to include humans, physically, along with the apes in the order *Anthropomorpha*. Although Linnaeus still maintained the intellectual and moral chasm separating humans from the other higher primates, his great authority encouraged naturalists and social thinkers to draw sharp distinctions among the various races of mankind. This ostensibly scientifically valid categorization of human races was, in fact, predicated upon a complex (confused?) juxtaposition of physical *and* cultural attributes. By the start of the nineteenth century it had already become usual to define the various human races in terminology that both sanctioned and facilitated the construction of racial stereotypes that now assumed the aura and authority of science. Evolutionism added a formidable component to the Victorian predilection for regarding human races in a hierarchy from superior down to inferior. Although the details would differ, evolutionary explanations of the origin of these races already bore the portentous imprint of classificatory criteria that could, and would, lend themselves readily to theoretical and practical ideas of conflict and control in an imperialist age (Lorimer 1997, 216–217; Stocking 1987).

"Scientific Racism"?

Ethnocentrism was nothing new to the western world. Aristotle had separated Hellenes from Barbarians in order to underscore that the latter, from the perspective of classical Greece, were obviously deficient in civlization and no better than brutes. Nearly two thousand years later, the eminent Scottish philosopher and historian David Hume (1711–1776) was equally explicit. "I am apt to suspect the negroes," he wrote in 1748, "and in general all the other species of men (for there are four or five kinds) to be naturally inferior to whites. There was never a civilized nation of any other complexion than white. . . . No ingenious manufactures amongst them, no arts, no science. . . . Such a uniform and constant difference could not happen, in so many countries and ages, if nature had not made an original distinction betwixt these breeds of men" (W. Coleman 1971, 98). Hume's explicit hierarchical racial categorization was not unique. By the close of the Enlightenment and throughout the course of the nineteenth century, this conflation of physical and cultural traits to define the various human races constituted the premise upon which the majority of European and North American thinkers developed the science of anthropology. The history of anthropology is extremely complex (Stocking 1987). During the Victorian era, anthropological speculation was marked by an emphasis upon linking the divergent physical racial characteristics (drawn from comparative anatomy and more overt traits such as skin color) to rankings along a cultural scale from the most to the least civilized.

In a popular 1854 textbook used in many schools in the United States, Josiah Nott and George Gliddon verbally and visually urged, on ostensibly scientific grounds, the near bestial condition of a distinct "Negro race." Nott and Gliddon marshaled evidence drawn from physical and cultural anthropology, in addition to sketches and other artifacts depicting humans and their near relations. The two authors concluded that a person would have to "be blind not to be struck by the similitudes between some of the lower races of mankind, viewed as connecting links in the animal kingdom, . . . with the Orang-Outan and Chimpanzee." In a famous illustration from the text, Nott and Gliddon depicted three skulls of a Caucasian male, a black male, and a young chimpanzee. The skulls, with artists' conceptions of the actual facial appearance of each specimen, were intended to leave no doubt in the students' minds that the chimpanzee looked more similar, and thus *was* more closely related on the evolutionary scale, to "African and Oceanic Negroes than the latter are to the Teutonic types" (Nott and Gliddon 1854). Nineteenth-century evolutionary theories reinforced cultural biases by providing the mantle of scientific authority to claims concerning racial differences. The very subtitle of *The Origin of Species*, "Preservation of Favoured Races in the Struggle for Life," seemed to afford license to those biologists, anthropologists, sociologists, and political theorists who sought to demonstrate that racial differences translated into scientifically sanctioned models for discrimination. Obviously, or so it appeared, the "fittest" races and nations survived the incessant struggle for existence. But what made one nation or race fitter than others? During the second half of the nineteenth century, social as well as natural scientists became obsessed with deploying the evidence of evolutionary biology to determine what the different human races actually signified. Their conclusions were based upon the real, or imagined, complex of associated hereditary mental and physical traits.

Walter Bagehot's *Physics and Politics, or, Thoughts on the Application of the Principles of "Natural Selection" and "Inheritance" to Political Society* (1872) was one of numerous books invoking evolutionary biology to reach sociopolitical conclusions. Bagehot (1826–1877), an economist, political analyst, and editor of *The Economist*, was one of the most influential journalists of the mid-Victorian period. In *Physics and Politics*, he sought to demonstrate that throughout history the strongest nations had always dominated or conquered weaker ones. He emphasized that the progress of civilization depended upon the operation of natural selection applied to national groups. Bagehot's polemical, and sometimes contradictory, interpretation of natural selection is less significant than the clear message of his book. It was the so-called liberal democracies of nineteenth-century Europe and North America, he proclaimed, that would produce the variations required for further, higher societal evolution. Competition in the world of ideas, as well as economic and military power, would ensure the ultimate success of superior ideas and practices. Bagehot linked race to national character in a forceful piece of propaganda for the validity and political utility of the new "scientific racism" (Bowler 1989, 289). Darwin was fascinated by Bagehot's bold mix of evolutionism and politics (Desmond and Moore 1991, 557–558).

The Victorian public was treated to a rich diet of books on scientific racism. Representative titles included the Reverend John George Wood's (1827–1889) *The Natural History of Man* (1868–1870); the botanist turned science popularizer Robert Brown's (1842–1895) *The Races of Mankind: Being a Popular Description of the Characteristics, Manners and Customs of the Principal Varieties of the Human Family* (1873–1876); and Augustus Henry Keane's (1833–1912) textbooks *Ethnology* (1896) and *Man: Past and Present* (1899). The cumulative result of this avalanche of print was to promote a culturally sanctioned set of racial stereotypes that afforded politicians and their constituencies guidelines for imperialist and colonialist attitudes toward non-European peoples. Evolutionism had become one of the more potent ingredients in an edifice of racist ideology that both epitomized a dominant strand of Victorian thought and left a powerful legacy for the racism of the twentieth century. As historian Douglas Lorimer has shown,

> in assessing the influence of science in effecting this regressive change in Victorian attitudes to race, it needs to be recognized both that the community of scientists inherited the racism of the common context and that, as science gained greater intellectual authority and institutional and professional status, it came to reshape that inheritance. . . . Insofar as science came to shape popular views of race, this influence occurred not [only] as a direct consequence of the controversies of the 1860s, but [also] as a result of changes in the external and domestic context between the 1880s and the outbreak of World War I. (Lorimer 1997, 218–229)

Francis Galton's descriptions of his own African explorations typify this strand of regressionist racial stereotyping. They are replete with crude and arrogant comments about indigenous peoples whom he likened to baboons, pigs, and dogs. While Galton's Eurocentric attitudes were extremely blunt, they nonetheless reflected the reactions and conclusions of many Victorian explorers who encountered exotic peoples and cultures in their fieldwork (Fancher 1998, 108–109, 112).

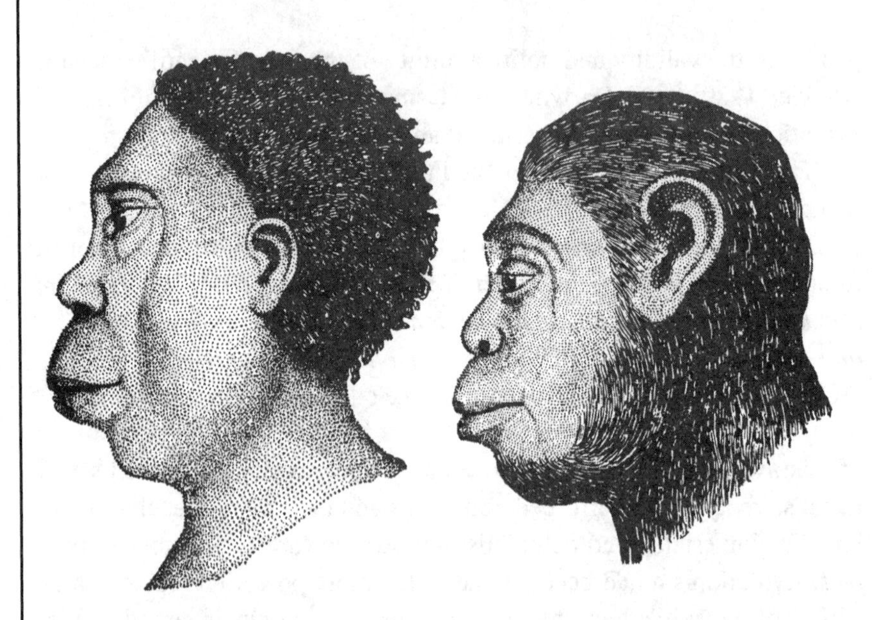

Fig. 4. "Comparison of female Hottentot with female gorilla." A blatant example from a late-nineteenth-century American anthropology text purporting to illustrate racial inferiority by exaggerating the supposed close physical similarities between blacks and primates. (From Alexander Winchell, *Preadamites* [1888])

Increasing Contact between Races

A fundamental question underlying the debate on racism was whether human races had descended from a common ancestor (monogenism) or from several radically distinct ancestral human species (polygenism). This debate was not merely one of objective scientific investigation, rather, it was situated within the problematic context of Victorian Europe's increasing contacts with non-European peoples. Ideologies of colonialism and imperialism, as well as the progressivist and hierarchical assumptions of certain strands of evolutionary theory, merged to produce a cultural climate in which pretensions to the scientific study of race became critical ingredients in political, economic, and military encounters between European and non-European peoples. Europe and its explorers, missionaries, and commercial and industrial entrepreneurs began an unprecedented experiment in bringing into close contact societies that had previously—before the age of railroads, the telegraph, and steamships—enjoyed relative isolation from each other. Policies and ideologies for guiding this integration of cultural identities and destinies became more overt and more contentious. Whether the so-called savages now swept into the European sphere of influence would be exploited or treated as joint partners in a global society was one of the major issues confronting Victorian culture (Brantlinger 1988; R. Stafford 1989; Stepan 1982).

As shown in chapter 3, Wallace attempted in his essay "The Origin of Human Races" (1864) to resolve the dispute between the monogenists and polygenists. The essay, which was regarded by Spencer, Lyell, Darwin, James, and most of the intellectual community in Britain and America as successfully settling the dispute, was not, however, restricted to the question of the origin of human races. Wallace extended the argument to include the issue of racial superiority. Even if most Victorians were satisfied that the various human races had evolved from a common ancestor, this by no means terminated debates regarding which races were superior and which inferior. Wallace suggested that those races that were exposed to harsher climatic conditions would become hardier, more provident, and more social than the races that lived in subtropical and tropical regions where food was more abundant and "where neither foresight nor ingenuity are required to prepare for the rigours of winter." Wallace appealed to history to support biology on this point. Claiming that all "the great invasions and displacements of races have been from North to South, rather than the reverse," he cited the successive conquests of the Indian peninsula by races from the northwest and the conquest of southern Europe by the "bold and adventurous tribes of the North" as proof that the inhabitants of temperate regions are always superior to the races of the tropics. The "great law of 'the preservation of favoured races in the struggle for life'" operated as inexorably in the human realm as it did throughout the rest of the natural world (Wallace 1864, clxiv).

Wallace's essay is notable for its statement of the racial superiority of Europeans. In terminology that he himself would soon abandon—but that was to become commonplace in the latter decades of the nineteenth century—Wallace asserted that natural selection "leads to the inevitable extinction of all those low and mentally undeveloped populations with which Europeans come in contact." The indigenous populations of North America, Brazil, Australia, Tasmania, and New Zealand succumbed "not from any one special cause, but from the inevitable effects of an unequal mental and physical struggle." As late as 1864, then, Wallace argued that the European race—and its descendants—would always conquer the savage races with which it came in contact "in the struggle for existence, and . . . increase at [their] expense, just as the more favourable increase at the expense of the less favourable varieties in the animal and vegetable kingdoms" (Wallace 1864, clxv). Despite Wallace's personal opposition to overt forms of racial discrimination, it was clear even from his early writings that evolutionary concepts and vocabulary could readily be appropriated by European and North American racist social theorists (Bannister 1979, 180–200; Haller 1971).

As Haller states, nineteenth-century evolutionism "rationalized and helped to justify the value system upon which the idea of racial inferiority rested in American thought." For intellectuals and the general public, the history of America revealed a sharp divide between the Caucasian and the "colored" races. This commonplace division was made by both liberals and conservatives of the nation:

> For many Americans who [overtly] shunned the stigma of racial prejudice, science became an instrument which "verified" the presumptive inferiority of the Negro and rationalized the politics of disenfranchisement and segregation into a social-scientific terminology that satisfied the troubled conscience of the middle class. To understand attitudes of racial inferiority in the context of nineteenth-century science and social science is a first step in fathoming the depth of race prejudice in our own day. Inferiority was at the very foundation of their evolutionary framework and, remaining there, rose to the pinnacle of "truth" with the myth of scientific certainty. To see racial prejudices in their scientific robes is to understand why, despite later conceptual changes in evolution and methodology, attitudes of racial inferiority have continued to plague western culture. (Haller 1971, x–xi; Feagin 2000)

Were Europeans Superior?

Wallace's background and prolonged contacts with other races in his lengthy travels predisposed him to question the moral superiority of Europeans to other human cultural groups. He could not deny that Europeans had been empowered by the scientific and technological developments of the eighteenth and nineteenth centuries, and thus placed in a position of undeniable global economic, industrial, and military supremacy. Spiritualist, theist, and, finally, socialist precepts gradually led Wallace to articulate increasingly unequivocal condemnations of the misuse of European and North American intellectual prowess. That Wallace read his paper at a meeting of the Anthropological Society of London on 1 March 1864, and published the article in that Society's *Journal*, takes on added interest since the controversies over race within the rival Anthropological and Ethnological Societies had erupted into full-scale war at that period. The Anthropologicals, a break-away group from the Ethnological Society led by the openly racist Robert Hunt, accused the Ethnologicals of being less scientific because of their attachment to the "rights of man mania" (Stocking 1987, 248–254). The Anthropologicals also advocated polygenesis as a theory that justified slavery as a policy. They thus differed from their parent society on ideological and political as well as scientific grounds (E. Richards 1989b, 373–436). Since Wallace's 1864 paper was intended, in part, to ameliorate the tense situation between the proponents of monogenism and polygenism, he did not see the opposition between the two societies as irreconcilable at this point. However, when it became apparent that the Anthropological Society was unrelenting in its doctrinaire racism and antiegalitarian politics, Wallace joined forces with Thomas Huxley, John Lubbock, and others over the next several years to limit and control the contributions of the Anthropologicals to the BAAS. Huxley's group wanted to give what they called "proper direction" to anthropology and to reunite the societies (Barton 1998, 439). These efforts paid off. Under Huxley's leadership, the two rival societies were merged in 1871 in the newly formed, and professionally more respectable, Anthropological Institute.

Respectability did not guarantee, however, that racist ideology vanished. Although a substantial number of Victorian evolutionists were opposed to stark forms of racial discrimination, a residue of Eurocentrism remained prevalent among even the most moderate thinkers. In 1866, the English intelligentsia was torn apart by the "Governor Eyre controversy." In 1865, Edward James Eyre, the governor of Jamaica, had ordered his troops to brutally crush a local peasant revolt. More than 1,000 blacks were either executed or flogged. British antislavery radicals and liberal politicians joined forces to form the Jamaica Committee to bring Eyre to justice. Darwin, Wallace, Huxley, Lyell, Spencer, and Tyndall contributed money and moral support to the prosecution effort. For Darwin, the issue was especially painful, since it reminded him of his bitter arguments against Robert FitzRoy's defense of slavery during the *Beagle* voyage (Desmond and Moore 1991, 540–541). Although the government recalled Eyre, the whole affair was a divisive one for Britain. The opponents of Eyre were matched by equally adamant defenders of the massacre. Hunt's racist clique maintained that English law did not apply to a "*naturally* wild . . . inferior race." In this vitriolic episode, the evolutionists' theories could easily backfire. The *Pall Mall Gazette* despicably pinned Huxley's "nigger" politics on to his "peculiar views" of human ancestry. The *Gazette* informed its readers that no one who saw a hairy chimp "as 'a man and a brother'" would balk at giving blacks the same "sympathetic recognition" as did Huxley (Desmond 1998, 352–353). Although England was not America, Victorian evolutionism yielded mixed messages on race (and gender) on both sides of the Atlantic. If Darwin, Wallace, Spencer, and Huxley themselves were able to extract inconsistent, even contradictory, hypotheses and opinions from their own evolutionary theories, it is hardly surprising that the general public could interpret, or misinterpret, those theories with relative ease.

Wallace's 1864 paper further implied that evolution accounted not only for the economic and military superiority of certain races but also for the preeminent status humans as a whole held within the animal kingdom. At that period when the human mind had become of greater importance than bodily structure, "a grand revolution was effected in nature—a revolution which in all the previous ages of the earth's history had had no parallel." Because humans could respond to changing environmental conditions by an advance in mental capabilities, Wallace declared that they were in "some degree superior to nature, inasmuch as [they] knew how to control and regulate her action." Wallace had always been convinced of humanity's unique status in the hierarchy of nature. His maturing evolutionary thought now encouraged him to declare that those who maintained that human attributes argued for a "position as an order, a class, or a sub-kingdom by [itself], have some reason on their side." Furthermore, humans were not merely at the summit of organic nature. The continued action of natural selection—and spiritual agencies—on their intellectual and moral characteristics destined *Homo sapiens* to an ever higher level of existence. Human physical well-being would be complemented by an ever-increasing mental and moral evolution, whose nature Wallace described ecstatically in the concluding paragraph of his essay (Wallace 1864, clxviii–clxx).

In one important respect, Darwin shared Wallace's views of the preeminent status of the human species. Although Darwin's first encounter—on his *Beagle* voyage—with the aboriginal inhabitants of the southern tip of South America at Tierra del Fuego was deeply shocking, it did not destroy his conviction that all humans constituted a single species. The sight of the naked savages who screamed from the cliffs was so strange that Darwin could scarcely reconcile their appearance with that of human inhabitants of the regions of the globe with which he was familiar. The native Fuegians seemed "like the troubled spirits of another world." Further contact with the natives, however, moderated Darwin's initial amazement. He came to realize that so-called primitive peoples could, with civilizing influences, measure up in all important respects to the nineteenth-century European. Darwin believed that all humans could progress culturally and, in time, form members of one human community. His steadfast commitment to naturalistic evolution guaranteed that the terms savage and civilized were relative ones, not absolutes. But Darwin's faith in the ultimate equality of all humans was predicated upon the bedrock of scientific naturalism. Though a vast gulf lay between civilized and uncivilized races, the possibility of bridging that gulf lay in the united, and wholly naturalistic, forces of biological and cultural evolution (Browne 1995, 239–250).

Wallace, in an ever sharpening contrast to Darwin's framework and worldview, came to conceive of human biological and cultural evolution as only partially autonomous forces. For Wallace, human evolution had, at certain specific stages, required providential intervention and guidance. The Darwin-Wallace difference with respect to human evolution and human destiny reflected a more general split among Victorians. It also foreshadowed twentieth-century debates about the role of nonmaterialist factors in human development (Turner 1974).

5

WALLACE AND DARWIN

THE MAJOR DIFFERENCES

Wallace's and Darwin's differing views on human evolution are as famous as their joint discovery of natural selection. The Wallace/Darwin divergence on various issues represents in microcosm the ambiguities evolutionary theories presented to Victorian culture generally. This chapter will focus on the contrasting paths taken by Wallace and Darwin in ferreting out what each believed were the most appropriate messages evolutionism held for illuminating that elusive concept, "human nature." The two men also disagreed on several more technical questions, including the origin of hybrid sterility, the origin of sexual dimorphism within animal and plant species, the inheritance of acquired characteristics, and the laws governing variations within species. These debates were not merely academic. They took place between very real human beings. From the 1860s onward, Darwin and Wallace developed enormous admiration for each other's intellectual vigor and genius as well as personal character. Because of these mutual feelings of sincere respect, it mattered a great deal to each to convince the other in their debates. To Darwin, for whom the opinions of those few he regarded as his intellectual peers were especially significant, the seemingly endless disagreements with Wallace over the scope and interpretation of natural selection proved to be quite distressing (Kottler 1985, 367–369, 424–427). Wallace, for his part, was disappointed by the unresolved differences with Darwin but appeared to be less personally tormented (Wallace 1905, vol. II, 14–22). If the two cofounders of natural selection theory could hold such radically opposed views on aspects of the "meaning" of evolutionism, it is little wonder that Victorian society at large was divided so strongly on the cultural implications of the nineteenth-century revolution in biology.

1869: Wallace Shocks the Darwinian Camp

Wallace's perceived public position as a foremost proponent of the thesis that human evolutionary history could be reconstructed solely on the basis of natural selection was altered abruptly in April 1869. In a review of two new editions of geological treatises by Charles Lyell, Wallace announced that human intellectual capacities and moral qualities—unique phenomena in the history of life—were not explicable by natural selection. Rather, these, as well as certain physical attributes of the human race, required the intervention at appropriate stages of "an Overruling Intelligence" that "guided the action of those laws [of organic development] in definite directions and for special ends" (Wallace 1869a, 359–394). Wallace's response to Lyell's tenth edition of *Principles of Geology* (1867–1868) is doubly significant. First, Wallace applauded Lyell's long-awaited public endorsement of evolutionism. Second, Wallace's announcement of his own explicit views parallels (though for different reasons) Lyell's frequently expressed extreme reservations concerning natural selection and human evolution. Darwin was, not surprisingly, disappointed with both Wallace and Lyell (Bartholomew 1973).

Darwin believed he had demonstrated that the great stumbling block for evolutionism, the origin and development of the higher human attributes, could be accounted for by natural selection. In *Origin*, he had suggested how *altruism*, or behavior that benefits others even if detrimental to the individual, would arise and persist among members of a given species. Darwin focused on the case of neuter, or sterile, insects. This example is highly relevant to the problem of altruism, because such insects represent the extreme case of the loss of individual reproductive fitness—the very essence of natural selection. Darwin showed that the contradiction was easily resolved. Insects that are born sterile are capable of devoting their energies not to reproduction but to work for the community—the colony or hive. Darwin argued that the sterility of certain castes of "worker insects" enhanced the fitness—and reproductive success—of the entire species. Sterility was thus a result of natural selection acting at the group level. By extending the scope of evolutionary analysis from the individual to group level, Darwin was able to explain the development of altruistic traits in higher animal species, such as intentional (rather than programmed) cooperation and sympathy, by natural selection (Rosenberg 1992, 19–28). Darwin did not publicly expand this argument to include the evolution of human moral and social behavior until the appearance of *Descent of Man* in 1871. But he had long believed that the higher human traits, such as sympathy and sacrifice, could be explained in the same manner as physical traits.

That it was Darwin's *principle of utility* that Wallace invoked to substantiate his own claims for the limits of natural selection only added to Darwin's discomfiture. In *Origin*, Darwin had argued that natural selection could never produce a structure harmful to an organism. Nor could it produce a structure that was of greater perfection than was necessary for an organism at that particular stage of its evolutionary development (Darwin 1859, 201–202). Citing the culture of the "lowest savages"—and, by implication, humans at more remote periods in their evolutionary history—Wallace maintained that the utility principle precluded natural selection as the agent responsible for four characteristic human features: the brain, the hand, the organs of speech, and the external form of the body. The brain of savages, Wallace noted, was of practically the same size and complexity as that of the average European. It could, under appropriate cultural conditions, be capable of the outstanding intellectual achievements of civilized man. Yet, the mental requirements of the lowest savages are "very little above those of many animals" (Wallace 1869a). Wallace concluded that the savage's, and early humans', highly developed brain must be regarded as an organ of greater perfection than necessary for survival. Natural selection, he argued, using the utility principle, would have provided early humans with an intellect only slightly superior to that of the apes. It cannot, Wallace declared, be the factor that explains the complexity of the early human brain.

The hand of the savage is, Wallace continued, similarly an organ of greater refinement than required and could not have been produced by natural selection alone. But humanity's highest civilized accomplishments—art, science, and technology—were dependent upon "this marvellous instrument." For Wallace, the savage's already perfect hand is evidence of provision by a higher intelligence of an organ that would be fully utilized only at a later stage in human development. The erect posture of the savage (and prehistoric man), "his delicate and yet expressive features, the marvellous beauty and symmetry of his whole external form," are additional examples of modifications Wallace claimed were of no physical use to their possessors. He argued that the absence of extensive body hair, compared to other higher primates, was disadvantageous for human evolutionary survival in harsh environments. This comparative human nakedness was, thus, inexplicable by natural selection. Wallace argued again for intelligent intervention and provision in the evolutionary process. "The supreme beauty" of the human form and countenance, though initially of no practical use, had (probably) been the cause of human aesthetic and emotional qualities, which Wallace believed could not have arisen if early humans had retained the appearance of an erect gorilla. Wallace further suggested that human nakedness, "by developing the feeling of personal modesty, may have profoundly affected our moral nature." The same arguments applied to the complex and delicate physical and mental apparatus responsible for human speech. Wallace regarded these as instruments developed in advance of the immediate needs of their possessors (Wallace 1869a).

Wallace's 1869 review concluded with the proposition that a "new standpoint [was now possible] for those who cannot accept the theory of evolution [by natural selection] as expressing the whole truth in regard to the origin of man." Wallace was careful to declare that the higher intelligence, whose action he had invoked to explain that which natural selection could not, was fully compatible with scientific methodology. Using the analogy of domestic variation—the same analogy he had criticized Darwin for using so extensively in *Origin*—Wallace stated that just as man had used the laws of variation and selection to produce fruits, vegetables, and livestock, so also "in the development of the human race, a Higher Intelligence has guided the same laws for nobler ends." In both cases, the "great laws of organic development" had been adhered to, not nullified. Natural selection had been supplemented by conscious selection. In human evolution, Wallace concluded, "an Overruling Intelligence has watched over the action of those laws so directing variations and so determining their accumulation, as finally to produce an organization sufficiently perfect to admit of and even to aid in, the indefinite advancement of our mental and moral nature" (Wallace 1869a).

Wallace and Spiritualism

A majority of historians of science have interpreted Wallace's 1869 announcements as representing a volte-face with respect to his previous conceptualization of evolution. This presumed radical shift is usually attributed to Wallace's growing involvement with, and ultimate conversion to, spiritualism in the period from 1865 to 1870. If, however, his thoughts and writings from 1845 to 1870 are analyzed within the broader framework of Wallace's holistic approach to human evolution, a different picture emerges. From the outset of his career, Wallace worked within a context that incorporated philosophical, ethical, sociopolitical, as well as biological interests and investigations. Wallace's modifications of certain causal explanations of human evolution are developments from, rather than repudiations of, his earlier and preliminary hypotheses. Using geological metaphor, Wallace was not an intellectual catastrophist but an intellectual uniformitarian.

If it is argued a priori that spiritualism and evolutionary biology represent mutually exclusive conceptual schemes, then of course Wallace's acceptance of spiritualism would necessarily be seen as the cause of his rejection of natural selection in explaining man's higher faculties. If, however, it is recognized that "the fundamental principles of Wallace's approach to the study of man/nature were set in his mind well before he finally [hit upon] natural selection, and given the fact that he repeatedly re-affirmed his belief in those principles in his writings over a span of seventy years . . . it is extremely difficult to believe that either natural selection or spiritualism had any profound effect on re-directing them. His relation of the two ideas is the product of his personal evolution of thought, not its cause." Quite simply, Wallace never thought that natural selection alone was competent to explain the evolution of human higher faculties. He, and many other Victorians, had always envisioned some additional explanatory model to fully resolve the question of human origins, their higher faculties, and their future evolution (Charles Smith 1992, 1, 19–20, 43–49).

Spiritualism and natural selection were not viewed by Wallace as mutually exclusive explanatory pieces of a larger evolutionary teleology. He certainly appreciated the fact that many of his scientific colleagues, notably Darwin, Huxley, Tyndall, and the eminent physiologist William Benjamin Carpenter, would regard them as such. Carpenter's *Principles of General and Comparative Physiology* (1839), which denied any miraculous tinkering in nature, was highly regarded by Huxley and other advocates of scientific naturalism (Desmond 1998, 180; Desmond 1989, 210–222). For tactical reasons, then, Wallace chose to emphasize the utilitarian objections against the total efficacy of natural selection in the 1869 review. He hoped they would be read as a scientifically more respectable analysis of the limitations of natural selection than an overtly spiritualist critique would have been. Wallace's review was aimed more at the professional scientific community than the general public. The Victorian public, in any event, was often sympathetic to accounts of phenomena drawn from phrenology and mesmerism, in addition to spiritualism and alternative medicine (Cooter 1984; Winter 1998; Bynum and Porter 1987).

In this respect, Wallace's strategy succeeded. A number of biologists, already dubious that natural selection was the sole or even primary mechanism of evolution, recognized the force of Wallace's utility critique (Bowler 1983, 28). As discussed previously, there were major rivals, such as neo-Larmarckism, to Darwinism in the Victorian period. Wallace also realized that certain, but not all, of his fellow scientists would be unresponsive to arguments drawn from the then controversial data of psychic phenomena, and he worded his 1869 review accordingly. For the next twenty years, to the dismay of the hard-core Darwinists, Wallace publicly held to the position that a utilitarian analysis was a major basis for his critique of the scope of natural selection in human evolution. But he gradually introduced theism, in addition to conventional spiritualism, to demonstrate the inadequacy of natural selection as the explanation for certain uniquely human characteristics. In one of the many ironies of Victorian evolutionism, Wallace remained throughout the rest of the century the staunchest advocate of the total sufficiency of natural selection as the agent of animal and plant species evolution. The increasingly teleological and theistic imprint upon Wallace's evolutionary theory was always most apparent in his views on human origins and development. In his major exposition of natural selection, *Darwinism* (1889), when Wallace conceded that natural selection could account for many of the *physical* features of man, he still rigorously exempted human moral and intellectual qualities from its sway. But this exemption had its roots in Wallace's earliest evolutionary speculations of the 1840s.

Wallace and the Evolutionary Theists

Wallace's theistic interpretation of certain parts of the evolutionary puzzle did not make him an anomaly in the British scientific community. Important groups existed within the ranks of professional scientists in the later

Victorian period who fully endorsed the notion that there was an integral religious dimension to science. Some were physicists. Noteworthy among them were the "North British" physicists, who included such titans as William Thomson, James Clerk Maxwell, and Peter Guthrie Tait. Many were biologists, including the Christian Darwinists. There were also eminent idealist natural philosophers, such as T. H. Green, F. H. Bradley, and Edward Caird, who appropriated science and evolutionary theory in constructing theistic metaphysical systems. Members of these groups shared certain of Wallace's interests in theistic and spiritualist matters and would not have considered it unscientific to do so. When the topography of later Victorian British science is viewed not merely from the perspective of Huxley and the scientific naturalists but from the broader intellectual landscape, theistic science is seen as a powerful paradigm in that era (Crosbie Smith 1998; Moore 1979; Den Otter 1996).

In the United States, prominent scientists such as Asa Gray, Joseph LeConte, and James Dwight Dana were widely regarded as providing scientific evidence that the elements composing human personality were "too numerous and too peculiar to have come in by slow increments," in other words, by the process of natural selection. Many American Protestant evolutionists maintained that the "biological solution does not exclude the theological." Wallace's emerging position in the 1870s and beyond was viewed as buttressing their contention that the hypothesis of evolutionary descent served to reinforce the conviction "that humanity bore the image of God." These American scientists regarded evolutionary theism as confirming the Christian belief "that the elements attesting to the special relationship between God and human beings resided in the fact that the human species possessed attributes—self consciousness, reason, the moral sense, free will, and religiosity—that were different in kind from those of all other animals" (Roberts 1988, 176–177).

The cult of science did not render theistic evolution less significant, even if it did redirect the manner in which religious beliefs were expressed. As noted historian Ronald L. Numbers has recently suggested, "theistic evolution was undergoing privatization more than elimination." Although overt references to the divine became less visible in the scientific literature toward the close of the Victorian period, many American evolutionists retained their religious views (Numbers 1998, 40). The American Protestant theologian James McCosh—a president of the College of New Jersey (later Princeton University) and prolific author of works on the harmony of science and Christian faith—was among those who regarded Wallace as having provided scientific arguments for limited divine intervention in the course of human evolution (McCosh 1890 [1888], ch. 6). In dissociating himself from a completely materialist evolutionary naturalism, Wallace was joining Lyell, Gray, and a substantial group of evolutionists who incorporated theistic and teleological elements into their scientific theories (Hull 1973, 64–65).

The 1869 review represents a public watershed in Wallace's career. It was, as Darwin noted, an "inimitably good" exposition of natural selection, but one that concluded with those few remarks on humans that made him "groan." Wallace, in turn, fully appreciated Darwin's (and others') reactions with "regard to my 'unscientific' opinions as to Man." He wrote Darwin that he would "look with extreme interest for what you are writing on Man [Darwin was preparing his *Descent of Man* for publication], and shall give full weight to any explanations you can give of his probable origin" (Marchant 1916, 200).

Wallace's depiction of his views as "unscientific" opinions is both intentionally ironic and historically significant. Wallace had never deemed unaided natural selection to be a sufficient mechanism for human evolution. His tracts on spiritualism in the several years prior to the Lyell review emphasized further the need to posit auxiliary agencies to explain human higher faculties. Wallace's theistic rendition of evolutionary theory is germane to current studies in the historiography of science and religion. Within the context of the rich, ambiguous, and contested Victorian philosophies of nature, a demarcation between the categories of science and religion was (and is) difficult, if not impossible, to fix precisely (Fichman 1997). Wallace was articulating the personal as well as metaphysical tensions induced by the striking rise of institutionally professionalized science during the late nineteenth century. His views on human evolution were made clearer still the following year with the publication of *Contributions to the Theory of Natural Selection* (1870).

"Limits of Natural Selection"

The final essay in *Contributions*, "The Limits of Natural Selection as Applied to Man," elaborated upon the arguments sketched in Wallace's 1869 review. Wallace made his philosophical commitment to an evolutionary teleology more explicit. In rejecting a completely materialistic version of evolution, he focused on two phenomena—the origin of consciousness and the development of man from the lower animals—to demonstrate that certain human characteristics cannot be explained by natural selection. Wallace started from the then widely accepted premise that the size of the brain, "universally admitted to be the organ of the mind" (Wallace 1891, 188), is proportional to mental capacity. Citing evidence from Huxley and the anthropologists Pierre Paul Broca and Sir John Lubbock, Wallace emphasized that the brain size of prehistoric man and many of the lowest savages is wholly comparable to that of modern Europeans. This "apparent anomaly," he continued, suggests the idea of "a surplusage of power—of an instrument beyond the needs of its possessor" (Wallace 1891, 190). The fact that all the higher intellectual (and moral) faculties do occasionally manifest themselves in the primitive state, Wallace argued, indicates their latency in the large brain of savages and early humans. He declared that the theory that the "savage's" large brain exceeds its actual requirements was substantiated by comparative anatomy. Wallace noted that certain of the higher animals, with far smaller brains, exhibit behavioral traits similar, if not identical, to those of early humans. Wallace included in this category the ingenuity of the jaguar in catching fish, the hunting in packs of wolves and jackals, and the placing of "lookout guards" by antelopes and monkeys.

Wallace and Darwin were now using identical data to support quite different evolutionary conclusions. The evidence of continuity in psychological and behavioral processes from the higher animals to early man had provided Darwin with some of the most crucial support for his theory of human evolution by natural causes only. In Wallace's hands, the data became powerful support for his own claim that the large brain of early humans was "prepared in advance, only to be fully utilised as he progresses in civilisation." The brain, Wallace concluded, "could never have been solely developed by any of those laws of evolution, whose essence is, that they lead to a degree of organisation exactly proportionate to the wants of each species, never beyond those wants." Aside from Darwin, Wallace had another target: Huxley's controversial article "On the Physical Basis of Life" (Wallace 1891, 186–193, 206–207, 212). Wallace sought to refute Huxley's assertion that "thoughts are the expression of molecular changes in that matter of life which is the source of our other vital phenomena" (Wallace 1891, 207). Wallace drew a distinction between the ontological status of life—"the name we give to the result of a balance of internal and external forces in maintaining the permanence of the form and structure of the individual"—and consciousness. He granted Huxley that life may conceivably be regarded as the result of "chemical transformations and molecular motions occurring under certain conditions and in a certain order." But, Wallace declared, no combination of merely material elements, no matter how complex, could ever produce the "slightest tendency to originate consciousness in such molecules or groups of molecules" (Wallace 1891, 209–210).

Wallace held matter and consciousness to be "radically unlike, exclusive, and incommensurable." Moreover, the presence of consciousness in "material forms is a proof of the existence of conscious beings, outside of, and independent of, what we term matter" (Wallace 1891, 207–210). In an adroit stroke, Wallace argued that Huxley's reductionism and adherence to traditional materialism was actually inconsistent with "the most recent [scientific] speculations and discoveries as to the ultimate nature and constitution of matter." Wallace invoked the theory—then current in physics—that what is commonly called matter is actually an arrangement of centers of attractive and repulsive force. Wallace asserted that the special properties of matter, (e.g., electrical, chemical, magnetic) can be explained on the basis of the interaction between these force centers. In repudiating the materialist solution to the problem of consciousness—that all matter is conscious—Wallace declared matter itself to be "essentially force, and nothing but force." Moreover, the various forces in nature, of which matter and consciousness are but different manifestations, may be ultimately reducible to "will-force; and thus the whole universe is not merely dependent on, but actually is, the WILL of higher intelligences or of one Supreme Intelligence" (Wallace 1891, 207–212).

Wallace's evolutionary theory was a complex construction. He was at once the most effective advocate of natural selection as the primary mechanism of species development (save in the case of humans) *and* a formidable opponent of a completely materialistic evolutionary naturalism. No aspect of Victorian evolutionism was more sensitive to the play of ideological forces than that which dealt with humans, particularly the evolution of moral and intellectual attributes. The intense public interest and controversy engendered by *Origin* could hardly have arisen if the question of human development from the lower animals was not perceived as an inextricable component of the theory of natural selection (Ellegard 1958, 332). Wallace's own views on human evolution could scarcely be ignored, and "Limits of Natural Selection" drew immediate criticism, as well as support.

Ironically, Wallace was chided by both Darwinians and their opponents. Darwinians objected to his spiritualist interpolations, although they could not effectively repudiate all of Wallace's utilitarian critiques of the complete sufficiency of natural selection in human development. Evolutionary explanations of the origin and development of the moral faculties remain a subject of debate among biologists in our own day (Wilson and Dugatkin 1992, 29–33; Cronin 1991). Opponents of evolutionary naturalism, while receptive to Wallace's position on human origins, felt that he still accorded too great a power to natural selection in the plant and animal kingdoms (Kottler 1974, 157–159). Darwin had, perhaps, the most accurate, and generous, assessment of his and Wallace's differing perceptions of human evolution. In a letter dated 20 April 1870—when Wallace's views had just been presented publicly—Darwin wrote to his friend and colleague, "I hope it is a satisfaction to you to reflect—and very few things in my life have been more satisfactory to me—that we have never felt any jealousy towards each other, though in one sense rivals. I believe that I can say this of myself with truth, and I am absolutely sure that it is true of you" (Marchant 1916, 206–207).

Darwin, Wallace, and Evolutionary Images of Gender

The notion of prudish, sexually repressed Victorians, who cautiously guarded themselves against any temptation, no matter how slight, is a familiar one. In one transparent sense this stereotype is accurate. A number of prominent thinkers of the period promoted such views for a variety of religious, social, political, and scientific reasons. However, recent scholarship of the actual as opposed to stereotypical behavior of Victorians has successfully questioned this conception and proven it fundamentally inaccurate. During the nineteenth century, even in seeking the apparently conservative happy ending of marriage, Victorian men and women were inevitably led to the consummation of their love. Sex and sexuality, then, were unavoidable issues for the Victorians. Evolutionary theories both shaped and mirrored the complex attitudes toward sex and gender. The evolution of sexual reproduction and of sex differences were central topics confronting biologists investigating the development and behaviors of animal and plant species. The human species was no exception.

It wasn't until the early 1900s that scientists connected sex chromosomes to sex-linked characteristics or that they discovered the workings of hormones. Thus, "We [begin] to see why for some forty years [prior to 1900] the exact nature of sex differentiation and its psychic accompaniment was a subject of intense, though inconclusive debate" (Conway 1970). The publication of *Origin* in 1859 was one factor prompting a scientific analysis of what exactly differentiated men from women mentally, emotionally, and physiologically. Why the human species evolved into two sexes was a fundamental question that both intrigued and confounded leading evolutionary theorists such as Herbert Spencer and Scottish biologist and sociologist Patrick Geddes (1854–1932) as well as Darwin and Wallace. What emerged most frequently from this inconclusive, but culturally potent, set of investigations and speculations was some version of a *dyadic*, or twofold/bipolar, model. According to the dyadic model, in addition to the obvious differences in sex organs and physical attributes, men were considered over women to be the dominant, active agents in society. Males, it was argued, expended energy while women were sedentary, storing and conserving energy. Many Victorian evolutionists believed that these feminine and masculine attributes traced back to the metabolic processes of the lowest forms of life. An evolutionary dichotomy was posited to define the categories "feminine" and "masculine" biologically. Females were thought to have evolved with a synthesizing and nurturing (*anabolic*) constitution, which conserved energy. Males, in contrast, were characterized by a *katabolic* physiological and psychological nature, whose essence was to release physical and intellectual energy in a variety of ways.

Such evolutionary speculations laid the groundwork for, and drew support from, the notion of separate spheres of thought and activity for men and women. According to the dyadic model, since men only concerned themselves with fertilization in the reproductive process, they could also spend energies in other arenas. This accounted, in Spencer's words, for "the male capacity for abstract reason . . . along with an attachment to the idea of abstract justice . . . [which] was a sign of highly-evolved life" (Spencer quoted in Conway 1970). Conversely, woman's prolonged role in pregnancy, menstruation (considered a time of illness, debilitation, and temporary insanity), and child rearing left her very little energy for other pursuits. As a result, these theorists announced that women's roles in Victorian society were the outcome of biological evolution. A (middle class?) woman had to stay at home in order to conserve her energy. Men, on the other hand, could and needed to go out and hunt and forage or, more typically in industrializing Europe, work in and direct factories, businesses, armies, and empires. Such evolutionary reasoning also provided justification for the presumed emotional and mental differences between men and women. This logic led Geddes, whose views were typical of many (male) evolutionary theorists, to assert that "Male intelligence was greater than female, men had greater independence and courage than women, and men were able to expend energy in sustained bursts of physical or cerebral activity. . . . Women on the other hand . . . were superior to men in constancy of affection and sympathetic imagination. . . . [They had] greater patience, more open-mindedness, greater appreciation of subtle details, and consequently what we call more rapid intuition" (Conway 1970, 47–62).

Having thus defined the sociocultural roles of men and women, Victorians still had to deal with the sexual act itself. The dyadic model held sway in this domain as well. In the earlier decades of the nineteenth century, women were usually considered the weaker, more innocent sex. According to conventional ideas, females had little to no sexual appetite, often capturing all the sympathy and none of the blame over indiscretions. Men represented the fallen, sinful, and lustful creatures, wrongfully taking advantage of the fragility of women. This situation was gradually, but strikingly, transformed in the latter half of the century.

According to the dyadic model of evolution, individual women had to be held accountable for sexual indiscretions, since this was not biologically "normal" for females. Men, slaves to their katabolic purposes and sexual appetites, could not really be blamed for their indiscretions. A double standard was being sanctioned by science. Women were frequently portrayed in the medical literature as either frigid or insatiable. A young lady was only worth as much as her chastity and appearance of complete innocence, for women were sexual time bombs just waiting to be set off. Once led astray, she was the fallen woman, and nothing could reconcile that until she died. This was, it must be emphasized, the emerging orthodoxy of evolutionary biology and medicine. Certain individuals or groups, most notably artists, poets, dramatists, and novelists, did not accept such strict roles in depicting men and women in either their sexualities or their contributions to sexual intercourse. The dyadic model for men and women permeated the age, but only served to try to encourage a specific ideal. In realistic life situations, as well as in art and fiction, Victorians could recognize the complexities of sex and gender and the broad spectrum of actual roles males and females played (Conway, Bourque, and Scott 1989; Fee 1974, 86–102; C. Parker 1995; Bashford 1998).

Darwin's *Descent of Man*

Both Darwin and Wallace explicitly drew gender implications from evolutionary theory. *The Descent of Man, and Selection in Relation to Sex*

Fig. 5. "Female Dentistry ("It's *nearly* out; but my Wrist is so Tired that I must really *Rest* a bit!"). An 1879 cartoon from the satirical magazine *Punch*, demonstrating a recurrent Victorian theme that the incongruity of fragile and attractive women tackling "masculine" jobs rendered women doctors and dentists, for example, comic or amusing. (From *Punch* 77 [1 November 1879]: 203)

(1871) is essentially a sequel to *Origin* in terms of expanding upon human evolution. In *Descent*, Darwin wrote, "The chief distinction in the intellectual powers of the two sexes is shown by man's attaining to a higher eminence in whatever he takes up, than can woman—whether requiring deep thought, reason, or imagination, or merely the use of the senses or hands." Darwin dismissed those aspects of intelligence traditionally attributed to women, such as intuition and imitation, as "characteristic of the lower races, and therefore of a past and lower state of civilization." The intellectual as well as physical differences between the sexes, he argued, were the product of the continued action of natural selection and its auxiliary, sexual selection. Darwin posited that during the long course of human evolution the male struggle for the possession of the female (sexual selection that emphasized male combat), as well as such activities as hunting and the defense of females and children, inevitably resulted in continually heightened intelligence. "Thus," Darwin concluded, "man has ultimately become superior to woman" (Darwin 1871, vol. II, 326–329).

The publication of *Descent* put Darwin's powerful authority behind the evolutionary ratification of patriarchal Victorian values. His emphasis on the "natural" supremacy of white, middle- and upper-class males found a highly receptive audience in the context of imperial expansion, economic transformation, urban and industrial unrest, and the emergence of socialist working-class movements all over Europe. Females who were agitating with increasing urgency for suffrage, access to higher education, and entrance to various professions also had to contend with the message of *Descent*. By asserting the instinctively maternal and inherently modest traits of the human female and the male's innate aggressive and competitive characteristics, Darwin's was a formidable voice in gender debates (E. Richards 1997). He was regarded by many as establishing naturalistic corroboration of woman's narrow domestic role and contemporary gender as well as social inequities. Most evolutionists followed and reinforced Darwin's lead. Spencer, for instance, opposed the extension of the franchise and higher education to women primarily on the grounds that they were less highly evolved than men, and correspondingly less fit to handle political and professional responsibilities. Darwin was known to be intimidated by intellectual women, whom he scorned as "bores." He, and many of his colleagues and disciples, had scant use for John Stuart Mill's emancipationist manifesto *The Subjection of Women* (1869), which argued the case for female suffrage. Darwin patronizingly declared that Mill "could learn some things from biology," most notably female inferiority and ineradicable sexual difference (Desmond 1998, 447; Desmond and Moore 1991, 572).

The "Woman Question"

Other prominent Darwinians such as George John Romanes, Francis Galton, and Geddes joined forces with anthropologists, psychologists, and gynecologists to forge a formidable body of biological determinist theory. They purported to show that women were constitutionally different from men in their anatomy, physiology, temperament, and intellect. Galton met, and was polite to, important females such as novelist George Eliot, social reformer Beatrice Potter Webb, and the legendary Florence Nightingale. But he clearly never regarded them in any way as intellectual equals (Fancher 1998, 99–115). By arguing on ostensibly objective scientific grounds that women were inferior to men, as well as equating women with the "lower" races, Darwin's reconstruction of human evolution reinforced pervasive Victorian racial and sexual stereotypes.

Put bluntly, women could never hope to match the intellectual achievements of men, nor obtain an equal share of socioeconomic and political power and authority. As a number of feminist scholars have documented, Victorian science—and evolutionary theory in particular—was strongly gendered. Opposition to scientifically sanctioned patriarchy, however, was also present. Recent studies have begun to focus on those individuals and institutional alliances, largely but by no means exclusively female, which sought to counter the prevailing set of ideas and practices about gender, sexuality, and science. Specific individuals, including a small number of female scientists, succeeded in making preliminary advances into white, middle-class territories, but their efforts were the exception rather than the rule (E. Richards 1997, 119–122).

A notable exception to the Darwinians' evolutionary endorsement of masculine superiority was Darwin's eloquent public defender and propagandist, Thomas Huxley. An otherwise stalwart Darwinian, Huxley refused to make women's biological "limitations" the basis for a discriminatory educational policy. He wondered why careers "open to the weakest and most foolish of the male sex should be forcibly closed to women of vigour and capacity." Huxley became a modest, but influential, advocate for higher education for women. He was instrumental in convincing the Senate of London University, at which he was a professor, to start granting degrees to women in 1874 (Desmond 1998, 447–450). But Huxley's feminism had limits. He and other members of the Darwinian-dominated Ethnological Society succeeded in establishing a policy that excluded women from its regular professional meetings, which were for "scientific" discussions. Women could be admitted on special occasions, but only at the larger, general meetings open to the public—and by invitation only.

Significantly, a number of women intellectuals endorsed the same biological determinist positions on the "woman question" as did the Darwinians. And they did so for the same evolutionary rationale: females were constitutionally unsuited to rival males in the quest for high academic and intellectual posts. Eliza Lynn Linton (1822–1898) was a successful journalist and ardent Darwinian. Yet she campaigned vehemently against birth control and women's higher education, suffrage, and entry into the emerging professions, including science (E. Richards 1997, 122,

127–128; Jann 1994, 287–306). Clearly, paradoxes abounded in the Victorian embrace of evolutionism, particularly with respect to gender issues (Shteir 1997, 236–255; Benjamin 1991).

Wallace and Female Sexual Selection

Wallace was one of the most prominent evolutionists to propose an alternative account of gender and sexual relationships. His break with the prevailing Victorian orthodoxy on women occurred later in his career, after he had become an advocate of socialism. As indicated previously, Wallace was deeply impressed with the writings of the American Edward Bellamy. A socialist, Bellamy's utopian novels *Looking Backward* (1888) and *Equality* (1897) depicted an egalitarian future state in which socialism served to eradicate the economic and cultural barriers that held Victorian women in check. Taking his cue from Bellamy, Wallace looked anew at the hypothesis of sexual selection in human evolution. He developed a version of sexual selection that emphasized the potential of female choice, rather than male aggressiveness, in bringing about a new social order. Wallace first proposed his views on gender equity in two articles published in the early 1890s. In them, he argued that in human society, female choice of reproductive partners could become the most effective way of ensuring evolutionary progress. Wallace believed that socialism, by eradicating the dangerous conditions of labor under capitalism, would significantly lower the rate of male mortality relative to females (since men were more generally employed in dangerous occupations). The observed excess of females in the general population during the ages of most frequent marriage (from twenty to thirty-five years)—despite the statistically higher percentage of male births—would be neutralized.

In the monogamous socialist state Wallace envisioned, female choice could become dominant. Female sexual selection, operating upon an excess of males (or at least not a minority) would, Wallace argued, be biologically decisive. The greater option of female celibacy (possible because of financial independence) would augment the rigor of sexual selection.

Wallace declared that those individuals "who are the least perfectly developed either mentally or physically . . . or who possessed any congenital deformity [or tendency to hereditary disease] would in hardly any case find partners, because it would be considered an offence against society to be the means of perpetuating such diseases or imperfections." Such individuals, he was careful to point out, would not be deprived of the ability to lead contented lives but only of the ability to transmit their defective traits to any offspring. For Wallace, such a society was not only desirable but biologically plausible. In his interpretation of human evolutionary history, the cultivation of sympathetic feelings "has improved us morally by the continuous development of the characteristic and crowning grace of our human, as distinguished from our animal nature" (Wallace 1890, 325–337; Wallace 1892, 145–159). A foe of eugenics, Wallace hoped that female selection of the healthiest reproductive partners would be accepted by both the successful and unsuccessful in marriage. He believed that the congenitally defective or otherwise less able members of society would *voluntarily* concur in their removal from the reproductive pool in the nobler interests of society's future. Wallace's utopian vision of female sexual selection under socialism clearly held the potential for certain types of discrimination. These reproductive issues have lost none of their highly controversial implications down to the present day.

Wallace's combination of socialism, female choice, and biology was not unique. Aside from Bellamy, a number of feminist writers, notably the American Charlotte Perkins Gilman, also utilized sexual selection within a socialist framework. In *Women and Economics* (1898), Gilman agreed with Darwin that sexual selection was a force in human evolution. But she declared that under capitalism it served not to improve the species (as Darwin held) but to weaken it. Gilman, like Wallace, no longer separated economic factors from biological ones. In the Victorian marriage market, woman's dependence on man for subsistence nullified any genuine, and thus biologically efficacious, free female choice. The solution to this dilemma was clear to Gilman: Send women to work in the world beyond the middle-class home. Economic equality and independence were the keys to terminating capitalism's sexual exploitation (in many senses) of women, thereby restoring to females the evolutionary potential to make, in Gilman's words, "their rightful contribution to the future of the race" (Russett 1989, 84–86). In a like vein, the socialist Eliza Burt Gamble argued in *The Sexes in Science and History: An Inquiry into the Dogma of Woman's Inferiority to Man* (1894) that under capitalism, women had become "economic and sexual slaves . . . dependent upon men for their support" and dispossessed of their "fundamental prerogative" of aesthetic choice. Gamble's book is considered a major nineteenth-century rebuttal of Darwinian arguments for the continuing inferiority of women. She, like Wallace, envisioned a noncapitalist future, when women would regain their rightful power of sexual selection. For Gamble, as for Wallace, women, through the transmission of their "more refined instincts and ideas" to their offspring, would usher in a "new spiritual age" (E. Richards 1983, 110, n. 155).

The major differences in key areas between Darwin and Wallace are clear. These differences, however, do not obscure their common goal of using the insights of evolutionary biology to bring about a society that would best meet the challenges posed by the profound scientific and cultural upheavals of the Victorian age. Darwin and Wallace had variant images of evolutionary progress. But both of them, as did so many of their contemporaries, looked to evolutionism for new answers to age-old human dilemmas.

6 EVOLUTIONARY ETHICS

Since evolutionism became part of the fabric of nineteenth-century culture, it necessarily was drawn into the domain of ethical discourse. At first sight, it would seem to make perfect sense to think that evolutionary biology has something to offer in the realm of ethics. If humans have evolved over long periods of time to become what they are partly through the force of natural selection and other biological influences, then our biological history played an important role in developing human nature. Since all ethical systems rest, either explicitly or implicitly, on conceptions of human nature, the very foundation of ethical thinking would then seem rooted, to some degree, in biology. At one level, this is scarcely controversial. Enlightenment thinkers such as Denis Diderot and Julien Offray de La Mettrie demonstrated that biological factors—health, illness, diet, and medication, among others—clearly affected an individual's moral perceptions. During the Victorian period, the growing evidence for human evolution rendered appeals to biology to clarify at least some aspects of the foundations of ethical systems even more compelling. To a growing segment of Victorian society, biology appeared as a useful tool with which to explore age-old questions of right and wrong. What was controversial was whether biology had anything truly novel or philosophically significant to say about ethics, beyond merely obvious corroboration of traditional notions of morality (Bradie, 1994, 11).

Darwin himself believed that evolution did provide a powerful and novel key for understanding humanity's ethical sense. Human morality, he argued, should be explained in ways analogous to the evolution of the hand or eye. All were evolutionary adaptations of developing humans to environmental pressures: products of natural selection. Although Darwin's own position is actually far more complex and nuanced, his was a powerful voice for those nineteenth-century—and later—proponents of evolutionary ethics. He declared that "morals and politics would be very interesting if discussed like any branch of natural history." With less modesty than was usually characteristic of him, Darwin noted that the central question of the origin of our moral sense itself "has been discussed by many writers of consummate ability; [but] . . . no one has approached it exclusively from the side of natural history." He attempted to rectify this shortcoming in his *Descent of Man* (1871), a work as controversial as *Origin of Species*. As suggestive and eloquently argued as *Descent* is, Darwin's attempt to construct a naturalistic account of the origin and diversity of ethical systems raised as many questions as it purported to solve. Humans' ill treatment of each other throughout history—war, aggression, murder, greed—found ready explanation in the struggle for existence. Human morality, in contrast, presented a far greater challenge to evolutionary theory. How did such traits as compassion, altruism, benevolence, self-sacrifice, and generosity emerge from the "law of the jungle"? Darwin proposed a number of persuasive explanations of such instances of higher moral conduct. But they were only persuasive, not conclusive. Accordingly, Darwin's posited solutions to the enormously difficult questions of ethics evoked considerable criticism (Cronin 1991, 325–326). They also raised many people's blood pressure. If Darwin was on the right track, what would become of the traditional answers to ethics provided by religion and philosophy?

The siren call of the "temptations of evolutionary ethics" (in historian Paul Farber's apt phrase) repelled as many Victorians as it seduced. The claim that biology holds the key to human nature and morality was an explosive one in the nineteenth century. More pointedly, the assertion by numerous writers in the Victorian era—as well as in our own day—that "the theory of evolution provides the foundation for a deeper analysis of human action than anything available in the humanities or social sciences" necessarily evokes both admiration and hostility. At stake is the crucial question of whether, and to what degree, biology helps us to better comprehend individual and social human motivation and responsibility. This question is part of a broader one: Can scientific knowledge aid us in constructing a viable moral or political vision (Farber 1994, 1–2)? This chapter will explore some of the most famous Victorian answers to such a notoriously complex set of issues. Focusing on Darwin, Wallace, Huxley, and Spencer, we will get a taste of the vigorous debates surrounding the concept of evolutionary ethics. Such a historical overview sheds light upon one of the most contentious of all the cultural applications of evolutionary theory in the Victorian era. As will become evident, nineteenth-century history has much to tell us about the debates currently raging over the ethical bombshells created by recent advances in genetic engineering, reproductive technologies, and increased human manipulation of the biosphere.

Ethics and Philosophy

During the Victorian age, apart from traditional religions, two major approaches characterized theoretical and practical ideas on ethics and morality: the *intuitive* and *inductive* schools. Both philosophical approaches assumed that there is a single and highest normative principle, or standard, for ethical judgments. They differed in the crucial area of whether we have knowledge of that principle intuitively (without appeal to experience), or inductively (through empirical experience and observation). The German philosopher Immanuel Kant (1724–1804) was the most influential representative of the intuitive school. John Stuart Mill most forcefully defended the inductive approach for the Victorians. Which approach would prove more hospitable to evolutionism? Or would evolution suggest a radically new model for attempting to resolve the eternal ethical dilemmas confronting humans, individually and collectively?

Brilliant and influential as Kant's idealist ethical system is in the history of philosophy, it was rejected by most Victorian advocates of evolutionary ethics. According to Kant, intuitive knowledge is superior in certainty to empirical or inductive knowledge. He argued that we do not just receive sensations passively from the external world. Rather, the human mind is an active power, which organizes received sensations to make them comprehensible and rational. In Kant's idealist metaphysics, the observed regularity of natural law is a necessity of the organizing process of the human mind, rather than any necessity imposed by nature itself. In the realm of ethics, Kant argued for an innate moral sense, or conscience, that makes our duty clear. This innate sense is independent of any prospect of pleasure or pain, or of reward or punishment (Bowler 1989, 105). Accordingly, the morality of an act must not be judged by its consequence but only by its motivation. Intention, for Kant, is the sole criterion for judging whether an action is good or evil. Ethical duty in his view, is—or ought to be—based on a universally innate principle that is inherently right in itself. As the ultimate moral principle, Kant restates the golden rule in logical form: "Act as if the principle on which your action is based were to become by your will a universal law of nature." This rule is called the *categorical imperative* because it is unqualified and a command arising from the structure of the human mind itself. Advocates of evolutionary ethics, for the most part, rejected Kant's intuitive, or a priori, metaphysics. It seemed to them to contradict the fundamental precept of evolutionism that morality, as all other aspects of human nature, arises from a natural process of development. Most Victorian evolutionists turned instead to some version of the inductive approach in their attempts to construct a viable system of empirically based ethics.

The major inductive philosophy relating to ethics was *utilitarianism.* In its most basic version, utilitarianism is the doctrine that what is useful is good. Consequently, the ethical value of human conduct is determined by the utility of its results. *Usefulness*, however, is a notoriously difficult term to define. What is useful, and therefore good, for one individual may be quite different from that which is regarded as useful by another. Thus, the term *utilitarianism* is more specifically applied to the proposition that the supreme objective of moral action is the achievement of the *greatest happiness for the greatest number*. This objective is also considered the aim of all legislation and the ultimate criterion for assessing the appropriateness of social institutions. The utilitarian theory of ethics is generally opposed to ethical doctrines in which some inner sense or faculty, such as conscience, is made the absolute arbiter of right and wrong. Utilitarianism is likewise at variance with the view that moral distinctions depend on the will of God or the teachings of organized religion.

J. S. Mill and Utilitarianism

Utilitarianism was enunciated in its most characteristic form by the British jurist and philosopher Jeremy Bentham in his *Introduction to the Principles of Morals and Legislation* (1789). Bentham introduced the notion of a "calculus of pleasures and pains" to provide an ostensibly objective guide for ethical decisions. But it was John Stuart Mill who provided the definitive exposition of Victorian utilitarianism. Mill's major contribution to the theory appeared in his *Utilitarianism* (1863), published just four years after Darwin's *Origin.* Mill asserted in his influential treatise that distinctions of quality, in addition to those of intensity, among pleasures was fundamental to a valid system of ethics. Bentham had maintained that the quality of all pleasure was equal. In a famous phrase, Bentham asserted that push-pin (a child's game) was as good as poetry. Mill rejected Bentham's simplistic interpretation of human happiness. He contended that "it is better to be a human being dissatisfied than a pig satisfied"; in other words, human discontent is better than animal fulfillment. Mill regarded the direct identification of the concept of happiness with pleasure and the absence of pain, and the concept of unhappiness with pain and the absence of pleasure as an inadequate foundation for ethics. During the course of the Victorian era, other advocates of utilitarianism modified the system further. Spencer and Sir Leslie Stephen, for example, substituted intelligence for pleasure, or happiness, both as the supreme moral value and as the most reliable method of achieving other desirable moral aims.

Retaining the terms happiness and unhappiness, however, as useful for ethical discourse, Mill elaborated upon the ways by which we could decide what acts and institutions were moral, and which were immoral or evil. His vigorous empiricism, based on sense experience, was in full force in *Utilitarianism.* For Mill, the method of determining which acts were most ethically valuable meant looking to the consensus of experienced observers to learn which actions and institutions actually promoted the greatest happiness of the greatest number. In applying utilitarianism, Mill asserted that the happiness of every individual counts equally. In ethical assessments, the happiness of any one person is to count for no more or less that the happiness of any other person. This egalitarian element in Mill's utilitarianism appealed to many Victorians.

In *Utilitarianism*, Mill outlined what he regarded as the most reliable and objective criteria by which we could determine whether specific actions or institutions are moral/virtuous or immoral/evil. For Mill, actions and institutions are to be evaluated by considering their likely contribution (or failure to contribute) to the greatest happiness of the human race. On the face of it, however, this is not the way ethical evaluations are usually made. For example, we usually decide whether an act is right or wrong by reference to traditional rules of morality—either legal or religious. But Mill declared, as did many Victorians, that traditional ethical codes were becoming less relevant to the new scientific and industrial order emerging in Europe and North America. Mill's *Utilitarianism* was an attempt to provide a better account for, or improve upon, traditional methods of making ethical evaluations. The principle of utility could, and Mill believed it should, replace those traditional methods, such as religious dogma, for ascertaining moral behavior. Mill further argued that the principle of utility involves an assessment of only an action's consequences. It expressly does not address the motives or character traits of the agent performing the action. Utilitarianism is thus a radical departure from classical virtue theories, whether religious or philosophical, which explained morality by abstract and a priori concepts (Mill 1863).

Mill recognized that a critic of utilitarian ethics might argue that besides happiness, there are other things, such as virtue or sympathy, which humans desire. Responding to this, Mill said that everything we desire becomes part of happiness. Thus, happiness is a broad and complex phenomenon composed of many factors, including virtue; love of money, power, and fame; and sensual satisfaction. Victorian critics of utilitarianism further argued that unlike the supposition of the Utilitarians, true morality is not based on consequences of actions. Instead, these critics asserted, morality is based on the fundamental concept and motivation of justice. Mill, however, saw in the very concept of justice a rationale for utilitarian ethics. If justice were universal or innate, as the intuitionist school maintained, then concepts of justice would not be ambiguous. According to Mill, however, there are disputes in the notion of justice when examining theories of punishment, fair distribution of wealth, fair taxation, and so on. Only by appealing to utility can these disputes be resolved. Mill concludes that justice is a genuine concept, but it must be seen as based on utility (Mill 1863).

Evolution and Morality

It was in the context of these debates over the origin, practice, and sanction of morality that evolutionary theory made its striking, and lasting, impact on ethical thought. Although Darwin, Spencer, Tyndall, G. H. Lewes, and other major figures often differed in points of detail from Mill, they endorsed his effort to establish an empirical basis for ethical philosophy. They and their followers attempted to establish a "moral science" on the foundations of evolutionary theory (R. Richards 1987, 253–330). The question of the moral sense—its origin, its psychological force, its relationship to particular criteria of morality—seemed directly related to speculation about human evolution. The Darwin-Wallace

Fig. 6. Photograph of Charles Darwin in later life. (Courtesy of the Library of Congress, Z62-52389)

theory of natural selection, because it sharpened the debate on issues such as the relationship between instinctual and acquired (learned) behavior or between individual and group welfare and survival among all species, intensified also the debates concerning morality. It focused attention on whether and how human morality might have evolved according to the same laws governing all other physical and mental traits. The connection between human individual moral sentiments and collective social behavior had come under the evolutionary microscope.

As codiscoverers of natural selection, it was inevitable that Darwin's and Wallace's pronouncements on the implications of that theory for ethics would be among the most eagerly anticipated by Victorians. Darwin, in keeping with his lifelong predilection for the tenets of scientific naturalism, provided Victorian culture with powerful arguments and images of the evolution of human morality from the social behavior of nonhuman ancestors. His most detailed views on ethics appear in *The Descent of Man* (1871) and *The Expression of the Emotions in Man and Animals* (1872).

In *Descent*, Darwin agreed that "there can be no doubt that the difference between the mind of the lowest man and that of the highest animals is immense." He declared that he fully subscribed to the view that of all the differences between humans and other species, "the moral sense or conscience is by far the most important." The entire thrust of *Descent*, however, was to demonstrate that despite the obvious chasm between the striking human moral and intellectual traits and achievements and those of even the higher primates, there was a direct evolutionary linkage. Darwin's *Descent* and *Expression of the Emotions* presented massive data showing that "the difference in mind between man and the higher animals, great as it is, certainly is one of degree and not of kind." Darwin's scientific and rhetorical expertise was brilliantly showcased in these two works focusing on human evolution. Many readers would have found it difficult not to be swayed by Darwin's assertion that "we have seen that the senses and intuitions, the various emotions and faculties, such as love, memory, attention, curiosity, reason, etc., of which man boasts, may be found in an incipient, or even sometimes in a well-developed condition, in the lower animals." Darwin's great authority lay behind his conclusion that "man has risen, though by slow and interrupted steps, from a lowly condition to the highest standard as yet attained by him in knowledge, morals and religion" (Darwin 1871 [1874 ed.], 110, 142–143, 165).

Wallace, too, believed that many human social traits had evolved from natural selection of biologically based behavioral adaptations of lower animals. But he insisted that the higher human faculties, including the moral sense, required the intervention of forces other than wholly naturalistic ones in the human evolutionary process. The origins, evolution, and justifications of morality were major points of disagreement between Darwin and Wallace (see chapter 5). If the codiscoverers of natural selection could differ so dramatically on this point, it comes as no surprise that many other evolutionists put forward their own explanations of evolutionary ethics. It was Herbert Spencer who attempted to provide a comprehensive theory of human mental and moral evolution that would mediate between the competing hypotheses. Although Spencer rejected certain aspects of utilitarian social theory, he agreed with Mill that the ultimate criterion of truth—whether in politics, religion, economics, or ethics—resided in empirical validation. Spencer thus embarked upon his goal of establishing a thoroughgoing evolutionary psychology as a foundation for a science of ethics (R. Richards 1987, 258, 274–280).

Spencer and the Evolution of Morality

Herbert Spencer envisioned the development of life from simple organisms, unconsciously immersed in their environment, into more highly evolved species with increasingly complex mental traits. As he traced it, the evolution of mental life led from simple reflex reactions to instinct, memory, reason and, finally, will, or the power to make ethical choices. Following the naturalistic tradition, Spencer argued that the higher mental and moral traits evolved by means of the increasingly sophisticated response of higher organisms to the demands of the environment. Convinced of the evolutionary relationship between mind and body, Spencer developed what he termed the "law of intelligence": As species became more highly evolved, social as well as physical requirements pushed organisms to more complex and sophisticated mental states. Spencer's evolutionary framework was permeated by a teleological outlook. He believed that evolution was a process moving toward moral goals. Spencer attempted to demonstrate scientifically that nature inexorably inched its way toward perfection. In the case of human evolution, Spencer interpreted perfection as a complete adaptation to the social and moral state. Humans, thus, had evolved, and hopefully would continue to evolve, toward an increasingly finely tuned biological and ethical order. In this higher stage, evil and stupidity, which Spencer considered inferior responses to societal demands, would be replaced, gradually, by individual and collective freedom and rationality. This state of higher adaptation to society, Spencer concluded, would result in the maximizing of human happiness. Despite the fact that happiness was (and remains) notoriously difficult to define uniquely or unambiguously, Spencer's evolutionary psychology enabled him to construct a comprehensive and persuasive theory of the evolution of human morality.

Spencer's goal-directed evolutionary scenario relied upon certain Lamarckian views of adaptation. But the publication of Darwin's *Origin* gave Spencer's post-1860 writings a somewhat more Darwinian tone. Spencer's later explanation of the mechanisms that propelled the biological and social progression toward human betterment incorporated some of Darwin's findings and interpretations. Indeed, the Darwin/Spencer link was reciprocal. As noted previously, it was Spencer's term "survival of the fittest" that Darwin, at Wallace's suggestion, adopted to characterize the mechanism of natural selection. Many Victorian thinkers, including Mill, Darwin, Joseph Hooker, and John Lubbock, were impressed with Spencer's ambitious attempt to establish a framework for an evolutionary science of ethics. In his synthesis of biology and culture, Spencer made one of the more compelling cases in the late Victorian period for the belief that moral conduct and evolution led to the same end: a life fully adapted to the social state. In Spencer's vision, humans would enjoy the greatest happiness possible because evolutionary adaptation constituted the most just life possible (R. Richards 1987, 282–294, 298–309).

The "Naturalistic Fallacy"

Spencer's was but one of many such systems that proliferated in the latter decades of the nineteenth century, as evolutionism became a dominant metaphor for studying all aspects of human behavior and society. Despite their success in demonstrating the plausibility of an evolutionary model for ethical behavior, however, most such biologically based explanations of the origin and development of the higher moral traits of altruism, justice, and compassion are subject to the charge of committing what is called the *naturalistic fallacy*. According to the critics of evolutionary ethics—and there were many in the Victorian period as there are today—such systems are guilty of equating, or confusing, what *is* with what *ought to be*. The criticisms of evolutionary ethics range from the religious to the political to the economic (J. Greene 1981; Ruse 1999).

It was inevitable that such bold excursions into new foundations for the ethical and social dimensions of human activity would provoke diverse reactions. What Spencer, Darwin, and the majority of evolutionary scientists/philosophers were doing was attempting to elaborate a secular ethical framework to supplement, or replace, the faiths by which men and women had lived for centuries. Many radical thinkers embraced the new naturalism as an opportunity for agitating for ethical as well as sociopolitical reforms. Secular and materialist philosophies of nature and humanity, in the Victorian era as in our own day, draw strength from evolutionist foundations. Particularly in its strict Darwinist interpretation, evolutionism supports the view that the universe—and human society—is the product of impersonal forces that move nature toward unpredictable but adaptive ends. Even thinkers such as William James, Charles Sanders Peirce, and the French moral philosopher Henri Bergson (author of *Creative Evolution* [1907]), who were not materialists or atheists by any means, still found comfort and inspiration in the notion that the lack of a preordained pattern of evolutionary development could expand the boundaries of human freedom and potential.

For many other Victorians, the idea of an evolutionary universe, governed by what they perceived to be a lack of purpose or meaning, was degrading and discomfiting. The various critics of evolutionary ethics shared a common aversion to what they regarded as the Darwinian repudiation of goal-directed nature. It must be emphasized that Darwin himself never definitely stated that evolution meant the absence of purposive behavior, especially with respect to humans. But many of his critics simply assumed that Darwinism encouraged a belief in what England's poet laureate Alfred Tennyson described as "Nature, red in tooth and claw." Some religious and theistic critics rejected evolution outright. More frequently, they situated evolution within a broader framework of spiritual guidance and divine purpose (see chapter 3). One of the most powerful condemnations of evolutionary ethics came, surprisingly, from the inner circle of the Darwinian clan. Huxley, in the 1893 Romanes Lecture and its publication the following year with an extended introduction (Prolegomena) as "Evolution and Ethics," argued that humans must, in fact, oppose the cosmic process of evolution (see chapter 2). Only then, he believed, would it be possible to construct and maintain moral principles and feelings that go beyond nature to establish a sphere of activity that defined our humanity (Bowler 1989, 237–245).

Huxley's Ethical Twist

Thomas Huxley was one of the most active and articulate participants in the religious and scientific upheaval that followed the publication of Darwin's *Origin*. He was also a harbinger of the sociobiological debates about the implications of evolution that are current today. Huxley's Romanes Lecture argued that the human psyche is at war with itself. Humans beings, he contended, are alienated in a cosmos that has no special reference to their needs. Nature, therefore, was no guide to ethics. More specifically, Huxley argued strongly against the extension of evolutionary ideas, especially natural selection, to the formulations of ethical standards. This was a shocking conclusion coming from Darwin's bulldog and one of the Victorians' most brilliant proponents of scientific naturalism. Yet Huxley emphatically refused to sanction nature as a reliable guide for moral conduct, as outlined and approved by the so-called Social Darwinists such as Leslie Stephen and William K. Clifford. Huxley's agnosticism required that he abandon Christianity as a rationally justifiable foundation for ethics. But he also believed that the attempt to substitute either utilitarian arguments, such as Mill's, or an evolutionary foundation were just as invalid. Huxley thus stands at a troubling cultural threshold in the tradition of evolutionary biology, once it began to explore the ethical dilemmas of humans in nature and society. A fervent evolutionist, he felt that evolution did not, and could not, provide any clear solution to the darker aspects of the human condition. In this sense, Huxley has been seen as foreshadowing certain twentieth-century writers like William Faulkner and Albert Camus (Farber 1994, 63–69). In a direct challenge to the optimistic ethical scenarios of key figures like Spencer, Huxley asserted that human moral aspirations are in conflict with the conditions imposed by nature. Seen in the light of current understanding of the mechanisms of evolution, this declaration remains as controversial today as it was when Huxley enunciated it. George Williams, a leading evolutionary biologist, has claimed that recent biological ideas and data justify a more extreme condemnation of the "cosmic process" than even Huxley proposed. Williams starkly denies that the evolutionary forces that got us to our present stage are capable of maintaining a viable world. James Paradis, a scholar of Victorian science and culture, has shown that "Evolution and Ethics" confronted Victorians with a forceful rejection of assumptions about the harmony of nature (Paradis and Williams 1989).

Huxley's words in the Romanes Lecture remain as compelling now as when they were written. He declared that

> Men in society are undoubtedly subject to the cosmic process. As among other animals, multiplication goes on without cessation, and involves severe competition for the means of support. The struggle for existence tends to eliminate those less fitted to adapt themselves to the circumstances of their existence. The strongest, the most self-assertive, tend to tread down the weaker. But the influence of the cosmic process on the evolution of society is the greater the more rudimentary its civilization. Social progress means a checking of the cosmic process at every step and the substitution for it of another, which may be called the ethical process; the end of which is not the survival of those who may happen to be the fittest, in respect of the whole of the conditions which obtain, but of those who are ethically the best. (T. Huxley 1893)

Huxley clarified why he believed that ethical behavior must be seen as separate from those many other aspects of human culture that *have* benefited from the evolutionary process. He suggested that

> The practice of that which is ethically best—what we call goodness or virtue—involves a course of conduct which, in all respects, is opposed to that which leads to success in the cosmic struggle for existence. In place of ruthless self-assertion it demands self-restraint; in place of thrusting aside, or treading down, all competitors, it requires that the individual shall not merely respect, but shall help his fellows; its influence is directed, not so much to the survival of the fittest, as to the fitting of as many as possible to survive. It repudiates the gladiatorial theory of existence. . . . Laws and moral precepts are directed to the end of curbing the cosmic process and reminding the individual of his duty to the community, to the protection and influence of which he owes, if not existence itself, at least the life of something better than a brutal savage." (T. Huxley 1893)

Arguing against Spencer, Clifford, Stephen, and others who sought to erect a foundation of ethics upon a literal interpretation of evolutionism, Huxley warned that it "is from neglect of these plain considerations that the fanatical individualism of our time attempts to apply the analogy of cosmic nature to society. The struggle for existence which has done such admirable work in cosmic nature, must, it appears, be equally beneficent in the ethical sphere. Yet if that which I have insisted upon is true; if the cosmic process has no sort of relation to moral ends; if the imitation of it by man is inconsistent with the first principles of ethics," then Huxley was forced to a surprising conclusion. "Let us understand, once for all," he proclaimed, "that the ethical progress of society depends, not on imitating the cosmic process, still less in running away from it, but in combating it." He recognized that his prescription would strike the many Victorians captivated by evolutionism as "an audacious proposal" that would "pit the microcosm against the macrocosm and set man to subdue nature to his higher ends." Huxley hoped, however, that his audience would agree that his skillful but blunt rejection of evolutionary ethics would pave the way for an alternative but more "solid foundation." For Huxley, those very evolutionary forces that had brought humanity to the level of nineteenth-century knowledge and power would now be transcended by a difficult but ultimately achievable ethical "enterprise [which] may meet with a certain measure of success." True to his personal and philosophical agnosticism, Huxley preferred to venture onto terrain by which humans might rise above certain aspects of the evolutionary process and thus achieve a fuller ethical potential (T. Huxley 1893).

In rejecting nature as capable of providing, by itself, a moral or social norm, Huxley—like Wallace, but for quite different reasons—moved further away from the early nineteenth-century tradition of natural theology than had Spencer, Darwin, and many evolutionists. He could no longer see the benign face of nature that Paley had portrayed, nor could he share Darwin's (qualified) confidence that natural selection worked ultimately toward continued improvement of species, including *Homo sapiens*. Rather, Huxley conceived of science and technology as tools that would enable human beings to forge a new ethical and social order that transformed important aspects of the natural world. Huxley, a leader of the emerging breed of professional scientists, felt that science, including evolutionary biology, would create a new nature—and one presumably more amenable to human control and aspirations (Turner 1993, 126–127). Huxley's vision was seductive, but it would be left to the twentieth century and beyond to reap the actual benefits as well as dangers of a scientifically and technologically altered nature. The issues he raised in "Evolution and Ethics" are eloquent, if ambivalent, testimony to the "brave new world" that evolutionary theory unveiled in the Victorian era. Evolutionary biology would soon extend human ability to transform nature from the physical to the biological and cultural realms. That prospect of remaking not only nature but life was at once exciting and frightening. Indeed, the remarkable successes in the twentieth and early twenty-first centuries in extending both our knowledge and applications of evolutionary biology and related sciences such as neurobiology and cognitive psychology have kept the lure of evolutionary ethics alive and well.

Despite the fact that the varied attempts to erect an ethical system on the basis of evolutionary foundations are vulnerable to potent philosophical and cultural critiques, a number of important philosophers and scientists persist in the effort to ground an understanding of morality in biology. The Human Genome Project is just the most famous—and lavishly funded—example of the biological approach to human nature. Even in anthropology, which for much of the twentieth century has consciously contrasted "cultural" factors with "biological" ones, some researchers in recent years have been going back to evolution as a starting point for theories of human cultural development. Important thinkers in each generation (since the Victorian era) continue the quest to demonstrate that ethical values can be derived legitimately, at least in part, from evolutionary biology. The contemporary view, particularly in the academic world, that humans "read" values into nature rather than discover them there would seem a strong disincentive to such a quest. Deconstructionists in literary criticism, postmodernists in history, and social constructivists in philosophy and sociology all emphasize the subjective and/or social origins of moral (and political) values. Yet, the evolutionary perspective in ethical theory refuses to move offstage. Darwin's and certain other Victorians' belief that evolution held a crucial key to unraveling the dilemmas of ethics has proven remarkably resilient, even if relatively unproductive thus far (Farber 1994, 8–9). What accounts for the perennial appeal to nature as a guide for ethical conduct?

The most obvious answer is that all members of the human species are united by their possession of a common species perceptual system and common emotional psychology. If our judgments of right and wrong have some basis in nature, then surely—as Darwin and like-minded Victorians argued—evolutionary biology must have something to tell us about the origin and development of human ethical standards. The main difficulty, as Darwin's critics and many contemporary thinkers have stressed, is that evolutionary ethics has serious philosophical inconsistencies. The fact/value, or, is/ought, gap is the most problematic. Moreover, even assuming a common ethical structure for human judgment were shown to be a product of our evolutionary history, how do we account for the vast array of (often contradictory) individual and cultural norms defining what is good/ethical and bad/immoral (Maienschein and Ruse 1999, 309–326)? These objections to evolutionary ethics are formidable and have yet to be overcome. The final words on the scope and limits of evolutionary theory to our understanding of morality are yet to be written. As our knowledge of human biology becomes ever more impressive, the proponents for and against rooting morality in biology become more vocal. Most likely, we will see a more sophisticated application of certain biological theories as *one* element in a broader approach to solving the ethical dilemmas of contemporary technoscientific society (Bradie 1994, 163–175). The questions posed by and to the Victorians concerning the relevance of evolution to morality constitute one of the most enduring, if elusive and contentious, consequences of the success of nineteenth-century evolutionism. The cultural controversies surrounding evolutionary biology have grown more, not less, complex and urgent since the Victorian era. The implications of genetic engineering, reproductive technologies, and humans' manipulation of the biological and physical environment are among the most pressing issues facing early twenty-first-century societies. This brief survey of the history of evolutionary ethics has shown both the appeal and weaknesses of attempts to use biology—and science and technology more generally—to solve cultural problems. History, of course, provides analysis and documentation of past approaches to uncovering the possible links between biology and morality. History also demonstrates the extent to which such approaches involve often facile readings into nature of various cultural values to claim scientific legitimacy and justification for the most diverse moral values. But historians should not try to predict the course of future events, such as the fate of evolutionary ethics, or they cease to be historians (Farber 1994, 175). Such professional strictures do not, however, apply to a new group of thinkers who came to prominence in the Victorian era: science fiction writers. Quite the reverse. It was the very audacity of the prophecies of science fiction writers that catapulted them to a position of prominence in so rapidly changing a world as that of the late nineteenth century.

Ethical Hopes and Anxieties: Victorian Science Fiction

Victorian science fiction offers additional insight into the ways in which scientific knowledge and practice were transforming the realities of the era. It also serves as an uninhibited vehicle for projecting the contrasting ethical visions of what continued scientific and technological advances might bring. Most crucially, science fiction reveals the ambivalence in popular attitudes toward scientific theories, invention, scientists and their professional institutions and growing authority, and social change. In an age when science and technology were becoming increasingly dominant cultural forces, works that sought to make sense of science's impact upon the present as well as future state of affairs for elite, middle-class, and working-class audiences (including women and children) had significant appeal. Commercial novelists, scientific popularizers, journalists, and amateur naturalists—as well as many scientists—drew upon narrative and didactic modes of discourse that had been present since the late eighteenth century (Cooter and Pumfrey 1994, 237–267).

Science fiction as a separate genre is generally considered to have emerged as a significant literary form in the latter decades of the nineteenth century. There are obvious precursors, notably the gothic scientific romances such as Mary Shelley's *Frankenstein* (1818). But the origins of modern science fiction, especially mass science fiction in the form of novels and magazines published on both sides of the Atlantic, lay in the 1880s and 1890s. Most definitions of science fiction include not only the obvious fact of depicting social change as initiated or mediated by science (and technology) but also suggesting novel scientific and technical possibilities and applications. In this respect, science fiction helps create expectations of change by providing a cognitive estrangement. This encourages the reader to ponder both the contingent nature of what passes as the received culture of any era and, equally crucially, the potential realities of unknown but plausible futures (Suvin 1983; Alkon 1994).

The strange new worlds opened to late Victorians included time travel, possibilities of intelligent life forms elsewhere in the universe (most usually centered on speculations about Mars), and, of course, future evolution. Scientists such as Darwin, Huxley, Spencer, and Wallace included speculations about the possible future course of human evolution in the main body of their works. But it was H. G. Wells (1866–1946) who made evolutionism a staple of late Victorian science fiction. Wells had his precursors, some of them obscure and others (still) well-known. Arthur Brookfield wrote *Simiocracy: A Fragment from Future History* (1884), in which future left-wing evolutionists succeed in gaining equal rights for orangutans and English humans end up subjugated. Robert Louis Stevenson's better-known

The Strange Case of Dr. Jekyll and Mr. Hyde (1886) involves chemically induced "devolution," an explicitly biological metaphor for original sin. Wells, like Darwin and Huxley, also wrote nonfictional accounts of possible evolutionary scenarios, including scientific speculations on human evolutionary progress, extinction, and degeneration. Many Victorian writers wrestled with a major dilemma posed by evolutionism: Would the future of the human race be graced by progress—ethically as well as technically—or, rather, scarred by degeneration and regression?

According to the poet Mathilde Blind, Darwinian evolution would produce "better, wiser, and more beautiful beings . . . in the ages to come . . . than we can now have any conception of." Others were less sanguine. They worried that regression was at least as likely in the natural and social orders as progress. Francis Galton's eugenics project, in particular, played to both optimistic and pessimistic evolutionary projections. If the "right kind" of people were encouraged to breed, future prospects for humanity seemed rosy. However, if the "wrong kind" of people reproduced, the future would be bleak. Numerous warnings that misguided albeit well-intentioned efforts to insulate the poor and infirm from the harsh but ultimately beneficial effects of struggle and natural selection abounded in the late Victorian era. In that confused cultural climate, fears that inferior segments of society were breeding prolifically while the middle and, especially, upper classes were displaying declining birthrates, percolated not only in Britain but throughout Europe and North America (Paul 1995). The devolutionary alarm had been sounded sharply by the time Huxley's protégé Ray Lankester's *Degeneration: A Chapter in Darwinism* appeared in 1880. It was Wells's brilliant series of science fiction novels of the 1890s, however, that most effectively dramatized for the Victorian public the ambivalent ethical forecasts to be drawn from evolutionary speculation (Fayter 1997, 256–280; Morton 1984).

Wells and Devolution

In *The Time Machine* (1895), Wells challenged progressionist readings of evolution. He undercut both communist and liberal bourgeois utopian pretensions by presenting a future of cannibalistic underground workers (the Morlocks) who feed upon, literally, the ostensibly carefree and beautiful gentrified class of Eloi. The Eloi cavort happily but naively on the surface of the earth. In this novel, the worst fears of devolution are depicted with terrifying realism. One year earlier, Huxley had presented his own dark moral vision of nature as a decaying garden in the "Prolegomena" (1894) to *Evolution and Ethics*. Huxley's preface reinforced a strain of cosmic pessimism which reflected the fin de siècle melancholy of many Europeans as they faced a very uncertain twentieth century. Huxley modified his own (and others') earlier optimistic views that the relentless but ultimately beneficent biological laws of nature guaranteed a future for humanity in which the fittest had survived and human social and physiological improvements reigned triumphant. Huxley suggested such might be the short-range outcome, but those same biological and physical laws could ultimately dictate inevitable decline and extinction of the human race (Paradis and Williams 1989).

Wells evoked similar concerns voiced by Galton's opponents by depicting a dark and alien underside of eugenics in *The First Men in the Moon* (1901). It was *War of the Worlds* (1898), however, that represented the most powerful dramatization of the outer limits of Darwinism. Victorian evolutionism had been permeated by a complex set of ideological metaphors that suggested positive as well as negative ethical images of the human species. Wells's *War of the Worlds* gave vivid expression to the ambivalent attitudes and assumptions about power, progress, and purpose at individual, national, and global levels. Given the imperialist context of late Victorian culture, it was inevitable that Wells would use his science fiction to probe the realities and rationalizations of invasion, conquest, colonization, and extermination of certain societies by other, technologically and militarily more powerful, industrialized states. *War of the Worlds* serves as a prototype for subsequent science fiction that deals with the possible consequences of contact between humans and extraterrestrial civilizations. In the opening paragraph of the novel, Wells described how complacent Victorians went about "their little affairs, serene in their assurance of their empire over matter. . . . At most [they] fancied there might be other men upon Mars, perhaps . . . ready to welcome a missionary enterprise. Yet across the gulf of space, minds that are to our minds as ours are to those of the beasts that perish, intellects vast and cool and unsympathetic, regarded this earth with envious eyes, and slowly and surely drew their plans against us" (Fayter 1997).

Wells's fictional account of the invasion of the earth by hostile Martians interweaves several moral and cultural themes drawn from Darwinian images of struggle and competition. First, it is a commentary on the wars of Europeans with "less civilized" peoples during the aggressive and rapid imperialist expansion of the 1880s and 1890s. Wells asked his readers to "remember what ruthless and utter destruction our own species has wrought, not only upon animals . . . but upon its inferior races." He was referring specifically to the Australian Tasmanians' extermination by European immigrants. Wells asked pointedly, "Are we such apostles of mercy as to complain if the Martians warred in the same spirit?" In *Origin*, Darwin had asserted that natural selection among and within species "almost inevitably induces extinction." In *War of the Worlds*, it is the English who are the Tasmanian aborigines, and the Martians who are the conquerors. Second, *War of the Worlds* is about the biological war taking place everywhere in nature. This is what Darwin (echoing Malthus) called the struggle for existence in the great battle of living organisms, in which (following Spencer) "only the fittest survive." Third, Wells's novel is about the kind of war that might occur if science were devoted to serving military interests by producing advanced weapons of mass destruction. The apocalyptic and, to its many readers, terrifying account of the Martian invasion of earth was (and remains) deeply disturbing. Wells grants the human species a reprieve at the novel's end. The Martians succumb to a terrestrial infection by bacteria to which they on their planet had never been exposed. But Wells warns that the reprieve is likely only temporary. In the unsettling epilogue to *War of the Worlds*, the narrator anticipates another invasion. "We cannot regard this planet as . . . a secure abiding place," he writes, "for in the larger design of the universe other potentially hostile threats await" (Fayter 1997, 270–273).

An Uncertain Evolutionary Future

Victorian science fiction, as much as Victorian anthropology, demonstrates how evolutionary science forced Europeans and North Americans to confront both their own and other cultures from a radically new perspective. Such texts were as much examinations and speculations on exotic or alien beings as they were commentaries on the ethical dilemmas confronting Victorian society itself. Science fiction had emerged as a major vehicle for bringing evolutionary ideas, assumptions, and ideologies to the broader public that was increasingly affected by evolutionist doctrines and practices (Fayter 1997, 263–264, 273–274; H. Kuklick 1992). It is worth noting that just as the late Victorian period witnessed an explosion of works about Mars, so, too, is our own day seeing a renewed fascination with the possibility of life on the red planet (Dick 1996).

Evolutionary theories had a profound impact upon Victorian conceptions of ethics, human history, nineteenth-century society, and future prospects for *Homo sapiens*. As the writings and actions of a multitude of scientists, politicians, philosophers, theologians, and literary figures attest, evolutionism raised as many questions as answers regarding central issues of power, class, gender, and race. By the close of the Victorian era, people from all ranks of society and from different nations could well wonder how evolutionary science and its associated ideologies would play out in the opening decades of the twentieth century.

7

EVOLUTION AND RELIGION

TENSIONS AND ACCOMMODATION

The conventional view that evolution and religion were locked into inevitable conflict during the Victorian era is now discredited. Although nineteenth-century evolutionism did pose challenges to traditional religious beliefs and institutionalized practices, the notion that individual scientists or members of the general public had to choose between evolution and religion is a gross simplification. The reality of Victorian culture allowed for a wide spectrum of individual and collective responses to the implications of evolutionary theories for concepts of God and divine activity. Traditional religious forces, such as the powerful Anglican Church, remained potent factors in individual and societal decisions and actions. But religious thinkers did have to acknowledge the significance of the growing scientific evidence for evolution. At the other extreme, atheists could enlist evolutionary arguments to buttress their own proclamations of divine nonexistence or irrelevance. For most Victorians, however, evolutionism fostered—at times demanded—a reevaluation of the relationship between science and religion. There was a crisis of faith during the late nineteenth century, but the outcome of that crisis, both individually and collectively, was most frequently an accommodation between some version of scientific naturalism and some type of theism. The metaphor of a battle between science and religion was utilized during the Victorian era, and continues to be adopted in certain circles in our own day. That metaphor, however, must be taken figuratively rather than literally. Evolutionism and religion were both complex domains and the interaction between the two redefined certain important conceptions of the sacred and the secular (Turner 1993, chaps. 1, 3).

Theistic Evolutionism

The careers of prominent figures in the history of evolutionary biology, such as Wallace, Asa Gray, Joseph LeConte, and St. George Mivart, provide clear examples of how theistic beliefs shaped aspects of the theories of a substantial number of Victorian scientists. Most evolutionists tended to emphasize scientific naturalism as a major metaphysical underpinning of their biological thought. Among them, significant groups—including Wallace, Gray, and like-minded biologists and philosophers on both sides of the Atlantic—had also been receptive to theistic frameworks. They regarded theism as both compatible with, and necessary for, a complete evolutionary explanation of humans and the natural world (Brand 1997). Commitments to theism must be viewed as part of the total picture of each individual evolutionist's worldview, which included explicit espousals of political and social ideologies ranging from capitalist imperialism to socialism. Gray reflected the desire of many who sought to provide a theistic, yet scientifically rigorous, rendering of evolutionary theory (Gray 1876). For Gray, as for Wallace and Mivart (author of *Gen-*

esis of Species [1871]), theism profoundly affected the specific content of their evolutionary hypotheses. The impact of theistic convictions was particularly evident in questions of human evolution. The evolutionary theists made specific claims regarding (1) the limitations of natural selection, (2) teleology, (3) intelligent design, and (4) mind/matter interactions (Gray 1880, 44, 99–103).

Evolutionary theism must be contrasted to *pantheism*, the philosophy that God is identical with the totality of nature or the laws of nature. Unlike pantheists, Victorian evolutionary theists generally maintained that the deity is a purposeful, eternal being who continues to sustain relations to the natural world. The theists believed that a supreme being affected the course of evolution in certain well-defined and nonmiraculous ways. For the most part, they did not embrace the concept of an "absentee God," or First Cause, who created the world and instituted universal laws requiring no further divine role. Many theists maintained that divine guidance was at work in a few, but immensely important, evolutionary developments such as the origin of life and consciousness (Mossner 1967, 326–336). In the terminology of modern theology, evolutionary theism treats divine action as an alternative/complement to scientific language, not as a competitor with it. As Ian Barbour remarks, the "cosmic drama can be interpreted as an expression of the divine purpose. God is understood to act in and through the structure and movement of nature and history." In certain respects, nineteenth-century evolutionary theism may also be seen as a precursor of twentieth-century process theology. In the process model, God is a creative participant in the cosmic community, however different from all other participants. Process theology is regarded as consonant with an ecological and evolutionary understanding of nature as a dynamic and open system, characterized by emergent levels of organization, activity, and experience. As such, it avoids the sharp dualisms of mind/body and humanity/nature (Barbour 1997, 231–239, 326, 331; Griffin 1989, 69–82).

The literature on the varieties of theism is vast (Quinn and Taliaferro 1997, 197–521). The definition provided by Robert Flint, a prolific Scottish scholar, in his *Theism* (first published in 1877 and reprinted in its eleventh edition in 1905) was a generally accepted one in late Victorian

Britain. Flint stated, "Theism is the doctrine that the universe owes its existence, and continuance in existence, to the reason and will of a self-existent Being, who is infinitely powerful, wise, and good" (Flint 1877, 18). In adopting Flint's definition, certain qualifications are helpful for clarifying evolutionary theism. A number of Victorian thinkers believed in a God, but avoided allegiance to any traditional confessional, doctrinal, or institutional position. Moreover, many Victorian evolutionists never asserted that God's existence is absolutely provable. They nonetheless belong to the ranks of those who, when taking "account of a sufficiently comprehensive range of data—not only the teleological character of biological evolution but also man's religious, moral, aesthetic, and cognitive experience"—argue that theism is the most probable worldview (Hick 1990, 26–28).

Atheism

At the other end of the spectrum from theism is atheism, the critique and denial of metaphysical beliefs in a god or gods. Most atheists do not merely lack belief in God, they make the positive assertion that God does not exist. In part, Victorian atheism was a reaction against traditional belief systems, such as Christianity, Islam, and Judaism. Atheists rejected, often on political and cultural as well as strictly theological grounds, the dogmatic assertions of most historical religions on the nature of deity, moral codes, rituals and sacraments, and the necessity for membership in an institutionalized religious community.

Atheism has a long history in Western thought and philosophy. Plato argued against it in *Laws*, while Democritus and Epicurus argued for it in the context of their materialist doctrines. The eighteenth century witnessed the emergence of a strand of atheism among Enlightenment thinkers influenced by the scientific revolution. They combined British analytic empiricism and the mathematical natural philosophy of Isaac Newton with René Descartes's mechanistic conception of the universe, but abandoned Newton's and Descartes's own theistic views. David Hume, in his *Dialogues Concerning Natural Religion* (1779), argued against the traditional proofs for the existence of God. He was a philosophical skeptic, rather than an atheist in the strict sense, but his emphasis upon human knowledge as fundamentally limited to ideas gained from actual sense experience undercut theology. Hume and his followers regarded the existence of God as a matter of faith, incapable of empirical validation. In an era increasingly permeated by a veneration for empiricism, they thus laid a philosophical basis for Victorian atheism. Atheism appealed to certain groups in all segments of British and European society. Upper-class, middle-class, and working-class radicals and freethinkers found atheism useful in their efforts to discredit the authority and power of traditional religion (Desmond 1997).

One of the most important proponents of nineteenth-century atheism was the German philosopher Ludwig Feuerbach (1804–1872). He put forward the argument that God is a projection of man's ideals. Feuerbach associated his denial of God's existence with the affirmation of human freedom. For Feuerbach, the assertion that God is mere projection liberates humans for their own self-realization and potential. Karl Marx drew on Feuerbach's thesis, and incorporated atheism into his socioeconomic and political thought. In a famous phrase, Marx referred to religion as the opiate of the masses. He hoped that societal progress, under socialism and communism, would free humans from their need to believe in a God.

Another strain in late Victorian atheism is the existentialist. Friedrich Nietzsche (1844–1900) proclaimed the "death of God" and the consequent loss of all traditional values. The only tenable human response, he argued, is that of nihilism; without God, there is no traditional answer to the question of purpose and meaning in life. In Nietzsche's view, the death of God freed humanity to fulfill itself and find its own essence, and thereby create its own values through free ethical choice (Berman 1988).

Most nineteenth-century atheists, having analyzed the diverse arguments for God's existence, concluded that there is simply no compelling evidence for the existence of a deity or deities. Although there is no inherent link between atheism and evolutionism, many Victorian atheists saw evolutionary theory as confirming their views. They were convinced that science "proved" that the universe, Earth, and its life forms evolved by wholly naturalistic processes. Victorian atheists saw no evidence or reason to think that the evolutionary process required intervention by any spiritual entity. They confidently maintained that ethical, political, and social systems can be developed without reference to any code of behavior of allegedly divine origin. Darwin, at different points in his life, made statements that led some of his contemporaries to label him an atheist. However, he repudiated that label for himself. There exists a lively body of literature debating the question of whether or not Darwin believed in God. The verdict is still out. Darwin did note that theism and evolution were clearly compatible. He cited Gray and Charles Kingsley as examples of those individuals who were both ardent theists and evolutionists. He never placed himself in that category, but did assert at the close of his life that he had "never been an atheist in the sense of denying the existence of a God." The Victorian crisis of faith did not leave Darwin untouched. He remained deeply uncertain on matters theological. When pressed, Darwin declared that if he had to define his religious stance, "an agnostic would be the most correct description of my state of mind" (Desmond and Moore 1991, 636, 657–658).

Agnosticism

In the perplexing climate of belief versus unbelief, agnosticism was a respectable and comforting position for a significant number of late Victorians. Agnostics often met with indignation from both theists and atheists. The former usually accused agnostics of knowing the "truth," but ignoring it. The latter charged that agnostics are atheists at heart but were afraid to admit it because of social and family criticism. Most theists and atheists claimed that agnostics belonged to one or the other of their two opposing camps. Few gave agnostics proper recognition by admitting that agnosticism was, in fact, a philosophically rigorous third possibility (Turner 1974). Historians have now demonstrated the significance of agnosticism among important segments of Victorian society (Lightman 1987). Because the term was coined by Thomas Huxley, agnosticism assumes a crucial role in assessing the Victorian debates on evolutionism and religion.

Agnosticism, which in contemporary usage often implies little more than passive disbelief, was a provocative and well-articulated concept in the later nineteenth century. The word itself was first used by Huxley in 1869, in private conversation at the Metaphysical Society. He intended it to denote "not a creed . . . but a method" (Desmond 1998, 567). Huxley felt that none of the standard terms for describing religious beliefs accurately reflected his own developing views. He sought a word that would characterize himself and like-minded thinkers, who did not consider traditional faith as providing adequate answers to the burning questions of the appropriate role of religion in an age of scientific naturalism. These questions had been debated for centuries. Does God exist? How can we know Him? Why would He create evil, and why would He allow the good to suffer and the wicked to flourish? Does He intervene miraculously in this world? Such perennial questions, and their various answers, had been given greater urgency in the Victorian period because of the advances in and increased prestige of natural science. Huxley maintained that neither he nor anyone else could answer those questions without resorting to a knowledge (or *gnosis*) that went beyond the legitimate reach of reason. As a leading advocate both of scientific naturalism and the professionalization of science, Huxley insisted on reason and the empirical method as the only properly scientific way of knowing this world. If reason could not answer questions such as God's existence, then the only honest course would be to admit that one could not answer such questions definitively. A master of language, Huxley invented the term *agnosticism* to express his conviction that knowledge of the divine was unattainable. For Huxley, traditional religious faith meant believing what is literally incredible, in other words, unreasonable, and thus was inappropriate for a scientist or any citizen in a scientific age.

Huxley always insisted that there was no such thing as organized agnosticism, and certainly no agnostic church. Initially, he used the term to describe only his own views. But when other prominent Victorians such as W. K. Clifford and Leslie Stephen began calling themselves agnostics, the term spread. The number of those who called themselves agnostics increased. The movement's first journal, the *Agnostic Annual*, appeared in 1883. Despite his own strong reservations about creating an institutionalized sect, Huxley recognized that the zeal to promulgate agnosticism on the part of its increasingly vocal practitioners could be extremely useful. It would advance his own agenda of portraying science, especially evolutionism, as a value-neutral and hence (ostensibly) objective method of knowledge acquisition and dissemination to a receptive public (Desmond 1998, 527–528). Agnosticism thus became an important adjunct to the secularization of various aspects of Victorian culture. Those who attacked Huxley and agnosticism tended to ignore the careful philosophical distinctions that he made. They lumped agnostics in with atheists, materialists, and other "infidels." Huxley's eloquent exposition of the basic agnostic tenets was ultimately effective. When coupled with the generally traditional and conservative morals of the first agnostics—who, for the most part, were careful to comport themselves like model middle-class Victorians, posing little threat and much promise to the institutions of Victorian culture and enterprise—agnosticism became an influential movement (Lightman 1987).

In 1889, Huxley put forward the classic definition of agnosticism:

> Agnosticism, in fact, is not a creed, but a method, the essence of which lies in the rigorous application of a single principle. That principle is of great antiquity; it is as old as Socrates; as old as the writer who said, "Try all things, hold fast by that which is good"; it is the foundation of the Reformation, which simply illustrated the axiom that every man should be able to give a reason for the faith that is in him; it is the great principle of Descartes; it is the fundamental axiom of modern science. Positively the principle may be expressed: In matters of the intellect, follow your reason as far as it will take you, without regard to any other consideration. And negatively: In matters of the intellect, do not pretend that conclusions are certain which are not demonstrated or demonstrable. That I take to be the agnostic faith, which if a man keep whole and undefiled, he shall not be ashamed to look the universe in the face, whatever the future may have in store for him. (T. Huxley 1889)

Huxley admitted that the agnostic approach would vary according to individual knowledge and capacity, and according to the general state of science. He was aware that what is unproved today may be proved, by the help of new discoveries, tomorrow. Huxley asserted that the only legitimate obligation in pursuing knowledge "is to have the mind always open to conviction." He wittily observed that agnostics who never fail in carrying out their principles are, I am afraid, as rare as other people of whom the same consistency can be truthfully predicated. But, if you were to meet with such a phoenix and to tell him that you had discovered that two and two make five, he would patiently ask you to state your reasons for that conviction, and express his readiness to agree with you, if he found them satisfactory. The apostolic injunction to "suffer fools gladly" should be the rule of life of a true agnostic. I am deeply conscious how far I myself fall short of this ideal, but it is my personal conception of what agnostics ought to be." (T. Huxley 1889, 245–247)

Huxley's interpretation of agnosticism demonstrates the tentative nature of the definition of this widely used term. Although Huxley has come to stand as the iconic Victorian agnostic scientific naturalist, the historical reality is more complex. Recent scholars have clarified the intricate, indeed somewhat ironic, nature of Huxley's and others' agnosticism. Far from spurning religion, many agnostics actually sought to set forth a serious new, nonclerical, religious synthesis. Historians have now located the origin of key aspects of agnostic epistemology within the thought of theologian H. L. Mansel. Thus, even for the scientific naturalists, the sharp secular/religious dichotomy is highly suspect. What is not in doubt, however, is that the scientific naturalists deployed evolutionary theory to advance the status of science and scientists in Victorian social, intellectual, and political spheres.

This strategy proved successful, and evolutionism became an essential part of the fabric of Victorian culture. But this success has tended to obscure the links between evolutionism, agnosticism, and religious epistemology in the late nineteenth century. More significantly, the tactical distinction drawn by Huxley between the empirical discourse of natural science and the vocabulary of religious thought and practice has become for many an absolute dividing line in contemporary culture. What is today presented as the value-neutral professional world of contemporary science is commonly perceived as more reliable and objective than the subjective concerns of religion and ethics (Lightman 1987; Turner 1993, 20–21; Paradis 1997, 166–167). This presumed clear-cut divide, however, between objective science and subjective human values is problematic, especially when we are confronted with profound ethical and environmental dilemmas involving the control of nature and of human behavior.

Agnosticism did serve as a comfortable vessel for navigating the muddy cultural waters of Victorian evolutionism. Similarly, the nondogmatic approach of a significant number of nineteenth-century evolutionary theists defused certain tensions between evolution and religion. This accommodating outlook stands in sharp contrast to the more strident positions taken by late twentieth and early twenty-first-century religious opponents of evolution. The current bitter debates over so-called creation science testify to the still powerful emotions generated by the findings of evolutionary science, especially regarding human evolution (see chapter 8). But controversy had already sharpened in the early decades of the twentieth century, most famously in the now legendary—in all senses of the word—Scopes trial.

The Scopes Trial

Until the 1990s, no trial in American history had attracted more attention—and been more misunderstood—than the 1925 trial in Dayton, Tennessee, of John Thomas Scopes, accused of violating a state law banning the teaching of human evolution. The Scopes trial has been the subject of scores of scholarly and popular books and articles. It has been immortalized in the famous Hollywood film *Inherit the Wind* (1960), based on a 1955 play by Jerome Lawrence and Robert E. Lee. The trial was, and remains, controversial. It has achieved mythological status in the annals of debates concerning evolution and religion. The Scopes trial demands our attention precisely because of the public interest it generated in 1925 and for its enduring historical significance. But the actual events of the trial, and its consequent interpretation for the history of science/religion interactions up to the present day, require careful analysis. The confrontation between prosecution and defense, which took place in a hot summer courtroom in Dayton, has become the subject of myth because of its ability to serve so many competing interests. The Scopes trial ranks with the Victorian confrontation between Huxley and Wilberforce as a rich source for legend-making about the nature of science, religion, and their mutual interactions (Numbers 1998, 76–77, 91). The courtroom antics in Tennessee represent, in microcosm, the collision of the diverse and powerful forces of evolutionism and antievolutionism. Scopes's trial was both a culmination of the evolution/religion debates of the Victorian era and a prelude to the vigorous debates of the late twentieth and early twenty-first centuries (Tompkins 1965).

As we have seen, Darwin's specific version of evolutionary theory, with its emphasis on random variations and the rigorously competitive process of natural selection, posed a crucial problem for many Victorians. For some conservative theologians and pious scientists, the ultimate challenge of Darwinism to a traditional Christian worldview was blunt: If nature reflected the character of its creator, then the God of a Darwinian world acted randomly and cruelly. It seems clear that Darwin himself did not embrace such a concept and sought refuge in agnosticism. Other scientists, including Wallace, Lyell, Gray, and St. George Mivart, propounded varieties of theistic evolutionism that accorded divine activity some role in the evolutionary process. But in the confrontational decade immediately following the publication of *Origin*, many church leaders raised an alarm against teaching evolution, particularly within seminaries and denominational colleges.

During the 1870s and 1880s, however, scientific developments seemed to cast some doubt on the validity of Darwinian mechanisms. Darwinism was criticized by a number of important biologists and physical scientists. Evidence from geology and geophysics suggested that the earth was not sufficiently old to allow enough time for Darwin's slight, random variations to accumulate and produce the current array of biological species, much less to generate life from nonlife. This was a serious scientific objection, and was not shown to be invalid until the twentieth century. By the 1920s, techniques of radioactive dating definitively proved the earth to be ancient enough to allow the great time span demanded by Darwinian evolution (Bowler 1989, 137–138, 206–207). But Darwin's nineteenth-century critics had a powerful weapon from physics in their arsenal.

A second serious scientific difficulty that plagued Darwinism was the lack of an adequate understanding of the mechanism of biological inheritance. Gregor Mendel's laws of genetics were discovered in 1865 but remained largely unknown or unappreciated until 1900. Without a knowledge of the precise means by which heritable differences are preserved, the small, gradual variations stipulated in Darwin's theory seemed to many biologists not to lead anywhere (Bowler 1989, 270–275). Like many naturalists working before the acceptance of Mendelian genetics, Darwin believed that the inherited traits of an offspring consisted of a blending of those possessed by its parents. Slight, random variation in an individual—no matter how much it helped that animal or plant survive—quickly would be swamped as that individual bred with others of its species. It appeared that gradually each succeeding generation would lose its distinctiveness. If organic evolution occurred (and by 1880 most naturalists believed that it did), then some mechanism must accelerate and direct variation. For the devoutly religious, this left a specific role for God in the evolutionary process.

Two major alternatives to Darwinism were discussed widely among European and American scientists during the final third of the nineteenth century. Gray and Wallace, among others, proposed theories of theistic evolution in which God or other higher, spiritual powers, directed variations into a pattern of progressive development. For some, this offered a way to reconcile religious faith with evolutionism. Other naturalists, led in the United States by Joseph LeConte, Clarence King, and Edward Drinker Cope, revived Lamarckian explanations to account for the speed and direction of evolution. According to these neo-Lamarckians, variations became purposeful and natural selection marginalized. These alternative theories of evolution might not fit completely with traditional Christian doctrines, but they certainly could be spiritual. Conservative Christians might dislike evolutionism on theological grounds, but few raised loud objections. Many liberal Christians wholly embraced an evolutionary creed (Larson 1997, 17–19).

Neo-Lamarckism and other non-Darwinian forms of evolutionary thought gained broad acceptance within the late Victorian scientific community, particularly in the United States. As one historian has observed, from "the high point of the 1870s and 1880s, when 'Darwinism' had become virtually synonymous with evolution itself, the selection theory had slipped in popularity to such an extent that by 1900 its opponents were convinced it would never recover. Evolution itself remained unquestioned, but an increasing number of biologists preferred mechanisms other than selection to explain *how* it occurred" (Bowler 1989, 246). With the "eclipse of Darwinism," as Huxley's grandson Julian later referred to this period in the history of biology (Bowler 1983), many conservative Christians toned down their rhetoric. The future prosecuting attorney in the Scopes trial, William Jennings Bryan, was not among them. He assured his audiences in the early twentieth century that he did "not carry the doctrine of evolution as truth as some do." Bryan then added sarcastically that "I do not mean to find fault with you if you want to accept the theory; all I mean to say is that while you may trace your ancestry back to the monkey if you find pleasure or pride in doing so, you shall not connect me with your family tree without more evidence than has yet been provided" (Bryan 1909, vol. II, 266–267). Bryan was clearly staking out a position that pleased his Fundamentalist friends. Influential as Bryan was, however, many highly orthodox Protestant theologians such as Princeton's James McCosh and Rochester seminary president A. H. Strong took the more conciliatory position that the eclipse of Darwinism offered. The theologian James Orr endorsed theistic evolution, asserting that in the more fluid biological climate of the early twentieth century "the Bible and science are felt in harmony" (Larson 1997, 20–21).

Science/Religion Debates Reignited

If orthodox Christians were not plotting to reignite the science/religion flames, who was? The highly significant point emerges that at the start of the twentieth century it was secular historians and essayists, rather than theologians and scientists, who were most keen on stirring things up. The secular historians were largely responsible for keeping alive the public perception of an inherent hostility between Christians and evolutionism. Later developments in biology in the 1920s and 1930s, particularly in such fields as population genetics and geographic variations within and between species, would eventually reestablish the Darwinian paradigm in what is now called the "modern evolutionary synthesis." These developments seriously eroded the explanatory status of neo-Lamarckian and other alternative evolutionary mechanisms that had readily supported theistic evolutionism (Bowler 1989, 307–318). But the early architects of the modern synthesis were concerned with biology, not theology. It was academic historians who trumpeted the military metaphor of a battle between science and religion. This battle ideology underlay the stridently polemical evolutionist and antievolutionist public crusades in the period directly preceding the Scopes trial. Two works in particular provided ammunition for those who viewed civilization as a battle between the progressive force of modern science and the reactionary force of traditional religion. John William Draper's *History of the Conflict between Science and Religion* (1874) was the first. The second was that of his fellow American, Andrew Dickson White. In 1876, White published *The Warfare of Science*, with a preface by John Tyndall. This was later expanded into the two-volume *History of the Warfare of Science with Theology in Christendom* (1896).

The conflict thesis is vivid, but grossly simplistic. Many, possibly most, scientists and religious thinkers did not *as individuals* see evolution and theology as intellectually or emotionally irreconcilable. The more contentious issues arose at the public rather than private level. Passions were aroused with respect to the public, collective perception of cultural authority. A major facet of late Victorian society was the shifting of cultural authority from traditional bastions, including religion, to emerging new centers of cultural power such as professionalizing science. That there was a competition for cultural authority in the late Victorian period is beyond dispute. The professionalizaton of science was a key element in this process. The expansion in the number of professional scientists, and the widespread dissemination of scientific ideas in educational institutions and among the general population, did tilt the tide in favor of science as a powerful voice of public authority. Huxley, as we have seen, was a skillful and successful player in this process. For epistemological as well as political and economic reasons, it was necessary for the professionalizing scientific community to propagandize in order to win greater governmental and public support. The image of a *public* conflict between science and religion assumed a polemical importance that the actual (private) views of many individuals, scientists as well as laypersons, did not necessarily reflect or justify (Turner 1993, 171–176).

Polemics, Politics, and Evolutionism

Polemics are crucial elements in the political process. Both Draper's and White's works were short on detailed historical accuracy, but laden with inflammatory prose. In the rapidly changing world that characterized the close of the Victorian era, the metaphor of a battle between science and religion was a potent tool in certain hands. Draper's and White's books were especially widely read in the United States. They recounted Roman Catholic attacks on Copernican astronomy, including the seventeenth-century trial of Galileo and execution of Giordano Bruno. Draper and White fostered the impression that religious critics of Darwinism threatened to rekindle the Inquisition. Their and other similarly biased historical accounts specifically failed to discuss the many areas of harmony between theologians and evolutionists at the turn of the nineteenth century. The conflict view of science and religion gained a wide following among secular scholars in the early twentieth century. It stiffened their resolve to defy the antievolution crusades of Bryan and certain Fundamentalist groups during the 1920s. In the years leading up to the Scopes trial, this reaction inspired an outpouring of academic books, articles, and essays discussing the conflict between science and religion, with an increasing focus on the seemingly pivotal issue of Darwinism (Larson 1997, 21–22; de Camp 1968).

Perhaps more than any single event, it was the discovery of the Piltdown Man (see chapter 2) that fueled the never dormant controversy over human origins. Word of the discovery became front-page news in December 1912 throughout the United States, where prominent creationists still publicly denounced the Darwinian theory of human evolution. Utilizing the recent advances in communications technology, the *New York Times* published a summary of Dawson and Woodward's initial presentation within hours of the event. "Paleolithic Skull Is a Missing Link," the *Times* headline proclaimed, "Bones Probably Those of a Direct Ancestor of Modern Man." The next day, the *Times* followed up with a telegraphic interview of Woodward. "Hitherto the nearest approach to a species from which we might have been said to descend that had been discovered was the cave-man," Woodward noted in this interview, "but the authorities constantly asserted that we did not spring direct from the cave-man. Where, then, was the missing link in the chain of our evolution? To me, at any rate, the answer lies in the Piltdown skull, for we came directly from a species almost entirely ape." Other American newspapers carried similar reports. Although the Piltdown fossil suggested a link in the record of human evolution, no single fossil discovery, on its own, could prove the Darwinian theory of human evolution. But the increasing number of other similar fossil discoveries was, cumulatively, making a stronger scientific case in Europe and North America for supporting a Darwinian view of human origins. The momentum of these scientific discoveries and interpretations helped set the stage in the early 1920s for a concerted crusade by certain fundamentalist groups against teaching evolution in public schools. It was only a matter of time before these debates would find their day in court (Numbers 1992, 37–71, 283–298; Nelkin 1982, 93–147; Viguerie 1981; Larson 1997, 13–14).

By 1925, the warfare model of science and religion had become part of the intellectual and cultural vocabulary of many secular Americans. Clarence Darrow, the leading defense counsel in the Scopes trial, learned it as a child from his anticlerical father who had read Draper, Huxley, and Darwin. Similar views characterized Scopes's other defenders. This erosion of the conciliatory approach to science/religion issues of the latter decades of the nineteenth century provoked, at least in the United States, an increasingly strident fundamentalist backlash against evolution by the 1920s. Recent historians have identified two different causes for the timing of this particular American antievolution crusade. First, Darwinism did not become a fighting matter for many fundamentalists until it began to influence their children's education in the 1920s. Second, the leeway afforded Christian biologists during the period from 1880 to 1920, when the mechanism of natural selection had come under increasing scientific attack and neo-Larmarckian and other alternatives flourished, became increasingly constricted (Larson 1997; Moore 1979). Theories and researches by experimental geneticists, biometricians, and field naturalists provided powerful new arguments for the primacy of random, inborn variations as the raw material of evolutionary change. Not all biologists became born-again Darwinians. Thomas Hunt Morgan, the preeminent American geneticist of the era, favored hypothetical large-scale mutations rather than Darwin's posited gradual, small-scale variations. The crucial point, however, is that by the 1920s, Darwinian mechanisms of random variation and natural selection had once again become dominant paradigms in biology (Moore 1979; Morgan 1932, 109–110).

This resurgence of specifically Darwinian mechanisms, coupled with the accumulating evidence supporting the hypothesis of human evolution from nonhuman ancestors, rekindled many of the Victorian debates over evolutionism. A major difference between the two eras, however, was that by the 1920s public education, particularly at the secondary level, had become widespread.

> The number of pupils enrolled in American high schools lept from about 200,000 in 1890, when the federal government began collecting these figures, to nearly two million in 1920. Tennessee followed this national trend, with its high school population rising from less than 10,000 in 1910 to more than 50,000 at the time of the Scopes trial in 1925. Commenting on this trend with respect to Tennessee, Governor Austin Peay—who signed the state's antievolution bill into law—boasted in his 1925 inaugural address, "High schools have sprung up throughout the state which are the pride of their communities." This was certainly true for Dayton, site of that year's Scopes trial, which opened its first public high school in 1906.

In line with the most recent developments in American science, these high schools explicitly included Darwinian concepts in their biology curricula (Larson 1997, 24; Hunter 1914).

A Civic Biology

By the start of the twentieth century, the more comprehensive subject of biology replaced traditionally separated fields such as botany and zoology in school curricula. More significantly, the Darwinian mechanism of natural selection figured prominently in textbooks as examples of which hypotheses the most recent scientific findings in biology tended to favor. One representative biology text featured a picture of Darwin and a subchapter titled "The Struggle for Existence and Its Effects" (Peabody and Hunt 1912). Another extolled Darwin for discovering "the laws of life," including the concept of organic evolution through natural selection (Hodges and Dawson 1918). George William Hunter's *A Civic Biology* was the best-selling biology textbook of the time. It was also the specific high school text at issue in the Scopes trial. Hunter summarized Darwin's evolutionary mechanism in a section entitled "Charles Darwin and Natural Selection." As Hunter explained, Darwin postulated new species "arising from very slight variations, continuing during long periods of years." This mechanism attributed these all-important variations to random individual differences inborn in the offspring rather than to Lamarckian vital forces or acquired characteristics (Hunter 1914, 253; Larson 1997, 16–17, 23). Such passages were bound to arouse concern in the religiously devout families, whose children were now exposed to evolutionary concepts that (once again) appeared to render teleological worldviews less tenable.

A further development spurred Bryan and other antievolutionists, namely, the rise of a strident version of Social Darwinism. The mechanism of natural selection was repugnant to Bryan on political as well as religious grounds. He saw its central tenet of competition as inextricably linked—validly or not—with unbridled capitalism. In the laissez-faire social evolutionism advanced by William Graham Sumner, a political scientist at Yale who propagated Darwinism in the social sciences, market forces determine the survival of the fittest. Powerful late Victorian captains of industry like Andrew Carnegie and John D. Rockefeller eagerly advocated this version of Social Darwinism. Bryan denounced that version as early as 1904 as "the merciless law by which the strong crowd out and kill off the weak." He also thought that the ruthless competition that he saw as the heart of Darwin's model of evolution was reflected in the militarism that led to World War I. As President Wilson's secretary of state, Bryan strove to avert this war by proposing several treaties that required nations to arbitrate their disputes. He resigned from that post when it became clear that Wilson no longer shared his goal. Many Americans did espouse the version of Social Darwinism that Bryan detested. They welcomed the association of natural selection with a survival-of-the-fittest mentality to justify their ideological embrace of laissez-faire capitalism, imperialism, and militarism.

During the years immediately preceding the antievolution crusade, another scientific-sounding form of these social doctrines—eugenics—gained widespread public attention. Following Galton, Hunter (in another of his popular textbooks) defined eugenics as "the science of improving the human race by better heredity." In the United States, many evolutionary biologists endorsed eugenics early in the century. The public campaign to impose eugenic restrictions on reproduction peaked in the 1920s. Consequently, the eugenics movement coincided with the antievolution crusade in many states. Typically justifying their actions on the basis of evolutionary biology and genetics, a number of American states enacted laws to compel the sexual segregation and sterilization of certain persons viewed as eugenically unfit. The legislatures included in this category the mentally ill and retarded, habitual criminals, and epileptics. Some antievolutionists denounced eugenics as the horrid consequence of Darwinian thinking that assumed that since humans evolved from beasts, they could be bred like cattle. Bryan decried the entire program as "brutal" and in Dayton offered it as another reason for not teaching evolution. Everywhere the public debate over eugenics colored people's thinking about the theory of human evolution (Paul 1995; Larson 1997, 26–28).

The specific event that triggered the Scopes trial (known then popularly as "The Monkey Trial") was the passing in March 1925 of a statute by the Tennessee legislature. This statute made it unlawful for any teacher in any educational institution supported by the public school funds "to teach any theory that denies the story of the Divine Creation of man as taught in the Bible, and to teach instead that man descended from a lower order of animals." The statute further stated that any teacher violating this section would be guilty of a misdemeanor and fined not less than $100 nor more than $500 for each offense. Tennessee was not the first state to explicitly attack evolution in the legal arena. Florida and Oklahoma already had laws on the books that criticized the promoting of evolutionary theory in schools. The Tennessee law, however, was the first to render the teaching of human evolution a crime that carried a legal punishment. In the spring of 1925, John Thomas Scopes was a twenty-four-year-old science teacher at Rhea County High School in Dayton. Among the discussions generated by the Tennessee legislative action were those that took place in Dayton. There, a small group of local citizens opposed to antievolution laws persuaded Scopes to openly violate the Tennessee statute in order to provide the legal pretext for a court test case. The American Civil Liberties Union (ACLU) was also actively interested in supporting such a test case. The resulting worldwide publicity that quickly developed surprised even the event's planners. With Scopes's arrest and indictment, the proposed trial took on national, and not merely local, proportions.

Darrow vs. Bryan

The lead prosecutor was William Jennings Bryan. He had been thrust into national prominence in 1896 by his "Cross of Gold" speech at the Democratic Convention in Chicago. The speech argued for a more liberal monetary policy and was directly responsible for Bryan winning the Democratic presidential nomination that year. Heading the defense was Clarence Darrow, America's most famous criminal lawyer and an agnostic. Darrow came to Dayton fresh from his success in the Chicago Loeb-Leopold murder trial of 1925. He had been contacted by the ACLU. Arthur Garfield Hayes and Dudley Field Malone were two other well-known lawyers of the day who aided in Scopes's defense. Hundreds of reporters descended on Dayton for the trial. Press coverage of the Scopes trial probably exceeded coverage of any event up to the time. One reason for this was that the recently invented radio was used to report the proceedings. Newspapers from all over the world sent not just reporters but editors to witness the courtroom battle. The *Baltimore Sun* sent a team of five writers, the best-known of whom was the sharp-penned wit and critic H. L. Mencken. It was estimated that 10,000 visitors from far and near overran the town daily. The trial began on July 10 and lasted eight days, in one of the hottest and driest Tennessee summers on record.

As a legal contest, the trial dealt with the specific issue of whether Scopes had violated the Tennessee statute in his teaching as a substitute biology teacher (physics was his main subject field). The public interest generated by the trial ensured, however, that the court proceedings would extend to more general questions. These included academic freedom, tolerance, bigotry, and the radically differing views on the relationship between evolution and religion. The jury convicted Scopes of violating the law, and the presiding judge fined him $100. Later, the Tennessee Supreme Court overturned the conviction on the technicality that the jury, rather than the judge, should have set the fine. Bryan died in his sleep on a Sunday afternoon five days after the trial. The aging Darrow returned to Chicago, where he lived until his death in 1938. Scopes drifted, by the fall of 1925, to the University of Chicago, where his study of geology led him to a career in the oil industry in Louisiana (Larson 1997, 170–193).

The complexities and subtleties in both the details of the trial and the actual views of the participants have tended to be obscured in the attempt by various groups to interpret the trial to suit their own agendas. For example, a number of historians have claimed that, despite Scopes's legal conviction, the trial actually represented a public-relations victory for the evolutionists. *Inherit the Wind* conveys that message. This popular and influential film did much to foster the myth that Bryan's testimony in Dayton, in which he admitted the antiquity of life on earth, destroyed his credibility with fellow Fundamentalists and brought about the demise of the antievolution movement. The available evidence, however, supports none of these claims. Many journalists did indeed review Bryan's performance in Dayton harshly, writing that he revealed his ignorance of both religion and science. But Darrow also received considerable criticism in the press: for disrespecting the judge, for treating Bryan rudely, and for trying to deny the people of Tennessee their democratic right to determine what should be taught in their tax-supported schools. In fact, Darrow became such a liability that the ACLU tried (unsuccessfully) to dump him from the defense team handling Scopes's appeal to the state supreme court (Larson 1997, 207–210, 240–246).

Legacy of the Scopes Trial

Overall, the Fundamentalists emerged from the trial flushed with a sense of victory and proud of the way Bryan had handled himself. The head of the World's Christian Fundamentals Association, which had invited Bryan to Dayton, praised him for his "signal conquest" on behalf of Fundamentalism: "He not only won his case in the judgment of the Judge, in the judgment of the Jurors, in the judgment of the Tennessee populace attending; he won it in the judgment of an intelligent world." Fundamentalist leaders would not have felt betrayed by Bryan's advocacy of an ancient earth. Except for the Seventh-day Adventist George McCready Price, they agreed with Bryan—and most other Americans—that life on earth long antedated Adam and Eve. The events at Dayton neither ended the antievolution crusade nor slowed it down. Nearly two-thirds of the antievolution bills introduced in state legislatures in the 1920s came after 1925. Far more significant than these, and later, laws, however, was the impetus given to turn efforts away from the courthouse to local schools. Fundamentalists sought to influence curricula regarding the precise manner in which evolution—particularly human evolution—should be taught and publicized. The debate between creationists and evolutionists ended neither with the publication of Darwin's *Origin* nor with the Scopes trial. The central issues involved, touching as they do on the extremely complex interactions between science and religion in different cultures and at different historical periods, simply do not admit of definitive resolution. The revival of contemporary scientific creationism, with attendant critiques, signals that the Victorian debates over the implications of evolutionism resonate with us still (Numbers 1998, 76–91).

The role of the ACLU in the Scopes trial and its continuing aftermath is significant. The trial brought the ACLU to the general public's attention for the first time. The organization moved to the forefront of the debates over the separation of church and state and, especially, the teaching of creationism in public schools. These debates are still raging today, and the ACLU continues to represent one major faction in the battle for religious liberty and the preservation of the First Amendment. The ACLU is committed to the Constitution's principle that keeping religion out of public schools is the surest way to preserve religious liberty for all students (Walker 1990). In 1998, the ACLU intervened in Seattle, Washington, where a ninth-grade science teacher was including creationism, under the guise of "intelligent design theory," in biology classes (see chapter 8). In 1997, in Charleston, West Virginia, the ACLU blocked a public school's plan to teach creationism alongside evolution. In 1996, the ACLU stepped in when authorities in Tennessee sought to compel public schools to teach creationism. In the same year, the ACLU took action in Lakewood, Ohio, to stop two physics teachers from teaching creationism to their students (Donohue 2000).

The Scopes trial has come to symbolize the broader controversy regarding the separation of church and state and their appropriate public roles in North American culture. Conceptions regarding the implications of human evolution are one focal point in this controversy. The most enduring lesson derived from the Scopes trial and its history and myths is that evolution and antievolution paradigms are far more than simply scientific, or pseudoscientific, models. They have become cultural forces that testify to the fallacy of conceiving of science as a value-neutral enterprise or disembodied conceptual scheme. The issue of the ethical and political neutrality of science was, as has been demonstrated, a central feature of the Victorian evolutionary debates. Every society and every culture creates its own (attempted) resolution of questions about the nature of science and the scope of scientific authority. In the case of evolutionary biology, its theoretical range, explanatory power, and practical consequences (such as genetic engineering) have become more powerful precisely because of the abundance of accumulating experimental evidence supporting evolution. Biologists today speak with greater authority than before, not only with respect to their technical discipline but also with respect to the ethical, social, and environmental issues arising from advances in biology. Not surprisingly, controversy erupts over the implications of the findings of biology and the pronouncements of biologists for cultural decision-making. These controversies have been fueled by the emergence of so-called creation science, to which we now turn.

8

THE CONTEMPORARY DEBATES

"CREATION SCIENCE"?

The Scopes trial was a dramatic example of the enduring public fascination with the implications of evolutionary theory, especially regarding humans. Such events do not provide, however, any definitive resolution of the issues involved. Nor can they be expected to do so. The emergence of a scientifically potent evolutionism by the end of the Victorian era had crucial implications for the course of the twentieth-century debates concerning human origins, human nature, and human culture. As the theoretical and experimental support for evolution continues to accumulate, opponents of an evolutionary worldview have been forced to develop more complicated strategies for attempting to confront evolution on scientific grounds. There is ample precedent for contesting specific evolutionary claims and hypotheses in the history of biology itself. The rise, fall, and subsequent vindication of the theory of evolution by natural selection are excellent illustrations of the manner in which science itself advances by debates within the scientific community. What sets evolution somewhat apart from other sciences is the immediacy of the relationship between evolutionary biology and human social, economic, political, and religious concerns. Evolution is controversial precisely because it confronts certain of the most deeply ingrained beliefs about human nature and human destiny. As evolutionary science becomes ever more firmly established, critics of evolutionism resort to increasingly more extreme measures in their attempt to discredit both its knowledge base and its cultural authority.

As in any other scientific domain, evolutionary biology has conceptual as well as practical consequences for human conduct at both the individual and social levels. Evolutionary biology, as much as nuclear physics, involves profound questions addressing the control of nature. Eugenics, as noted above, is a particularly striking attempt to apply—correctly or not—evolutionary ideas to human breeding. Clearly, evolutionism has emerged as one of the Victorian era's most potent, and malleable, legacies for the social context of science. The growing prestige and authority of science (and technology) during the twentieth century has compelled those who would contest evolutionary biology to adopt strategies that overtly address the current power relations of science and its institutions. Hence the rise, in the latter decades of the twentieth century, of what has been termed *creation science* (Strahler 1999). This movement differs in one important respect from previous creationist crusades. Creation science is explicit in its attempt to depict its ideological core as secular and scientific (McKown 1993). Although the basic motives of creation scientists are religious, it is this scientific patina that arouses the passionate defense and attacks regarding the movement (Montagu 1984; Ruse 1988; Eve and Harrold 1991). The struggle in the political and educational arenas, as well as in the public media, is a struggle for the power to influence the manner in which individuals and governments regard, and use, the concepts and findings of evolutionary biology (Scott 1997, 263–289).

Varieties of Creation Science

A wide variety of individuals and groups term themselves creation scientists. That rubric, however, covers several different categories. By far, the majority of creation scientists and their advocates come from the Fundamentalist Protestant wing of Christianity. The defining link that unites them is a belief in an inerrant Bible. They hold the Bible to be literally correct in all its original writings on all subjects, including its description of the Creation, Adam and Eve, and Noah's flood. As in any political and religious movement there are several schools of creationist thought, separated by doctrinal differences in their interpretations of the Bible. The three major schools are (1) the *day-age faction*, (2) the *gap theorists*, and (3) the *strict creationists*. According to one source, in 1984 there were no less than twenty-two national creationist organizations in the United States, and at least fifty-four state and local organizations (Numbers 1992, x–xiii).

The day-age faction of creationism argues that the days referred to in Genesis are really symbolic of enormous stretches of time, and not twenty-four-hour days. Perhaps the best-known of the day-age groups are the Jehovah's Witnesses. The gap theorists argue that there is an unmentioned lapse of time between the first and second verses of Genesis, and that the six-day creation event did not happen until after a long period of time had already passed. Many of the televangelists are gap theorists. Finally, the strict creationists assert that creation happened as described in Genesis, and that the universe and all life was created within six days, several thousand years ago. The first two factions, the day-age and the gap, accept the geological evidence of a very ancient earth (but not the evidence of evolution). They are usually referred to collectively as the *old-earth creationists*. The strict creationists, in contrast, assert that the entire universe is just 6,000 to 10,000 years old. Hence, they are referred to as *young-earth creationists*.

It is crucial to distinguish at the outset between theistic science and the various schools of creation science (Frye 1983). Theistic evolutionists—from the time of Gray, Wallace, LeConte, and many others in the Victorian era—argue that evolutionary science is wholly compatible with a theistic worldview. Nearly all mainstream religious denominations, as well as many scientists, are supporters of theistic evolution. They can be considered creationists in the sense that they do assert that the universe was made, or at least influenced, by divine or spiritual forces. But theistic evolutionists adhere to the methodology and findings of evolutionary biology. They are, consequently, viewed by Fundamentalists as opponents in league with the scientific naturalists and materialist evolutionists. It is more accurate to designate fundamentalist creationists as antievolutionists, since the one conviction that unites them all is the belief that evolutionary theory is contrary to the tenets of Christianity (Ecker 1990). Judged by this fundamentalist criterion, theistic evolutionism is as misguided and odious as materialistic evolutionism.

Young-Earth Creationists and "Balanced Treatment" Politics

It is the young-earth creationists who dominate the creation science movement and who head most of the major creationist organizations. Their specific viewpoints are those that set the tone of the various antievolution or *balanced treatment* policies sought by the creationists. The Arkansas Balanced Treatment Act (1981), for instance, defines creation science in terms of young-earth creationism:

> "Creation-science" includes the scientific evidences and related inferences that indicate: (1) Sudden creation of the universe, energy, and life from nothing; (2) The insufficiency of mutation and natural selection in bringing about development of all living kinds from a single organism; (3) Changes only within fixed limits of originally created kinds of plants and animals; (4) Separate ancestry for men and apes; (5) Explanation of the earth's geology by catastrophism, including the occurrence of a world-wide flood; and (6) a relatively recent inception of the earth and living kinds. (LaFollette 1983, 5)

The original impetus for young-earth creationism derived primarily from George McCready Price, a fundamentalist Seventh-Day Adventist who personally accepted the literal truth of the Bible. In 1923, Price published *The New Geology*, in which he argued that all of the geological features we see today were the result of Noah's flood and not the slow geological processes described by scientists. The "geological column," Price asserted, was nothing more than the deep sediments deposited by the flood while all of the various fossils were merely the dead bodies of organisms that had drowned in the deluge. Price declared conventional geology—by which he meant the generally accepted uniformitarian geological doctrines of the late nineteenth century—to be a fraud. He claimed it was fostered upon an unsuspecting public by scientists who were doing the work of the devil: "Some of the tricky methods used by the Great Deceiver to befuddle the people of the last days" (Numbers 1992, 137).

While most geologists dismissed Price as a crank and ridiculed *The New Geology* as being riddled with error and distortion, the book caused a sensation among religious Fundamentalists. They cited it as the first book to use science to show that the Bible is literally correct. This claim itself is historically inaccurate. Numerous treatises published in the eighteenth and early nineteenth centuries used geological arguments to attempt to prove that biblical accounts of catastrophic events such as major deluges were valid. Even by the close of the Victorian era, such biblical geologies were still being refuted by scientists of renown (M. Greene 1982, 165–173; Gillispie 1951). The tired and discredited thesis of *New Geology* detracted not at all from its enthusiastic reception among Fundamentalists. Price (who was not a geologist) was even cited during the Scopes trial as a scientific expert (Numbers 1992, 146). Much of Price's flood geology can be found, nearly intact, in the writings of modern creationists.

In 1935, Price helped to form the Religion and Science Association (RSA), the first nationwide creationist organization. The RSA had as its acknowledged purpose that of using scientific data to support the Bible. Shortly after it was formed, however, the RSA was torn by an internal feud between those who accepted Price's flood geology and those who rejected it (Numbers 1992, 112). In 1941 a new creationist organization was founded, the American Scientific Affiliation (ASA). Unlike the RSA, which was more concerned with theology than science, the ASA required all of its members to have legitimate scientific credentials. It also, revealingly, required all members to sign an oath of membership, swearing: "I believe the whole Bible, as originally given, to be the inspired Word of God, the only unerring guide of faith and conduct. Since God is the Author of this Book, as well as the Creator and Sustainer of the physical world about us, I cannot conceive of discrepancies between statements in the Bible and the real facts of science" (Numbers 1992, 159). This tactic of limiting membership to scientists who already agreed to the literal truth of Genesis would later be repeated. In effect, by using scientific knowledge as an apologetic for biblical truth, the ASA became the first creation science organization. Somewhat ironically, it was the ASA's insistence on a semblance of scientific respectability that proved problematic to some of its members. Once again, flood geology was at the center of the dispute. Dr. J. Laurence Kulp, a chemist and geologist, rejected flood geology and pointed out that it was demonstrably untrue. To insist upon it as biblically inspired, he warned, would make a laughingstock out of creationism. Kulp, aided by the biologist J. Frank Cassell, argued that the ASA's entire attitude on evolution had to change if it was to maintain any scientific respectability. They urged the ASA to adopt an attitude of theistic evolution. This effort was partially successful. Today, the ASA takes no official position on the question of creation science, and most of its members are theistic evolutionists (Numbers 1992, 167, 174–175).

HENRY MORRIS

Not surprisingly, the young-earthers vigorously defended their "science" against the attacks of Kulp and Cassell. During the 1953 ASA annual convention, Henry Morris presented a paper entitled "The Biblical Evidence for a Recent Creation and Universal Deluge." Morris, a staunch biblical literalist and young-earth creationist, had deliberately chosen to major in hydraulic engineering and minor in geology (at the University of Minnesota), so he could study the effects that flood waters would have on the earth. At the 1953 ASA convention, Morris first met John C. Whitcomb Jr., a theologian with an interest in flood geology and young-earth creationism. In 1957, Whitcomb finished a Th.D. dissertation entitled "The

Genesis Flood," which presented a detailed defense of the historicity and geological affects of Noah's flood. Shortly afterwards, he decided to publish the thesis as a book, *The Genesis Flood*, with Morris as coauthor (Whitcomb and Morris 1961). Financed by a number of religious Fundamentalists, *Genesis Flood* was published in February 1961 but dismissed by professional geologists as merely an updated version of McCready Price's scientifically discredited *New Geology*.

Genesis Flood was also criticized by old-earth creationists, who argued that the very idea of a global flood was not supported by any geological evidence. In response, Whitcomb and Morris answered simply that Genesis said there had been a global flood, therefore there must have been one. "The real issue," they declared, "is not the correctness of the interpretation of various details of the geological data, but simply what God has revealed in His Word concerning these matters" (Whitcomb and Morris 1961, xxvii). To the ASA *Journal*, which was vocal in its criticism of the book, Morris wrote, "The real crux of the matter is 'What saith Scripture?'" (Numbers 1992, 208). The dispute within the ASA over flood geology soon convinced the young-earthers that the ASA was getting "soft on evolution." In late 1961, the plant breeder Walter Lammerts, who had long been affiliated with creationist organizations, joined with Morris and Duane Gish to form an "antievolution caucus" within the ASA. Lammerts was an extremist even for a creationist. Unlike most young-earthers, who accepted a limited form of evolution within "created kinds," Lammerts rejected even this and asserted that no speciation of any sort was possible.

In 1963, Morris, Gish, and Lammerts formed a breakaway creationist organization called the Creation Research Committee. The Committee later changed its name to the Creation Research Society (CRS), the name by which it is still known. The major purpose of the CRS is "to publish research evidence supporting the thesis that the material universe, including plants, animals and man are the result of direct creative acts by a personal God" (Nelkin 1982, 78). Morris had by this time decided that scientific data could be used as an effective tool for bringing people to Christ. He began to point to his flood geology model as an "alternative science," one that proved the literal correctness of the Bible. Morris also began to explore the possibility of using the state legislatures to have balanced treatment acts passed, mandating equal treatment of what he termed "evolution science" and "creation science" in biology classrooms.

Clearly, the purpose of the CRS had less to do with scientific investigation than with proselytizing fundamentalist biblical literalism. In fact, a large number of creationists objected to the use of science at all. They felt that the religious message was weakened and cheapened by attempting to use scientific data to "prove" the act of creation (Numbers 1992, 230–231, 235, 246, 408). In 1978, Walter Lang, the editor of the creationist *Bible Science Newsletter*, echoed the sentiments of many creationists who felt that scientific justification for creation was unnecessary and detracted from the spiritual message. He noted that only "about five percent of evolutionists-turned-creationists did so on the basis of the overwhelming evidence for creation in the world of nature" (Numbers 1992, 233). Indeed, Lammerts, Gish, and Morris had all been committed creationists before they had gained any scientific experience. Morris, however, was completely committed to his strategy of using creation science to win a place for teaching the book of Genesis in American science classrooms. He made calculated steps to present creationism as a scientific, not a religious, outlook. Morris explained, confidently, that "creationism is on the way back, this time not primarily as a religious belief, but as an alternative scientific explanation of the world in which we live" (Morris 1974a, 16). Morris's *Scientific Creationism* (1974) was intended to be the definitive book on creation science, suitable for use in public school biology courses (Morris 1974b).

Institute for Creation Research

In 1970, Morris and Christian Fundamentalist preacher Tim LaHaye (of the Moral Majority Inc.), working with the Scott Memorial Baptist Church, raised money and set up the Christian Heritage College in San Diego, California. In its 1981 academic catalogue, the college (which had no accreditation from either religious or secular California educational authorities) advertised several courses in science. All were taught, according to the catalogue, in a "consistently creationist and Biblical framework." As for evolutionary theory, the catalogue states, "Biblical criteria require its rejection as possible truth" (LaFollette 1983, 107). Morris himself was teaching a course in creation science at the college. Morris also founded the Institute for Creation Research (ICR) as a scientific laboratory for the Christian Heritage College, with the avowed purpose of attempting to scientifically prove the literal validity of Genesis. Morris and his allies attempted the formidable, if not impossible, task of depicting their biblical science research as a purely secular enterprise. ICR today attempts to maintain the fiction that it is a scientific institute with no religious connections. However, most ICR staffers, including Morris and Gish, are still adjunct professors at the Christian Heritage College.

The ICR carries out no field research in any of the life sciences. Moreover, despite its claim to be purely scientific, ICR maintains its tax-exempt status with the Internal Revenue Service on the grounds that it is a religious institution carrying out "non-scientific research." It is responsible for much of the creationist literature that is available today. ICR also produces creationist films in conjunction with "Films for Christ." Somewhat miraculously, then, the ICR remains one of the preeminent institutional forces of the creation science movement. The ICR trumpets its "scientific credentials." Members of the ICR, it proudly declares, are required to have an advanced degree in at least one of the sciences. They usually fail to mention, however, that, like the CRS, all of its members must sign an oath affirming their belief in a literal interpretation of Genesis and their acceptance of Jesus Christ as their Lord and Savior. Muslims, Jews, Buddhists, and any other non-Fundamentalist creationists are not allowed membership in the ICR unless they renounce those beliefs and sign the ICR's oath of biblical infallibility. One of the ICR's favorite pamphlets is entitled "Twenty-One Scientists Who Believe in Creation." The pamphlet lists a number of holders of doctorates and master's degrees in various scientific disciplines who assert the literal correctness of Genesis. Of the twenty-one individuals showcased by ICR, however, only a small number hold a degree in any of the life sciences. Three hold doctorates in education, two are theologians, five are engineers. The remainder include a physicist, a chemist, a psycho-linguist, and a "food scientist" (Nelkin 1982, 78–83; Numbers 1992, 283–290). This interdisciplinary roster is not in itself problematic. Many emerging scientific fields, such as information technology, artificial intelligence, space science, and a number of environmental areas, of necessity must draw upon the combined expertise of individual scientists from a broad range of disciplines. Since creation science purports to deal primarily with biological phenomena, however, it is curious that so few of its leading exponents possess advanced degrees in the life sciences. Indeed, the ICR no longer finances or carries out field research of any sort. Its primary method of "scientific" research consists of scouring the published works of evolutionary theorists for quotations that can be pulled out of context and used to bolster creationist beliefs (Numbers 1992, 291–294, 302–303).

This paucity of authentic scientific research notwithstanding, the ICR has published what is referred to as the "scientific model" of creation. The ICR prefaces its account of that paradigm by noting that "creationism can be studied and taught in any of three basic forms, as follows: (1) 'Scientific creationism' (no reliance on Biblical revelation, utilizing only scientific data to support and expound the creation model); (2) 'Biblical creationism' (no reliance on scientific data, using only the Bible to expound and defend the creation model); and (3) 'Scientific Biblical creationism' (full reliance on Biblical revelation but also using scientific data to support and develop the creation model)." (All quotations are from ICR Impact No. 85, "The Tenets of Creationism," Henry Morris, July 1980.) The second and third of these three forms, of course, rely explicitly on Christian religious doctrines. They are therefore illegal to teach in public schools in the United States, under the Establishment Clause of the First Amendment. The first form, presumably, is the one that creation scientists are investigating and defending. It is also the one that they would like to have taught in science classrooms as an "alternative" to the accepted scientific model of evolution. On closer examination, however, it is evident that these three forms are essentially synonymous.

The legality of teaching any of these three forms, of course, constitutes the crux of most of the debates surrounding creation science. At least six federal court cases have ruled that scientific creationism is not science but religious doctrine dressed up as science, and hence illegal to teach in public schools. The creationists are fully cognizant of these legal strictures, and counter that scientific creationism and biblical creationism are the same doctrines, *but* differ according to their audience. Hence, in churches and Sunday schools, where teaching religious doctrine is perfectly acceptable, the ICR recommends teaching biblical creationism. But in public schools, where openly religious instruction is illegal, the ICR advocates teaching these religious doctrines in the guise of creation science. It is this attempt to defuse or divert attention from their real objectives that has, justifiably, angered both evolutionary scientists and the general public with respect to the pretensions of creation science (Numbers 1992, 241–251).

Tenets of Creation Science

What, then, are the main propositions of creation science? As set out in the ICR's "Tenets of Creationism" they include:

> (1) The physical universe of space, time, matter and energy has not always existed, but was supernaturally created by a transcendent personal Creator who alone has existed from eternity; (2) The phenomenon of biological life did not develop by natural processes from inanimate systems but was specially and supernaturally created by the Creator; (3) Each of the major kinds of plants and animals was created functionally complete from the beginning and did not evolve from some other kind of organism; (4) The first human beings did not evolve from an animal ancestry, but were specially created in fully human form from the start. Furthermore, the "spiritual" nature of man (self-image, moral consciousness, abstract reasoning, language, will, religious nature, etc.), is itself a supernaturally created entity distinct from mere biological life; (5) The record of earth history, as preserved in the earth's crust, especially in the rocks and fossil deposits, is primarily a record of catastrophic intensities of natural processes, operating largely within uniform laws, rather than one of gradualism and relatively uniform process rates. There are many scientific evidences for a relatively recent creation of the earth and universe, in addition to strong scientific evidence that most of the earth's fossiliferous sedimentary rocks were formed in an even more recent global hydraulic cataclysm; and (6) Since the Universe and its primary components were created perfect for their pur-

> poses in the beginning by a competent and volitional Creator, and since the Creator does now remain active in this now decaying creation, there do exist ultimate purposes and meanings in the universe. Teleological considerations, therefore, are appropriate in scientific studies whenever they are consistent with the actual data of observation, and it is reasonable to assume that the creation presently awaits the consummation of the Creator's purpose.

The above basic tenets represent an amalgam of religious, philosophical, and scientific assertions. Certain ICR tenets do in fact superficially reflect current controversies within evolutionary biology. The rate of geological processes is a legitimate area of investigation and competing hypotheses within the scientific community. Similarly, teleological considerations inform some important contemporary studies in both evolutionary biology and cosmology. The ICR has buried these kernels of scientific inquiry, however, under a dominant Fundamentalist religious edifice. It is evident that the proponents of creation science have an agenda other than an objective assessment of the strengths and weaknesses of prevailing hypotheses in both the life and physical sciences. One may deconstruct the tenets of creation science to reveal a program that is preeminently political and ideological. As we have seen, political and ideological forces are inseparable from the nature and functioning of the scientific enterprise. This was patently evident in the Victorian era and is equally evident in the contemporary period. Eugenics, reproductive technologies, and the Human Genome Project are but the most egregiously transparent domains of biology with respect to ideological and political contexts. What differentiates creation science, however, is its profound deficit of authentic scientific research and experimental validation coupled with an overt mandate to proselytize on behalf of specific religious goals under the rubric of science (Pennock 2001).

Recently, creation scientists have attempted to paint their outlook as purely secular and scientific, in an attempt to have it legislated into the classroom. Within their own ranks, however, they have all along been quite candid about the religious basis for their "science." Creation scientists have never concealed their belief that, wherever science and the Bible conflict, science must be rejected a priori. As articulated by Fundamentalist author and educator Donald Chittick, "Creation science begins with wholly Biblical presuppositions and interprets data from all of reality, including science, within that framework" (Rohr 1988, 156). Morris, the most prolific of the creationist writers, constantly reiterates the basic authority of religion over science. In his, and the vast body of creation science writings, biblical infallibility reigns. The Bible can't be wrong; therefore any and all scientific facts must be interpreted to fit into the biblical framework. And if there are any apparent conflicts between the biblical stories and the data of modern science, then it is the Bible that has priority. But Morris and his associates were fully cognizant of the shift in cultural authority from sacred to secular models that has been so marked a feature of Western societies since the Victorian era. Keeping up with the times—but with a quite different end in mind—Morris and the ICR have attempted to construct a creation science that would ostensibly provide a valid critique of evolutionary theory. "The creation model of origins and history," Morris writes, "may be used to correlate the facts of science at least as effectively as the evolution model." Morris's *Scientific Creationism* attempted to present a "scientific model" of creation which was, ICR argued, suitable for use as a textbook in a public school classroom. "The purpose of *Scientific Creationism* (Public School Edition)," Morris explained, "is to treat all of the more pertinent aspects of the subject of origins and to do this solely on a scientific basis, with no references to the Bible or to religious doctrine" (Morris 1974b, iv).

Morris's statement that the creation science literature did not use the Bible as its authority is devious. Both his own *Scientific Creationism* and Duane Gish's book *Evolution? The Fossils Say No!* (1972) were actually published by ICR in two separate versions. One, the "General Edition," contained all of the standard biblical citations and religious arguments against evolutionary theory. The other, the "Public School Edition," had the religious references edited out in a transparent attempt to make them suitable for public school use by deleting their obvious religious content and aims. In a letter to members of the ICR, Morris openly proclaims the real purpose behind the Institute for Creation Research, which he refers to as a "three-fold ministry": "We are convinced that this is the most effective way in which recognition of God as a sovereign Creator and Savior can be restored in our modern world, especially for the multitudes of young people in our schools who have been indoctrinated for so long with the false and harmful philosophy of evolutionary humanism" (LaFollette 1983, 24). In numerous letters and publications sent to ICR members, Morris makes it clear that the ICR's "research" is closely tied with its evangelical work. In the June 1995 "Acts and Facts" newsletter, John D. Morris, vice president of ICR's Outreach Ministries (and son and heir apparent to his father's creationist kingdom), informed readers that with "the rise of evolution and naturalism, 'science' has become the enemy of Christianity, but true science 'declares the glory of God' (Psalm 19:1). ICR desires to return science to its proper, God-glorifying, position" (John Morris, "Acts and Facts," June 1995). Ironically, John Morris is a better historian of science than he is a creation scientist. From the Newtonian age to the Victorian era and through the twentieth century, there have been numerous outstanding scientists who see compatibility rather than contest as the appropriate mode of interaction between science and religion. But the solid scientific foundation upon which Wallace, Isaac Newton, Asa Gray, Albert Einstein, and other figures have expounded the potential harmony of scientific and religious approaches to the study and understanding of nature is glaringly absent in creation science.

The aim of the Creation Science Research Center, by its own declaration, was "to reach the 63 million schoolchildren of America with the scientific teaching of Biblical creationism" (LaFollette 1983, 49; Nelkin 1982, 79). Such statements as this were, of course, meant for internal consumption only; they were not intended for the public. Publicly, the ICR and other creation science groups have continued to insist that religious belief has nothing at all to do with their efforts to force creationism into the classroom. Their own statements, however, clearly demonstrate that the goal of creation science is not, as they claim, to present a "balanced view" or an "alternative scientific viewpoint." Rather, it is an attempt to use the legislative power of the state to force children to be exposed to the Fundamentalists' own particular religious interpretation of Genesis. As Arkansas Judge William Overton commented, in ruling that state's balanced treatment law unconstitutional, creation science "is a religious crusade, coupled with a desire to conceal this fact. It was simply and purely an effort to introduce the Biblical version of Creation into the public school curricula" (LaFollette 1983, 45–73). The cumulative import of the "research," the "technical journals," and the public fora organized by the advocates of creation science may be characterized most accurately as a powerful, concerted effort to proselytize their Fundamentalist religious views. The chosen vehicle for this enterprise is the introduction of creation science teaching into the educational system. Accordingly, the most significant assessment of the success, or failure, of that enterprise is the record of legal battles that have marked the recent history of creation science.

Creation Science: A Legal History

After the Scopes trial, several states, including Arkansas, Mississippi, and Tennessee, kept their antievolution laws on the books but made little effort to enforce them. Then, in 1957, the Soviet Union launched its *Sputnik* satellite, shocking the United States out of its intellectual complacency and dramatically illustrating the inadequacy of science education in America (Nelkin 1982, 39–53; Bruner 1966; Numbers 1992, 238–240). In response to the new space race, Congress passed a number of laws, such as the National Defense Education Act, instituting a crash program to bring American science education up to par. One of these programs was the Biological Sciences Curriculum Study (BSCS), begun in 1959, to produce new, up-to-date biology textbooks. Written by professional experts in the several specialized areas of contemporary biology, the BSCS texts prominently featured evolutionary theory as the foundation of all the biological sciences. Within a few years, nearly half the high schools in the country were using BSCS biology textbooks, despite the fact that antievolution laws were still on the books in a number of states (Numbers 1998, 4–5).

In 1961, certain members of the Tennessee state legislature campaigned to repeal the 1925 Butler Act, the antievolution law under which John Scopes had been tried. The repeal failed after an acrimonious debate. Tempers in Tennessee ran as high in 1961 as they had thirty-six years earlier. In 1961, it was still technically a criminal offense to teach evolution as a valid scientific theory in state-funded schools. One legislator equated evolutionists with communists: "Any persons or any groups who assist in any way to undermine faith in the teachings of the Bible are working in harmony with communism" (Dykeman and Stokely 1971, 72). In 1967, teacher Gary Scott of Jacksboro, Tennessee, was fired for violating the Butler Act. He fought his firing in court and won. The Butler Act was finally ruled unconstitutional by the federal courts. Shortly afterward, Arkansas biology teacher Susanne Epperson filed a court challenge to the Arkansas antievolution law. When the Arkansas Supreme Court upheld the law, Epperson appealed to the U.S. Supreme Court. The Court ruled in 1968 that all state antievolution laws were unconstitutional on the grounds that they served to establish a state-supported religion and eroded the separation of church and state. The antievolution laws, the Court decided, were nothing more than "an attempt to blot out a particular theory because of its supposed conflict with the Biblical account, taken literally" (LaFollette 1983, 5–6, 76, 84). Despite the Supreme Court's ruling, the legal battles were about to escalate.

In 1973, the Tennessee state legislature passed a replacement for the Butler Act. The new law stated, "Any biology textbook used for teaching in the public schools, which expresses an opinion of, or relates a theory about origins or creation of man and his world shall [give] . . . an equal amount of emphasis on . . . the Genesis account in the Bible" (LaFollette 1983, 80). Within two years, this law was also struck down by the federal courts. The federal justices ruled that the Tennessee law was "a clearly defined preferential position for the Biblical version of creation as opposed to any account of the development of man based on scientific research and reasoning. For a state to seek to enforce such preference by law is to seek to accomplish the very establishment of religion which the First Amendment to the Constitution of the United States squarely forbids" (LaFollette 1983, 34–35). In response to this ruling and the earlier Epperson Supreme Court decision, creationist groups made a major tactical decision. They agreed to downplay the religious aspects of creationism, and to argue that creationism could be supported solely through scientific evidence, without any reference to God or the Bible. Thus, as noted above, creation science became the new mantra of Fundamentalists in the effort to gain entry into the public schools of the nation. These efforts have been challenged by evolutionists at the local as well as national level. Biology teachers have formed grassroots networks to counter the tactics of the creationists, recognizing that creation science is more a political than a scientific problem, and requires political as well as scientific attack (Park 2000, 349–370).

In 1981, Arkansas passed a law (Act 590) mandating that creation science be given equal time with evolution in public schools. Supported by legal help from the ACLU, a dozen or so clergymen of differing denominations sued. They argued that creation science was nothing more than Fundamentalist biblical literalism pretending to be science. Creationists from the Creation Research Society argued to the court that their viewpoint was a scientific model and not based at all on religion. Judge William Overton, after listening to both sides, was unconvinced by the creationists' arguments. He ruled that creation science was not a science but rather an attempt to introduce religious beliefs into the public school system. Under the United States Constitution, it is illegal for the federal government or for any state to pass a law that establishes government support for any religious view, or that serves to advance any particular religious view. The balanced treatment law, Judge Overton concluded, violated this Establishment Clause. The Arkansas antievolution law was ruled unconstitutional and was thrown out (LaFollette 1983, 81–82).

The creationists, however, remained adamant. The state representative who sponsored Act 590 told the newspapers, "If we lose, it won't matter that much. If the law is unconstitutional, it'll be because of something in the language that's wrong. . . . So we'll just change the wording and try again with another bill. . . . We got a lot of time. Eventually we'll get one that is constitutional" (*Washington Post*, December 7, 1981). On the same day that Judge Overton ruled the Arkansas law unconstitutional, the Mississippi state legislature passed a similar balanced treatment bill by a vote of 48–4. These persistent efforts of the proponents of creation science finally brought the issue to the U.S. Supreme Court. In June 1987, the Court, by a vote of 7–2, ruled that the purpose of creation science was "to restructure the science curriculum to conform with a particular religious viewpoint" (U.S. Supreme Court, *Edwards* v. *Aguilard*, 1987). Such acts, the Court decreed, "violate the Establishment Clause of the First

Amendment because they seek to employ the symbolic and financial support of government to achieve a religious purpose." As a result of this decision, all existing balanced treatment laws were thrown out.

Following this defeat, the creation scientists once again changed their tactics. First, they moved their focus from attempting to pass state laws mandating the teaching of creation science to attempting to pressure textbook committees and local school boards (where their highly organized and well-financed political machines can exert tremendous influence) into voluntarily granting equal time for creation science (Nelkin 1978; Nelkin 1982, 93–95). Second, they began changing the thrust of their arguments. Instead of claiming that creationism is a science and should therefore be taught in public schools, they began to admit that creationism really is religion. But, they assert, so is evolution. Evolution, they now say, is nothing more than the "religion" of "secular humanism," and therefore evolution should not be taught in public schools either. This argument has yet to be tested in the Supreme Court, but it has already failed in a number of federal courts.

Textbook Battles

The most recent creationist tactic has been to lobby state textbook committees to either drop mention of evolutionary biology altogether or to add a disclaimer to public school texts. The disclaimer the creationists dearly want would inform the student that evolution is "just a theory." On January 16, 1998, the Washington state senate introduced a bill (sponsored by senators Harold Hochstatter, Val Stevens, and Dan Swecker) that would require all biology textbooks to contain a printed disclaimer. All such textbooks purchased with state money would, if this bill were enacted into law, henceforth have the following notice placed prominently in them:

> A MESSAGE FROM THE WASHINGTON STATE LEGISLATURE
>
> This textbook discusses evolution, a controversial theory some scientists present as a scientific explanation for the origin of living things, such as plants, animals, and humans. No one was present when life first

> appeared on earth. Therefore, any statement about life's origins should be considered as theory, not fact. The word "evolution" may refer to many types of change. Evolution describes changes that occur within a species. (White moths, for example, may "evolve" into gray moths.) This process is microevolution, which can be observed and described as fact. Evolution may also refer to the change of one living thing to another, such as reptiles into birds. This process, called macroevolution, has never been observed and should be considered a theory. Evolution also refers to the unproven belief that random, undirected forces produced a world of living things. (Senate Bill 6058, State of Washington, 57th Legislature, 2001 Regular Session)

The disclaimer concludes with the legitimate assertion that

> There are many unanswered questions about the origin of life which are not mentioned in your textbook, including: 1) Why did the major groups of animals suddenly appear in the fossil record (known as the "Cambrian Explosion")?, 2) Why have no new major groups of living things appeared in the fossil record for a long time?, 3) Why do major groups of plants and animals have no transitional forms in the fossil record?, and 4) How did you and all living things come to possess such a complete and complex set of "Instructions" for building a living body?" (Senate Bill 6058)

Contemporary evolutionary biologists and geologists do in fact debate details of some of these experimental and conceptual issues. Such debates are a normal aspect of the institutional character of science and the professional scientific community. But creationists, such as the cosponsors of Washington State Senate Bill 6058, are confounding legitimate scientific controversy within a discipline—which is one of the most characteristic features of modern science, as the Victorian evolution debates amply demonstrate—with the conclusion that no consensus can be achieved. The creationists want to co-opt the normal, and necessary, function of scientific debate to make the categorical assertion that no single theory can ever rank as the most valid. This latest creationist twist on uncertainties in science would then render evolution merely one of several equally plausible theories. The essence of scientific inquiry, however, is to subject competing theories to the rigorous criteria of logical consistency and empirical validation. That theory, in any discipline, which best fits these criteria is deemed by the scientific community to be the accepted scientific paradigm for a given field. Science is an ongoing process. But the process yields theories that gain the status of being the best explanation of the phenomena in question. Evolution, after more than a century and a half of exhaustive philosophical, methodological, and experimental scrutiny, has emerged as the accepted basis of biological science. The crux of the creationist crusade, therefore, is to distort the consensual (but rigorously validated) methods and conclusions of modern science in an incessant, and deeply flawed, critique of evolutionary theory (Pennock 1999). Such a strategy, while politically and emotionally effective on certain occasions and to certain individuals and groups, parodies the very process by which modern science is able to construct valid theories about the natural world.

In April 1994, the Tangipahoa School Board in Louisiana passed a policy mandating that a disclaimer be presented before any discussion of evolutionary theory. A number of parents in the school district, including Herb Freiler, filed suit. In the *Freiler* v. *Tangipahoa Board of Education* case, the federal district judge ruled that the disclaimer was an unconstitutional establishment of religion. This decision was upheld on appeal by the federal circuit court. In its opinion upholding the appeal, the circuit court writes, "We conclude that the primary effect of the disclaimer is to protect and maintain a particular religious viewpoint, namely belief in the Biblical version of creation," and noted that the stated purpose of the disclaimer, to "exercise critical thinking," was "a sham" (U.S. Circuit Court, 5th Circuit, *Freiler* v. *Tangipahoa Board of Education*, August 19, 1999. No. 97–30879 and No. 98–30132). The *Freiler* ruling renders it highly likely that various courts will continue to reject the use of such disclaimers on constitutional grounds. But this appears to have had the result not of dampening creationist ardor but rather of shifting it more forcefully to the level of local school boards. This not unexpected focus on local school boards and local communities serves to emphasize that the current creationist organizations fully recognize their goal is political as well as religious. As such, the contemporary creationist attacks on evolution are more accurately seen not as scientific controversies but as the manifestation of deeply held convictions regarding the nature of civil society and the role to be accorded religious belief and authority therein. Creation science is one element in a larger Fundamentalist religious movement in the United States. Contextualized within this broader religious/political reality, it becomes clear why there are overt ties between advocates of creation science and the Christian Fundamentalist political spokespersons in the American public arena.

Creation Science and the Politics of Religion

In addition to providing financial support for creationist organizations, Christian Right televangelists advertise and promote creationist books and literature. "Experts" from the ICR and CRS are frequent guests on Fundamentalist television and radio programs. Jerry Falwell and his Moral Majority Inc. have become particularly close to the creationists. In late 1981, Falwell telecast an appeal for money to help defend the antievolution laws in Arkansas and Louisiana—using as the backdrop for his appeal the same Dayton, Tennessee, courthouse in which the original Scopes trial was held. As Philip Kitcher astutely points out, both the creationists and the Fundamentalists gain benefits from this partnership. "Jerry Falwell's Old Time Gospel Hour offers a forum for broadcasting creationist ideas. On the other hand, Falwell needs concrete issues around which to build his movement" (Kitcher 1982, 2). The televangelists recognize the creation scientists as powerful voices to bring new people into the Christian political movement. In turn, creation scientists have come to depend upon the Religious Right as a powerful political and economic ally. Moral Majority cofounder Tim LaHaye has close ties to the creationists. In his influential Fundamentalist manifesto *Battle for the Mind* (1980), LaHaye puts the fight against evolution squarely in the middle of the evangelical Christian worldview. The basic enemy of the Christian Right is something they refer to as "secular humanism." This term has a respectable lineage dating back at least as early as the Victorian era. But in the hands of LaHaye and his followers, secular humanism has become a catch-all phrase for any outlook or philosophy they find religiously offensive—everything from pornography to feminism to socialism to evolutionary science. The creationists and the Religious Right share a worldview that gives pride of place to eradicating the supposed evils of evolutionary science. Both groups see evolution as a major pillar that supports "Satanic secular humanism," and both are determined to do away with that pillar and substitute a religious outlook instead, in other words, creationism (Numbers 1992, 86–87, 290, 336–339).

But discrediting evolution is by no means their only goal. Paul Weyrich, a cofounder of Moral Majority and director of the Fundamentalist Committee for the Survival of a Free Congress, declares, "We are no longer working to preserve the status quo. We are radicals, working to overturn the present power structures of this country" (Kater 1982, 7). As perennial presidential candidate Pat Robertson echoes, "We have enough votes to run the country. And when the people say 'We've had enough,' we are going to take over" (Boston 1996, 29). Robertson told his supporters that his presidential bid was a direct command from God: "I heard the Lord saying, 'I have something else for you to do. I want you to run for President of the United States'" (Boston 1996, 39). In a fundraising letter for the Christian Coalition in July 1991 (Robertson founded the coalition and serves, along with his son, as one of the four members of the board of directors), Robertson asserts, "We at the Christian Coalition are raising an army who cares. We are training people to be effective—to be elected to school boards, to city councils, to state legislatures and to key positions in political parties. . . . By the end of this decade, if we work and give and organize and train, the Christian Coalition will be the most powerful political organization in America" (Boston 1996, 85). Ralph Reed, who serves as one of Robertson's front men in the Christian Coalition, says, "What Christians have got to do is take back this country, one precinct at a time, one neighborhood at a time and one state at a time. . . . I honestly believe that in my lifetime we will see a country once again governed by Christians." The Christian Right, with its supporters from the ranks of creation science, have formed alliances in order to carry out their goals of "overturning the power structures" and "taking over every area of life" and becoming "the most powerful political organization in America" (Boston 1996, 90).

As the Fundamentalists reiterate, one of the most important areas in which they must impose "Biblical moral law" are the local school districts. They make it clear that creation science is the key issue that provides them with the opportunity to do this. LaHaye declared, "The elite-evolutionist-humanist is not going to be able to control education in America forever" (LaHaye 1980, 3). Robertson stated that "humanist values are being taught in the schools through such methods as 'values clarification.' All of these things constitute an attempt to wean children away from biblical Christianity." Other fundamentalist apologists are just as blunt about their ultimate goals for public education. One spokesperson asserted,

> There are 15,700 school districts in America. When we get an active Christian parents' committee in operation in all districts, we can take complete control of all local school boards. This would allow us to determine all local policy; select good textbooks; good curriculum programs; superintendents and principals. . . . The Christian community has a golden opportunity to train an army of dedicated teachers who can invade the public school classrooms and use them to influence the nation for Christ." (LaFollette 1983; Boston 1996)

A portrait of this Fundamentalist "utopia" readily emerges from the cumulative proclamations cited above. The Christian Right has spent enormous time and effort in pushing for legislation to protect the "traditional family," as they define it according to biblical terms. In Indiana and Washington, Fundamentalists have sued in an attempt to have all state child-abuse and spouse-abuse laws repealed on the grounds that such laws violate their religious freedom by interfering with the biblical right of a father to "have dominion" over his wife and children and by abridging the father's "divine right" to discipline.

Another area that receives considerable Fundamentalist attention is homosexuality. Fundamentalists have made massive efforts to oppose and roll back civil rights for gays and lesbians, in an effort to marginalize them and eventually ghettoize them (Numbers 1992, 289). As extreme and offensive as these views may be, they nonetheless reflect the beliefs and aspirations of significant segments of the United States' population. The crucial point for this book is that evolutionism and antievolutionism both reflect and influence widely held cultural assumptions concerning "humans' place in nature." This was true in the Victorian era and remains true today.

Intelligent Design

A recent entry in the evolution/creation debates is the intelligent design (ID) movement. Intelligent design theorists began to make headlines in the 1990s with their dramatic fiats concerning scientific method and their claims of having found indisputable evidence of God in nature. As demonstrated in previous chapters, there is a long history of attempts to accommodate evolutionary science and religious beliefs. Eminent Victorian naturalists such as Wallace, Gray, and Mivart made compelling arguments for incorporating theistic elements into the broad framework of evolutionary thought. We have seen that their judicious, though at times controversial, efforts were influential and respected by members of both the scientific and religious communities, as well as the general public. Debates took a turn for the worse with the strident crusades of Fundamentalist creation science. Where do the ID advocates fit in?

The first ID book to reach a wide audience, and one of the first explicitly to adopt the intelligent design slogan, was Dean H. Kenyon and Percival Davis's *Of Pandas and People: The Central Question of Biological Origins* (1989). The book was intended for use as a high-school text, with its authors claiming to be writing from a Darwinian perspective. However, the authors found that Darwinian explanations frequently "required" an ID gloss. They defined intelligent design creationism as a framework that "locates the origin of new organisms in an immaterial cause: in a blueprint, a plan, a pattern, devised by an intelligent agent." In an appended "Note to teachers," Kenyon and Davis emphatically declared that their approach differed sharply from religious Fundamentalism and creation science. Most evolutionists remained highly skeptical, dubbing ID as a "creationist alias." In fact, until the mid-1990s no major academic or trade press had published a work supporting intelligent design creationism or creation science. The situation changed in 1996, when the Free Press of New York published Michael J. Behe's *Darwin's Black Box: The Biochemical Challenge to Evolution* (Numbers 1998, 15–17).

According to its architects, such as Behe, intelligent design theory—unlike creation science—does not require the vast majority of the scientific community to be in error. Intelligent design proponents argue that their theory is consistent with, not subversive of, scientific experimentation and data. The old age of the earth and the universe is not challenged by ID, nor is radioactive dating of fossils. More directly, ID advocates maintain that their theory challenges neither common ancestry nor the massive evidence for evolutionary descent with modification. They agree that the fossil record provides powerful data indicating a continual and gradual increase in complexity of organisms in time. ID differs sharply with fundamentalist creation science in that no issue is taken with the vast majority of the conclusions drawn by biologists, biochemists, microbiologists, and other researchers in fields related to evolution. Intelligent design, its advocates maintain, leaves most of evolutionary theory intact (Pennock 2001).

The basic tenets of intelligent design include the following: (1) The information needed for life is contained in a molecule known as DNA. This information can be analyzed with a field of science called information theory. (2) The complexity of life is a measure of the information in its DNA. Information and complexity are synonyms. (3) Natural selection does not create information. It only modifies existing information. Thus, new information may be created by genetic drift, or random changes to DNA. Or new information may be built into the evolutionary process by some designing agent. (4) The odds associated with events in the past (like the origin and evolution of life) can be accurately determined using information and probability theory. What the ID theorists do challenge, however, is the conclusion, drawn by scientists such as Richard Dawkins and Daniel Dennett, that naturalistic processes fully explain the origin and complexity of life, and that, as a result, there is no need for a designer of the complex evolutionary patterns observed. Intelligent design theorists claim that scientists such as Dawkins and Dennett represent only one faction within the scientific community. The ID theorists argue that those scientists willing to look beyond strict naturalism generally find that design is the best and most logical explanation for life's evolutionary complexity (Behe 1996). In this respect, ID can be regarded as a modern incarnation of William Paley's "argument from design," which held that God's existence could be proved by examining His works. Paley's celebrated example of divine design, as we have seen, was the vertebrate eye—a powerful metaphor that Victorian naturalists could scarcely ignore (Scott 1997, 279–280).

Intelligent design theory, in this sense, centers on the notion of teleological explanations in evolutionary biology. At least since the time of Aristotle, it has been clear that there is something distinctive about biological explanation. Living forms, with their exquisite adaptations to the environment and their complex functioning, seem to defy the notion that they are the products of random chance (Keller and Lloyd 1992, 324–333). Before Darwin, naturalists such as Georges Cuvier argued that there was some central organizing principle that served to coordinate the individual parts of a given organism and integrate them toward the well-being and fulfillment of the whole organism. Central to teleological explanation is this notion of forward-looking metaphor and language. One must legitimately ask questions about the function and purpose of organic structures and their future use in order to explain their nature and existence. Nineteenth-century evolutionary theory was deeply infused with teleological thinking. Darwin and Wallace, and their many peers, pondered the origin and evolution of complex structures such as the human eye or a bird's wing (Darwin 1859, 186–194).

Many nineteenth- and twentieth-century biologists followed Darwin's and Wallace's commitment to the teleologists' initial premise that living organisms are design-like, whatever the ultimate cause of this design. For most evolutionary biologists, the distinctive mark of the organic world is its forward-looking functionality (Buller 1999). Organisms *are* adapted to their particular environments and, hence, there is a teleological facet to explaining their origin and subsequent evolution. For Darwin and Wallace, natural selection provided the main, but not exclusive, explanatory tool for understanding how and why species evolved as they did. Darwin and Wallace differed on what additional factors other than natural selection were required for a complete explanation of the evolutionary process. These biological debates continue to the present day. If teleology is regarded at one level as a metaphor, then what we have is the undeniable fact that for most people, biologists as well as nonbiologists, organisms are treated as if they were designed. Whether this design is owing to God or some conscious intelligence, or whether it is simply the product of wholly naturalistic and random forces (the "blind watchmaker" of Dawkins's arresting phrase) is, of course, a central issue of debate in contemporary evolutionary science (Ruse 2000). With the discovery of the structure and functioning of the genetic material DNA in the 1950s, the debates were pushed to an even more fundamental level of biological analysis. It is at this point that contemporary ID theory enters the picture.

Michael Behe, a Catholic biochemist at Lehigh University, argued that biochemistry had "pushed Darwin's theory to the limit . . . by opening the ultimate black box, the cell, thereby making possible our understanding of how life works." The "astonishing complexity of subcellular organic structure," he asserted, led him to conclude "not from sacred works or sectarian beliefs" but from scientific data, that intelligent design had been operating in evolution. Behe declared that his result "is so unambiguous and so significant that it must be ranked as one of the greatest achievements in the history of science. . . . The discovery [of intelligent design] rivals those of Newton and Einstein, Lavoiser and Schroedinger, Pasteur and Darwin" (Behe 1996, 15, 33, 193, 232–233). Needless to say, not everyone agreed with Behe's self-assessment. Among his more vehement critics was the eminent biologist-philosopher Richard Dawkins, author of the famous book *The Blind Watchmaker* and one of the most adamant contemporary opponents of theistic science. Dawkins publicly rebuked Behe on television for lazily relying on intelligent design when he should have gone looking for scientifically acceptable explanations of the biochemical data. But others hailed Behe. The influential evangelical magazine *Christianity Today* gave its "Book of the Year" award for 1997 to *Darwin's Black Box.* By the mid 1990s, individual ID theorists joined forces to organize conferences, establish a well-funded Center for the Renewal of Science and Culture, and begin publishing a journal, *Origins & Design.* The goal of the ID advocates was to articulate a framework that included nonnaturalistic factors as legitimate elements of modern scientific explanations of phenomena. Not unexpectedly, the ID faction encountered criticism on many fronts. Their critics include theistic evolutionists (who are wary of invoking overtly nonnaturalistic factors in scientific theories), the Fundamentalist creation scientists (who disliked the ID theorists' failure to include scriptural references), and ardent naturalistic evolutionists like Dawkins and Daniel C. Dennett. The latter group saw ID as a "wedge," whose ultimate goal was to overturn Darwinism and scientific naturalism.

Dennett had achieved a degree of notoriety with his *Darwin's Dangerous Idea* (1995). In that book, Dennett approvingly announced that Darwinism is "a universal solvent, capable of cutting right to the heart of everything in sight," particularly in dissolving religious beliefs. Stephen Jay Gould, among other major evolutionary biologists, in turn criticized Dennett for his atheistic materialist rendering of evolution. Gould, a self-described agnostic, regretted the stridency which the extreme creationists, at one end of the spectrum, and the dogmatic scientific naturalists, at the other end, had injected into the long-standing attempt to find some sort of scientifically rigorous but religiously sensitive framework for evolutionary theory. Given that nearly 40 percent of the American population subscribes to some version of theistic evolutionism, Gould's conciliatory approach is broadly reflective of current attitudes—both within the scientific community and the general culture (Numbers 1998, 12–14, 18–21). The Victorian debates on the scientific, theological, social, and political implications of evolutionary theory were profound and exposed the deep and divisive impact evolutionism had upon traditional cultural assumptions. As the ID controversy demonstrates, those debates continue in full force today. The questions that the Victorians were forced to confront because of the overwhelming success of evolutionism as a scientific explanation of a vast range of phenomena have not gone away. If anything, advances in evolutionary science, genetics, and controversial areas such as reproductive technologies, genetically modified foods, and stem cell research, promise to keep the societal fires burning well into the twenty-first century. Whether ID is a new chapter in the creationist/evolution saga or merely an updated twist on creation science is the subject of considerable contemporary scholarly dispute (Pennock 2001).

There are two main reasons why ID is regarded seriously by some segments of the scientific community, even if it is still deemed a deeply flawed theory. Most of the leading ID scientists, as we have seen, pointedly dissociate themselves from creation scientists or scientific creationism. Specifically, ID theorists claim (whether rightly or wrongly is, to be sure, open to debate) that "intelligent design is a strictly scientific theory devoid of religious commitments." According to the mathematician-philosopher William Dembski,

> Whereas the Creator underlying scientific creationism conforms to a strict, literalist interpretation of the Bible, the designer underlying intelligent design is compatible with a much broader playing field. To be sure, the designer is compatible with the Creator-God of the world's major monotheistic religions like Judaism, Christianity, and Islam. But the designer is also compatible with the watchmaker-God of the deists, the demiurge of Plato . . . and the divine reason . . . of the ancient Stoics. One can even take an agnostic view about the designer, treating specified complexity as a brute unexplainable fact. (Dembski 1999, 247, 252)

Second, and perhaps more important, is that ID literature is far more sophisticated than creation science literature and is generally directed toward university audiences, rather than grade schools or the general public (though this may change). The quality of argumentation and the scientific credentials of its leading proponents, such as Behe and Dembski, are clearly superior to those of creation science. The most activist ID proponents, moreover, are associated with secular institutions of higher learning rather than sectarian organizations such as the ICR. Behe, Dembski, and like-minded colleagues promote ID as activist academics, not revivalist ministers. Consequently, their works now are seriously discussed at the university and college level, while the aggressive antievolutionist works by Morris and his followers are regarded primarily as curiosities. Although the ideological and religious subtext of ID is apparent (if often downplayed or denied), its leading advocates focus upon challenging certain of the fundamental assumptions of methodological materialism in evolutionary explanations, and modern science more generally (Scott 1997, 280–285). Debates surrounding ID remind us once again that the Victorian questions concerning the presuppositions and limitations of scientific naturalism continue to inform our contemporary concepts of the theoretical and practical consequences of science's powerful role in society and culture.

THE VICTORIAN DEBATES CONTINUE

Since the Victorian era, evolutionism has become an integral part of the debates regarding the appropriate place of science in the broader culture. The relationship between evolutionary science and religion is a critical aspect of these wide-ranging debates. It should be emphasized that many of the major world religions—including most of mainline Protestantism, as well as Catholicism, Judaism, Hinduism, and Buddhism—have embraced some form of evolutionism (either theistic or pantheistic) and rejected or allegorized the account of origins in Genesis. Fundamentalist factions, such as creationists, some sects of Orthodox Jews, and increasing numbers of Islamic sects, in contrast, regard evolution as anathema. The controversies that greeted the publication of *Origin* almost 150 years ago resonate with undiminished fervor today (McIver 1992). The issues raised so sharply in the Victorian era concerning the nature of the relationship between evolutionary science and religious beliefs and practices persist (Burkhart 1997). These issues involve not only differing concepts of knowledge and belief but competing views of cultural authority. The rhetoric has been transformed somewhat and the balance of cultural authority shifted in the past 150 years. When *Origin* appeared in 1859, evolutionists were in the minority. Huxley, Darwin, Spencer, and their associates had to wage an uphill battle to influence individual and institutional attitudes in government, education, industry, and society. At the start of the twenty-first century, a similar effort is being mounted by the antievolutionists.

Although their crusade—and that is the operative metaphor—will undoubtedly fail, their tactics and their influence over certain individuals and institutions are unfortunate. Creation science presents obstacles to a fuller understanding of the role evolutionary science will, and should, play in our culture. The critical questions regarding the power of science to alter both the natural and human environments have become even more urgent than they were in the Victorian era. The theory of evolution by natural selection put forward by Darwin and Wallace, with its subsequent enrichment by the findings of genetics, is one of the most powerful forces transforming the relationship between humanity and nature in the modern world. Creation science distorts the framework by which we seek to balance the potential power of biology to affect human evolution with the need to critically assess the psychological, environmental, and societal impacts such potential changes might have in a rapidly changing world. Evolutionary science is an inextricable feature of our changing world. The challenge is not, as the creationists would have it, to reject evolution. The challenge is to respect, and monitor, the efforts of those who accept the incontestable evidence for evolution and seek to define an appropriate and balanced place for evolutionary thought in the complex, and often troubling, sociopolitical and environmental realities of contemporary culture. Part of this challenge is to develop an adequate evolutionary biology that incorporates elements of a worldview involving personal freedom, purposive causation, and possibly a naturalistic theism (Griffin 2000, 309–310).

Evolutionary theory, from the Victorian period to the present, is ecumenical. It is not, as the creationists would have it, a recipe for atheism or cultural determinism. Rather, the findings of evolutionary biology—and the processes it shows to be constantly at work—provide us with one powerful basis for approaching fundamental questions involving human nature. Viewed thus, the relationship between evolutionism and religion has often been challenging and difficult; it has also been, and still is, stimulating and fruitful. What have frequently seemed at first to be firm barriers to evolutionary and religious perspectives coexisting within a broader framework, prove on closer examination to be precisely the point where advances can be made and understandings can be achieved (Ruse 2001, 216–219). Our conceptions of politics, philosophy, morality, and the ever-present battle between aggressive and cooperative tendencies—to name but a few of the many flash points of culture—benefit immensely from accurate knowledge of evolutionary biology. Only by an open and informed debate that critically evaluates both the beneficial and detrimental potential of evolutionary science can we hope to deal with, as Darwin termed it, the light evolutionary knowledge and practice will shed on human origins, our history, and our future (Sloan 2000; Ruse 1996; Eldredge 2000; Lewontin 2000; Keller 1995; Sapp 1999). The Victorian debates on the cultural meaning of evolution were profound, but ultimately inconclusive. What is incontestable, however, is that the theory of evolution has forever altered the study and interpretation of human nature and human society.

BIBLIOGRAPHY

Adams, Mark B., ed. 1990. *The Well-Born Science: Eugenics in Germany, France, Brazil, and Russia.* Oxford: Oxford University Press.

Alic, Margaret. 1986. *Hypatia's Heritage: A History of Women in Science from Antiquity to the Late Nineteenth Century.* London: Women's Press.

Alkon, Paul K. 1994. *Science Fiction before 1900: Imagination Discovers Technology*. New York: Twayne.

Allen, David. 1994. *The Naturalist in Britain: A Social History.* Princeton, N.J.: Princeton University Press.

Alter, Peter. 1987. *The Reluctant Patron: Science and the State in Britain: 1850–1920.* Oxford: Berg.

Bagwell, Philip S., 1988. *The Transport Revolution.* London: Routledge.

Baigrie, Brian S., ed. 1996. *Picturing Knowledge: Historical and Philosophical Problems Concerning the Use of Art in Science.* Toronto: University of Toronto Press.

Bannister, Robert C. 1979. *Social Darwinism: Science and Myth in Anglo-American Social Thought.* Philadelphia: Temple University Press.

Barbour, Ian G. 1997. *Religion and Science: Historical and Contemporary Issues.* San Francisco: Harper.

Barrett, Paul H., et al., eds. 1987. *Charles Darwin's Notebooks: 1836–1844: Geology, Transmutation of Species, Metaphysical Inquiries.* London: British Museum (Natural History) and Cambridge University Press.

Barrow, Logie. 1986. *Independent Spirits: Spiritualism and English Plebeians, 1850–1910.* London and New York: Routledge & Kegan Paul.

Bartholomew, Michael. 1973. "Lyell and Evolution: An Account of Lyell's Response to the Prospect of an Evolutionary Ancestry for Man." *British Journal for the History of Science* 6: 261–303.

Barton, Ruth. 1998. "Huxley, Lubbock, and Half a Dozen Others: Professionals and Gentlemen in the Formation of the X Club, 1851–1864." *Isis* 89: 410–444.

Basalla, George. 1988. *The Evolution of Technology.* New York: Cambridge University Press.

Bashford, Alison. 1998. *Purity and Pollution: Gender, Embodiment, and Victorian Medicine.* New York: St. Martin's Press.

Beddall, Barbara G. 1988. "Darwin and Divergence: The Wallace Connection." *Journal of the History of Biology* 21: 1–68.

Beer, Gillian. 1983. *Darwin's Plots: Evolutionary Narrative in Darwin, George Eliot, and Nineteenth-Century Fiction.* London: Routledge.

Behe, Michael J. 1996. *Darwin's Black Box: The Biochemical Challenge to Evolution.* New York: Free Press.

Ben-David, Joseph. 1991. *Scientific Growth: Essays on the Social Organization and Ethos of Science.* Berkeley: University of California Press.

Benjamin, Marina, ed. 1991. *Science and Sensibility: Gender and Scientific Enquiry, 1780–1945.* Oxford: Basil Blackwell.

Berman, David. 1988. *A History of Atheism in Britain: From Hobbes to Russell.* London and New York: Croom Helm.

Blinderman, C. 1986. *The Piltdown Inquest.* Amherst, N.Y.: Prometheus Books.

Bodde, Derk. 1991. *Chinese Thought, Society, and Science: The Intellectual and Social Background of Science and Technology in Pre-modern China.* Honolulu: University of Hawaii Press.

Boston, Rob. 1996. *The Most Dangerous Man in America? Pat Robertson and the Rise of the Christian Coalition.* Amherst, N.Y.: Prometheus Books.

BIBLIOGRAPHY

Bowler, Peter J. 1983. *The Eclipse of Darwinism: Anti-Darwinian Evolution Theories in the Decades around 1900*. Baltimore: Johns Hopkins University Press.

———. 1985. "Scientific Attitudes to Darwinism in Britain and America." In *The Darwinian Heritage*, ed. D. Kohn. Princeton, N.J.: Princeton University Press.

———. 1989. *Evolution: the History of an Idea,* rev. ed. Berkeley: University of California Press.

———. 1990. *Charles Darwin: The Man and His Influence*. Oxford: Basil Blackwell.

———. 1993. *Biology and Social Thought: 1850–1914.* Berkeley: University of California at Berkeley, Office for the History of Science and Technology.

Bradie, Michael. 1994. *The Secret Chain: Evolution and Ethics.* Albany: State University of New York Press.

Brand, Leonard. 1997. *Faith, Reason and Earth History: A Paradigm of Earth and Biological Origins By Intelligent Design.* Berrien Springs, Mich.: Andrews University Press.

Brantlinger, Patrick. 1988. *Rule of Darkness: British Literature and Imperialism: 1830–1914.* Ithaca, N.Y.: Cornell University Press.

Briggs, Asa. 1989. *Victorian Things*. Chicago: University of Chicago Press.

Broad, William, and Nicholas Wade. 1982. *Betrayers of the Truth.* New York: Simon and Schuster.

Brooke, John, and Geoffrey Cantor. 1998. *Reconstructing Nature: The Engagement of Science and Religion.* Edinburgh: T&T Clarke.

Brooks, John L. 1984. *Just Before the Origin: Alfred Russel Wallace's Theory of Evolution.* New York: Columbia University Press.

Browne, Janet. 1983. *The Secular Ark: Studies in the History of Biogeography.* New Haven, Conn.: Yale University Press.

———. 1992. "A Science of Empire: British Biogeography before Darwin." *Revue d'histoire des Sciences* 45: 453–475.

———. 1995. *Charles Darwin: Voyaging*. Princeton, N.J.: Princeton University Press.

Bruner, Jerome. 1966. *Toward a Theory of Instruction.* Cambridge, Mass.: Harvard University Press.

Bryan, William Jennings, ed. 1909. *Speeches of William Jennings Bryan*. Vol. 2. New York: Funk & Wagnalls.

Buller, David J., ed. 1999. *Function, Selection, and Design.* Albany: State University of New York Press.

Burchfield, J. D. 1975. *Lord Kelvin and the Age of the Earth.* New York: Science History Publications.

Burkhardt, F. H., S. Smith, et al., eds. 1983–. *The Correspondence of Charles Darwin.* Cambridge: Cambridge University Press.

Burkhardt, Richard. 1981. "Evolution" and "Inheritance of Acquired Characters." In *Dictionary of the History of Science*, ed. W. F. Bynum, E. J. Browne, and Roy Porter. Princeton, N.J.: Princeton University Press.

Burkhart, James B. 1997. *Redeeming Culture: American Religion in an Age of Science.* Chicago: University of Chicago Press.

Butt, Nasim. 1991. *Science in Muslim Societies.* London: Grey Seal.

Butts, Robert E. 1993. *Philosophical Essays: Historical Pragmatics.* Dordrecht: Kluwer Academic Publishers.

Bynum, W. F., and Roy Porter, eds. 1987. *Medical Fringe & Medical Orthodoxy: 1750–1850.* London: Croom Helm.

Camerini, Jane R. 1993. "Evolution, Biogeography, and Maps: An Early History of Wallace's Line." *Isis* 84: 700–727.

———. 1996. "Wallace in the Field." In *Science in the Field*, ed. Henrika Kuklick and Robert E. Kohler. *Osiris*, 2d ser., 11: 44–65.

———. 1997. "Remains of the Day: Early Victorians in the Field." In *Victorian Science in Context*, ed. Bernard Lightman. Chicago and London: University of Chicago Press.

Charlton, D. G. 1984. *New Images of the Natural in France: A Study in European Cultural History, 1750–1800.* Cambridge: Cambridge University Press.

Clark, W. E. LeGros. 1964. *The Fossil Evidence for Human Evolution.* Chicago and London: University of Chicago Press.

Cohen, I. Bernard. 1985. *Revolution in Science.* Cambridge, Mass.: Harvard University Press.

———, ed. 1994. *The Natural Sciences and the Social Sciences: Some Critical and Historical Perspectives.* Dordrecht: Kluwer.

Coleman, D. C. 1992. *Myth, History and the Industrial Revolution.* London: Hambledon Press.

Coleman, William. 1971. *Biology in the Nineteenth Century: Problems of Form, Function and Transmutation.* New York: Wiley.

Collini, Stefan. 1991. *Public Moralists: Political Thought and Intellectual Life in Britain, 1850–1930.* Oxford: Clarendon Press.

Conway, Jill K. 1970. "Stereotypes of Femininity in a Theory of Sexual Evolution." *Victorian Studies* 14: 47–62.

Conway, Jill K., Susan C. Bourque, and Joan W. Scott, eds. 1989. *Learning about*

Women: Gender, Politics, and Power. Ann Arbor: University of Michigan Press.

Cooter, Roger. 1984. *The Cultural Meaning of Popular Science: Phrenology and the Organisation of Consent in Nineteenth-Century Britain.* Cambridge: Cambridge University Press.

Cooter, Roger, and Stephen Pumfrey. 1994. "Separate Spheres and Public Places: Reflections on the History of Science Popularization and Science in Popular Culture." *History of Science* 32: 237–267.

Croce, Paul Jerome. 1995. *Science and Religion in the Era of William James.* Chapel Hill: University of North Carolina Press.

Cronin, Helena. 1991. *The Ant and the Peacock: Altruism and Sexual Selection from Darwin to Today.* Cambridge: Cambridge University Press.

DaCosta Kaufmann, Thomas. 1992. *The Mastery of Nature: Aspects of Art, Science, and Humanism in the Renaissance.* Princeton, N.J.: Princeton University Press.

Darwin, Charles. 1859. *On the Origin of Species. A Facsimile of the First Edition [1859].* Cambridge, Mass.: Harvard University Press, 1964.

———. 1871. *The Descent of Man, and Selection in Relation to Sex.* 2 vols. London: John Murray. 2d ed. New York: A. L. Burt Co., 1874.

———. 1872. *The Expression of the Emotions in Man and Animals.* London: John Murray. Facsimile reproduction with an introduction by Konrad Lorenz. Chicago: University of Chicago Press, 1965.

Darwin, Charles, and Alfred Russell Wallace. 1858. "On the Tendency of Species to Form Varieties; and on the Perpetuation of Varieties and Species by Natural Means of Selection." *Journal of the Linnean Society of London, Zoology* 3: 45–62.

Daston, Lorraine, and Katharine Park. 1998. *Wonders and the Order of Nature, 1150–1750.* New York and Cambridge, Mass.: Zone Books/MIT Press.

De Beer, Gavin. 1963. *Charles Darwin: Evolution by Natural Selection.* London: Nelson.

de Camp, L. Sprague. 1968. *The Great Monkey Trial.* Garden City, N.Y.: Doubleday.

Dembski, William A. 1999. *Intelligent Design: The Bridge between Science and Theology.* Downers Grove, Ill.: InterVarsity Press.

Den Otter, Sandra M. 1996. *British Idealism and Social Explanation: A Study in Late Victorian Thought.* Oxford: Clarendon Press.

Desmond, Adrian. 1982. *Archetypes and Ancestors: Paleontology in Victorian London: 1850–1875.* London: Blond and Briggs.

———. 1989. *The Politics of Evolution: Morphology, Medicine, and Reform in Radical London.* Chicago: University of Chicago Press.

———. 1998. *Huxley.* London and New York: Penguin Books.

Desmond, Adrian, and James Moore. 1991. *Darwin.* New York: Warner Books.

Diamond, Sara. 1990. *Spiritual Warfare: The Politics of the Christian Right.* Montreal: Black Rose Books.

Dick, Stephen J. 1996. *The Biological Universe: The Twentieth-Century Extraterrestrial Life Debate and the Limits of Science.* Cambridge: Cambridge University Press.

Donohue, William A. 2000. *Twilight of Liberty: The Legacy of the ACLU.* New Brunswick, N.J.: Transaction Publishers.

Draper, John William. 1874. *History of the Conflict between Religion and Science.* New York: Appleton.

Dykeman, W., and J. Stokely. 1971. "Scopes and Evolution—The Jury Is Still Out." *New York Times Magazine,* 12 March 1971, p. 72.

Ecker, Ronald L. 1990. *Dictionary of Science and Creationism.* Amherst, N.Y.: Prometheus Books.

Eldredge, Niles. 2000. *The Triumph of Evolution.* New York: W. H. Freeman.

Ellegard, Alvar. 1958. *Darwin and the General Reader: The Reception of Darwin's Theory of Evolution in the British Periodical Press, 1859–1872.* Goteborg, Sweden: Elanders.

Eve, Raymond A., and Francis B. Harrold. 1991. *The Creationist Movement in Modern America.* Boston: Twayne.

Fancher, Raymond. 1998. "Biography and Psychodynamic Theory: Some Lessons from the Life of Francis Galton." *History of Psychology* 1: 99–115.

Farber, Paul Lawrence. 1994. *The Temptations of Evolutionary Ethics.* Berkeley: University of California Press.

Fayter, Paul. 1997. "Strange New Worlds of Space and Time: Late Victorian Science and Science Fiction." In *Victorian Science in Context,* ed. Bernard Lightman. Chicago and London: University of Chicago Press.

Feagin, Joe R. 2000. *Racist America: Roots, Current Realities, and Future Reparations.* New York: Routledge.

Fee, Elizabeth. 1974. "The Sexual Politics of Victorian Social Anthropology." In *Clio's Consciousness Raised: New Perspectives on the History of Women,* ed. Mary Hartman and Lois Banner. New York: Harper.

Fichman, Martin. 1981. *Alfred Russel Wallace.* Boston: Twayne.

———. 1984. "Ideological Factors in the Dissemination of Darwinism in England, 1860–1900." In *Transformation and Tradition in the Sciences: Essays in Honor of I. Bernard Cohen*, ed. Everett Mendelsohn. Cambridge: Cambridge University Press.

———. 1997. "Biology and Politics: Defining the Boundaries." In *Victorian Science in Context*, ed. B. Lightman. Chicago and London: University of Chicago Press.

———. 2001. "Science in Theistic Contexts: A Case Study of Alfred Russel Wallace on Human Evolution." *Osiris* 16: 227–250.

Fisch, Max H. 1964. "Was There a Metaphysical Club in Cambridge?" In *Studies in the Philosophy of Charles Sanders Peirce*, ed. Edward Moore and Richard Robin. Amherst: University of Massachusetts Press.

Flint, Robert. 1877. *Theism: Being the Baird Lecture for 1876.* Edinburgh: W. Blackwood.

Frye, Roland M., ed. 1983. *Is God a Creationist? The Religious Case against Creation-Science.* New York: Charles Scribner's Sons.

Galton, Francis. 1889. *Natural Inheritance.* London: Macmillan.

———. 1909. *Essays in Eugenics.* London: Eugenics Education Society.

Gasman, Daniel. 1971. *The Scientific Origins of National Socialism: Social Darwinism in Ernst Haeckel and the German Monist League.* New York: American Elsevier.

———. 1998. *Haeckel's Monism and the Birth of Fascist Ideology.* New York: Lang.

Gates, Barbara T. 1997. "Ordering Nature: Revisioning Victorian Science Culture." In *Victorian Science in Context*, ed. Bernard Lightman. Chicago and London: University of Chicago Press.

Gavin, William Joseph. 1992. *William James and the Reinstatement of the Vague.* Philadelphia: Temple University Press.

Gillham, Nicholas W. 2001. *A Life of Sir Francis Galton: From African Exploration to the Birth of Eugenics.* Oxford: Oxford University Press.

Gillispie, Charles C. 1951. *Genesis and Geology: A Study in the Relations of Scientific Thought, Natural Theology, and Social Opinion in Great Britain, 1790–1850.* Reprint, New York: Harper Torchbooks, 1959.

Glick, Thomas, ed. 1974. *The Comparative Reception of Darwinism.* Austin: University of Texas Press.

Golinski, Jan. 1992. *Science as Public Culture: Chemistry and the Enlightenment in Britain, 1760–1820.* Cambridge: Cambridge University Press.

Goodman, David, and Colin A. Russell, eds. 1991. *The Rise of Scientific Europe: 1500–1800.* Milton Keynes: Open University.

Goodman, Jordan, and Katrina Honeyman. 1988. *Gainful Pursuits: The Making of Industrial Europe, 1600–1914.* London: Edward Arnold.

Gordon, Scott. 1991. *The History and Philosophy of Social Science*. London and New York: Routledge.

Graham, Loren R. 1993. *Science in Russia and the Soviet Union: A Short History*. Cambridge: Cambridge University Press.

Gray, Asa. 1876. *Darwiniana: Essays and Reviews Pertaining to Darwinism*. New York: D. Appleton.

———. 1880. *Natural Science and Religion: Two Lectures Delivered to the Theological School of Yale College*. New York: Charles Scribner's Sons.

Greene, John C. 1981. *Science, Ideology, and World View: Essays in the History of Evolutionary Ideas*. Berkeley: University of California Press.

Greene, Mott T. 1982. *Geology in the Nineteenth Century: Changing Views of a Changing World*. Ithaca, N.Y.: Cornell University Press.

———. 1992. *Natural Knowledge in Preclassical Antiquity*. Baltimore: Johns Hopkins University Press.

Griffin, David Ray. 1989. *God and Religion in the Postmodern World: Essays in Postmodern Theology*. Albany: State University of New York Press.

———. 2000. *Religion and Scientific Naturalism: Overcoming the Conflicts*. Albany: State University of New York Press.

Haines, Valerie. 1991. "Spencer, Darwin, and the Question of Reciprocal Influence." *Journal of the History of Biology* 24: 409–431.

Haller, John S. 1971. *Outcasts from Evolution. Scientific Attitudes of Racial Inferiority, 1859–1900*. Urbana: University of Illinois Press.

Hankins, Thomas. L. 1985. *Science and the Enlightenment*. Cambridge: Cambridge University Press.

Hartshorne, Charles, and Paul Weiss, eds. 1958–1965. *Collected Papers of Charles Sanders Peirce*. 8 vols. Cambridge, Mass.: Harvard University Press.

al-Hassan, Ahmad L., and Donald R. Hill. 1986. *Islamic Technology*. Cambridge: Cambridge University Press.

Hawkins, Mike. 1997. *Social Darwinism in European and American Thought, 1860–1945*. Cambridge: Cambridge University Press.

Helfand, Michael. 1977. "T. H. Huxley's 'Evolution and Ethics': The Politics of Evolution and the Evolution of Politics." *Victorian Studies* 20: 159–77.

Hick, John H. 1990. *Philosophy of Religion*. 4th ed. Englewood Cliffs, N.J.: Prentice-Hall.

Hodges, Clifton, and Jean Dawson. 1918. *Civic Biology*. Boston: Ginn.

Hull, David L. 1973. *Darwin & His Critics*. Cambridge, Mass.: Harvard University Press.

Hulme, Peter, and Ludmilla Jordanova, eds. 1990. *The Enlightenment and Its Shadows*. London: Routledge.

Hunt, Bruce J. 1997. "Doing Science in a Global Empire: Cable Telegraphy and Electrical Physics in Victorian Britain." In *Victorian Science in Context*, ed. Bernard Lightman. Chicago and London: University of Chicago Press.

Hunter, George William. 1914. *A Civic Biology*. New York: American.

Huxley, Leonard, ed. 1901. *Life and Letters of Thomas H. Huxley*. 2 vols. New York: D. Appleton.

Huxley, Thomas Henry. 1863. *The Evidence as to Man's Place in Nature*. London: Williams & Norgate.

———. 1889. "Agnosticism." In *Collected Essays*. Vol. 5, *Science and Christian Tradition*. New York: D. Appleton, 1894, pp. 209–262.

———. 1893. "Evolution and Ethics." In *Collected Essays*. Vol. 9. London: Macmillan, 1893–94.

Jacob, Margaret C. 1988. *The Cultural Meaning of the Scientific Revolution*. New York: Alfred A. Knopf.

James, William. 1865. "Rev. of Alfred Russel Wallace 'Origin of Human Races.'" *North American Review* 101: 261–263.

Jann, Rosemary. 1994. "Darwin and the Anthropologists: Sexual Selection and Its Discontents." *Victorian Studies* 37: 287–306.

Jardine, Nicholas, James Secord, and Emma Spary, eds., 1996. *Cultures of Natural History*. Cambridge: Cambridge University Press.

Jones, Greta. 1980. *Social Darwinism and English Social Thought: The Interaction between Biological and Social Theory*. Sussex: Harvester Press.

Karlin, Daniel, ed. 1999. *Rudyard Kipling*. Oxford and New York: Oxford University Press.

Kater, John L. 1982. *Christians on the Right: The Moral Majority in Perspective*. New York: Seabury Press.

Keller, Evelyn Fox. 1995. *Refiguring Life*. New York: Columbia University Press.

Keller, Evelyn Fox, and Elisabeth A. Lloyd, eds. 1992. *Keywords in Evolutionary Biology*. Cambridge, Mass.: Harvard University Press.

Kevles, Daniel J. 1985. *In the Name of Eugenics: Genetics and the Uses of Human Heredity*. New York: Knopf.

Kingsley, Charles. 1885. *The Water-Babies: A Fairy Tale for a Land-Baby*. London: Macmillan.

Kitcher, Philip. 1982. *Abusing Science: The Case against Creationism*. Cambridge, Mass.: MIT Press.

Kofahl, Robert E., and Kelly Seagreaves. 1975. *The Creation Explanation: A Scientific Alternative to Evolution.* Wheaton, Ill.: Harold Shaw Publishers.

Kohn, David, ed. 1985. *The Darwinian Heritage.* Princeton, N.J.: Princeton University Press.

Kottler, Malcolm Jay. 1974. "Alfred Russel Wallace, the Origin of Man, and Spiritualism." *Isis* 65: 145–192.

———. 1985. "Charles Darwin and Alfred Russel Wallace: Two Decades of Debate over Natural Selection." In *The Darwinian Heritage,* ed. David Kohn. Princeton, N.J.: Princeton University Press.

Kuklick, Bruce. 1977. *The Rise of American Philosophy: Cambridge, Massachusetts, 1860–1930.* New Haven, Conn.: Yale University Press.

Kuklick, Henrika. 1992. *The Savage Within: The Social History of British Anthropology, 1885–1945.* Cambridge: Cambridge University Press.

LaFollette, Marcel, ed. 1983. *Creationism, Science and the Law: The Arkansas Case.* Cambridge, Mass.: MIT Press.

LaHaye, Tim F. 1980. *The Battle for the Mind.* Old Tappan, N.J.: Revell.

Larson, Edward J. 1997. *Summer for the Gods: The Scopes Trial and America's Continuing Debate over Science and Religion.* Cambridge, Mass.: Harvard University Press.

Lewontin, Richard C. 2000. *The Triple Helix: Gene, Organism, and Environment.* Cambridge, Mass.: Harvard University Press.

Lightman, Bernard. 1987. *The Origins of Agnosticism: Victorian Unbelief and the Limits of Knowledge.* Baltimore: Johns Hopkins University Press.

———, ed. 1997a. *Victorian Science in Context.* Chicago and London: University of Chicago Press.

———. 1997b. " 'The Voices of Nature': Popularizing Victorian Science." In *Victorian Science in Context*, ed. Bernard Lightman. Chicago and London: University of Chicago Press.

Lindberg, David C. 1992. *The Beginnings of Western Science: The European Scientific Tradition in Philosophical, Religious, and Institutional Context, 600 B.C. to A.D. 1450.* Chicago: University of Chicago Press.

Lindroth, Sten. 1973. "Carl Linnaeus." In *Dictionary of Scientific Biography*, vol. 8, ed. Charles C. Gillispie. 16 vols. New York: Charles Scribner's Sons.

Lorimer, Douglas A. 1997. "Science and the Secularization of Victorian Images of Race." In *Victorian Science in Context*, ed. Bernard Lightman. Chicago and London: University of Chicago Press.

Lyell, Charles. 1863. *The Geological Evidences of the Antiquity of Man.* 2d ed. London: John Murray.

Lyell, [K. M.]. 1881. *Life, Letters and Journals of Sir Charles Lyell.* 2 vols. London: John Murray.

Lynch, Michael, and Steve Woolgar, eds. 1990. *Representation in Scientific Practice.* Cambridge, Mass.: MIT Press.

MacKenzie, Donald. 1981. *Statistics in Britain, 1865–1930: The Social Construction of Scientific Knowledge.* Edinburgh: Edinburgh University Press.

Maienschein, Jane, and Michael Ruse, eds. 1999. *Biology and the Foundation of Ethics.* Cambridge and New York: Cambridge University Press.

Mandler, Peter, Alex Owen, Seth Koven, and Susan Pedersen. 1997. "Cultural Histories Old and New: Rereading the Work of Janet Oppenheim." *Victorian Studies* 41: 69–105.

Maniquis, Robert M. 1992. *The Encyclopedia and the Age of Revolution.* Boston: G. K. Hall.

Marchant, James. 1916. *Alfred Russel Wallace: Letters and Reminiscences.* New York: Harper & Brothers. Reprint, New York: Arno Press, 1975.

Martin, Julian. 1992. *Francis Bacon, the State, and the Reform of Natural Philosophy.* Cambridge: Cambridge University Press.

Mathias, Peter, and John A. Davis, eds. 1991. *The First Industrial Revolutions.* Oxford: Basil Blackwell.

Mayr, Ernst. 1991. *One Long Argument: Charles Darwin and the Genesis of Modern Evolutionary Thought.* Cambridge, Mass.: Harvard University Press.

Mazumdar, Pauline. 1992. *Eugenics, Human Genetics and Human Failings: The Eugenics Society, Its Source and Its Critics in Britain.* London: Routledge.

McCosh, James. 1890. *The Religious Aspect of Evolution.* Rev. ed. New York: Charles Scribner's Sons, 1888.

McIver, Tom. 1992. *Anti-Evolution: A Reader's Guide to Writings before and after Darwin.* Baltimore: Johns Hopkins University Press.

McKinney, H. Lewis. 1972. *Wallace and Natural Selection.* New Haven, Conn.: Yale University Press.

McKown, Delos Banning. 1993. *The Mythmaker's Magic: Behind the Illusion of "Creation Science."* Amherst, N.Y.: Prometheus Books.

Mill, John Stuart. 1863. *Utilitarianism.* Ed. Roger Crisp. Oxford and New York: Oxford University Press, 1998.

Montagu, Ashley, ed. 1984. *Science and Creationism.* New York: Oxford University Press.

Moore, James R. 1979. *The Post-Darwinian Controversies: A Study of the Protestant Struggle to Come to Terms with Darwin in Great Britain and America, 1870–1900.* Cambridge: Cambridge University Press.

———, ed. 1989. *History, Humanity and Evolution: Essays for John C. Greene.* Cambridge: Cambridge University Press.

Morgan, Thomas Hunt. 1932. *The Scientific Basis of Evolution.* New York: Norton.

Morland, David. 1997. *Demanding the Impossible? Human Nature and Politics in Nineteenth-Century Social Anarchism.* London and Washington: Cassell.

Morrell, J. B. 1990. "Professionalization." In *Companion to the History of Modern Science*, ed. Robert C. Olby, G. N. Cantor, J. R. R. Christie, and M. J. S. Hodge. London: Routledge.

Morris, Henry M. 1974a. *The Troubled Waters of Evolution.* San Diego, Calif.: Creation-Life Publishers.

———. 1974b. *Scientific Creationism.* Gen. ed. San Diego, Calif.: Creation-Life Publishers.

———. 1984. *A History of Modern Creationism.* San Diego, Calif.: Master Book Publishers.

Morton, Peter. 1984. *The Vital Science: Biology and the Literary Imagination, 1860–1900.* London: George Allen and Unwin.

Mossner, Ernest Campbell. 1967. "Deism." In *The Encyclopedia of Philosophy*, ed. Paul Edwards. New York: Macmillan Company and Free Press.

Myers, Greg. 1990. *Writing Biology: Texts in the Social Construction of Scientific Knowledge.* Madison: University of Wisconsin Press.

Nelkin, Dorothy. 1978. *Science Textbook Controversies and the Politics of Equal Time.* Cambridge, Mass.: MIT Press.

———. 1982. *The Creation Controversy: Science or Scripture in the Schools.* New York: W. W. Norton & Co.

Nitecki, Matthew, ed. 1988. *Evolutionary Progress.* Chicago: University of Chicago Press.

Noakes, Richard J. 1998. *Cranks and Visionaries: Science, Spiritualism, and Transgression in Victorian England.* Ph.D. dissertation, Cambridge University.

Nott, Josiah Clark, and George R. Gliddon. 1854. *Types of Mankind: or, Ethnological Researches, Based upon the Ancient Monuments, Paintings, Sculptures, and Crania of Races, and upon Their Natural, Geographical, Philological and Biblical History; Illustrated . . . by Additional Contributions from Prof. L. Agassiz.* Philadelphia: Lippincott, Grambo.

Numbers, Ronald L. 1992. *The Creationists.* New York: Alfred A. Knopf.

———. 1998. *Darwinism Comes to America.* Cambridge, Mass.: Harvard University Press.

Numbers, Ronald L., and John Stenhouse, eds. 1999. *Disseminating Darwinism:*

The Role of Place, Race, Religion, and Gender. Cambridge: Cambridge University Press.

Oldroyd, David, and Ian Langham. 1983. *The Wider Domain of Evolutionary Thought*. Dordrecht: D. Reidel Publishing Co.

Oppenheim, Janet. 1985. *The Other World: Spiritualism and Psychical Research in England, 1850–1914*. Cambridge: Cambridge University Press.

Owen, Alex. 1990. *The Darkened Room: Women, Power and Spiritualism in Late Victorian England.* Philadelphia: University of Pennsylvania Press.

Paradis, James G. 1997. "Satire and Science in Victorian Culture." In *Victorian Science in Context.* Ed. Bernard Lightman. Chicago and London: University of Chicago Press.

Paradis, James G., and George C. Williams. 1989. *Evolution and Ethics: T. H. Huxley's "Evolution and Ethics," with New Essays on Its Victorian and Sociobiological Context*. Princeton, N.J.: Princeton University Press.

Park, Hee-Joo. 2000. "The Politics of Anti-Creationism: The Committees of Correspondence." *Journal of the History of Biology* 33: 349–370.

Parker, Christopher, ed. 1995. *Gender Roles and Sexuality in Victorian Literature.* Aldershot, England: Scolar Press.

Parker, Kelly A. 1998. *The Continuity of Peirce's Thought.* Nashville and London: Vanderbilt University Press.

Parsons, Timothy. 1999. *The British Imperial Century, 1815–1914: A World History Perspective.* Lanham, Md.: Rowman & Littlefield.

Paul, Diane B. 1995. *Controlling Human Heredity, 1865 to the Present.* Amherst, N.Y.: Humanity Books.

Peabody, James E., and Arthur E. Hunt. 1912. *Elementary Biology.* New York: Macmillan.

Peel, J. D. Y. 1975. "Herbert Spencer." In *Dictionary of Scientific Biography*, vol. 12, ed. Charles C. Gillispie. New York: Charles Scribner's Sons.

Peirce, Charles Sanders. 1906. "Rev. of [Alfred Russel] Wallace's *My Life*." *Nation* 82: 160–161.

———. 1997. *Pragmatism as a Principle and Method of Right Thinking: The 1903 Harvard "Lectures on Pragmatism,"* ed. Patricia Ann Turrisi. Albany: State University of New York Press.

Pennock, Robert T. 1999. *The Tower of Babel: The Evidence against the New Creationism.* Cambridge, Mass.: MIT Press.

———, ed. 2001. *Intelligent Design Creationism and Its Critics: Philosophical, Theological, and Scientific Perspectives.* Cambridge, Mass. MIT Press.

Pittenger, Mark. 1993. *American Socialists and Evolutionary Thought, 1870–1920.* Madison: University of Wisconsin Press.
Porter, Theodore M. 1990. "Natural Science and Social Theory." In *Companion to the History of Modern Science*, ed. Robert Olby et al. London: Routledge.
Potter, Vincent G. 1996. *Peirce's Philosophical Perspectives*, ed. Vincent M. Colapietro. New York: Fordham University Press.
Powell, Baden. 1855. *Essays on the Spirit of the Inductive Philosophy, the Unity of Worlds, and the Philosophy of Creation.* Reprint, Westmead, England: Gregg International, 1969.
Price, George McCready. 1923. *The New Geology.* Mountain View, Calif.: Pacific Press.
Pyenson, Lewis, and Susan Sheets-Pyenson. 1999. *Servants of Nature: A History of Scientific Institutions, Enterprises and Sensibilities.* New York and London: W. W. Norton & Company.
Quinn, Philip L., and Charles Taliaferro, ed. 1997. *A Companion to Philosophy of Religion.* Oxford: Blackwell Publishers Ltd.
Raposa, Michael L. 1989. *Peirce's Philosophy of Religion.* Bloomington: Indiana University Press.
Rehbock, Philip F. 1983. *The Philosophical Naturalists: Themes in Early Nineteenth-Century British Biology.* Madison: University of Wisconsin Press.
Reingold, Nathan, and Marc Rothenberg, eds. 1987. *Scientific Colonialism: A Cross-Cultural Comparison.* Washington, D.C.: Smithsonian Institution Press.
Richards, Evelleen. 1983. "Darwin and the Descent of Woman." In *The Wider Domain of Evolutionary Thought*, ed. David Oldroyd and Ian Langham. Dordrecht: D. Reidel Publishing Co.
———. 1989a. "Huxley and Woman's Place in Science: The 'Woman Question' and the Control of Victorian Anthropology." In *History, Humanity and Evolution*, ed. James Moore. Cambridge: Cambridge University Press.
———. 1989b. "The 'Moral Anatomy' of Richard Knox: The Interplay between Biological and Social Thought in Victorian Scientific Naturalism." *Journal of the History of Biology* 22: 373–436.
———. 1997. "Redrawing the Boundaries: Darwinian Science and Victorian Women Intellectuals." In *Victorian Science in Context,* ed. Bernard Lightman. Chicago and London: University of Chicago Press.
Richards, Robert J. 1987. *Darwin and the Emergence of Evolutionary Theories of Mind and Behavior.* Chicago: University of Chicago Press.
———. 1992. *The Meaning of Evolution: The Morphological Construction and*

Ideological Reconstruction of Darwin's Theory. Chicago: University of Chicago Press.

Ritvo, Harriet. 1997. "Zoological Nomenclature and the Empire of Victorian Science." In *Victorian Science in Context,* ed. Bernard Lightman. Chicago and London: University of Chicago Press.

Roberts, Jon H. 1988. *Darwinism and the Divine in America: Protestant Intellectuals and Organic Evolution, 1859–1900.* Madison: University of Wisconsin Press.

Rohr, Richard. 1988. *The Great Themes of Scripture.* Cincinnati: St. Anthony Messenger Press.

Rose, Steven, R. C. Lewontin, and Leon J. Kamin. 1990. *Not in Our Genes: Biology, Ideology and Human Nature.* London: Penguin Books.

Rosenberg, Alexander. 1992. "Altruism: Theoretical Contexts." In *Keywords in Evolutionary Biology*, ed. Evelyn F. Keller and Elisabeth A. Lloyd.

Rousseau, George S. 1980. *The Ferment of Knowledge: Studies in the Historiography of Eighteenth-Century Science.* Cambridge and New York: Cambridge University Press.

Rubino, Carl A. 1997. "Journeys, Maps, and Territories: Charting Uncertain Terrain in Science and Literature." *Intertexts* 1: 118–130.

Rudwick, Martin J. 1985. *The Great Devonian Controversy: The Shaping of Scientific Knowledge among Gentlemanly Specialists.* Chicago: University of Chicago Press.

Rupke, Nicolaas. 1994. *Richard Owen: Victorian Naturalist.* New Haven, Conn.: Yale University Press.

Ruse, Michael, ed. 1988. *But Is It Science? The Philosophical Question in the Creation/Evolution Controversy.* Amherst, N.Y.: Prometheus Books.

———. 1996. *Monad to Man: The Concept of Progress in Evolutionary Biology.* Cambridge, Mass.: Harvard University Press.

———. 1999. *The Darwinian Revolution: Science Red in Tooth and Claw.* 2d ed. Chicago: University of Chicago Press.

———. 2000. "Teleology: Yesterday, Today, and Tomorrow?" *Studies in History and Philosophy of Biological and Biomedical Sciences* 31: 213–232.

———. 2001. *Can a Darwinian Be a Christian? The Relationship between Science and Religion.* Cambridge and New York: Cambridge University Press.

Russett, Cynthia Eagle. 1976. *Darwin in America: The Intellectual Response, 1865–1912.* San Francisco: W. H. Freeman.

———. 1989. *Sexual Science: The Victorian Construction of Womanhood.* Cambridge, Mass.: Harvard University Press.

Sapp, Jan. 1999. *What Is Nature? Coral Reef Crisis.* New York: Oxford University Press.

Scott, Eugenie C. 1997. "Antievolution and Creationism in the United States." *Annual Review of Anthropology* 26: 263–289.

Seagreaves, Kelly. 1975. *Sons of God Return.* New York: Pyramid Books.

Searle, G. R. 1976. *Eugenics and Politics in Britain, 1900–1914.* Leiden: Noordhoff International Publishing.

Secord, James A. 2000. *Victorian Sensation: The Extraordinary Publication, Reception, and Secret Authorship of "Vestiges of the Natural History of Creation."* Chicago and London: University of Chicago Press.

Seitz, Frederick. 1992. *The Science Matrix.* New York: Springer-Verlag.

Sheets-Pyenson, Susan. 1989. *Cathedrals of Science: The Development of Colonial Natural History Museums in the Late Nineteenth Century.* Montreal: McGill-Queens University Press.

Shor, Elizabeth Noble. 1974. "Othniel Charles Marsh." In *Dictionary of Scientific Biography*, vol. 9, ed. Charles C. Gillispie. New York: Charles Scribner's Sons.

Shteir, Ann B. 1997. "Elegant Recreations? Configuring Science Writing for Women." In *Victorian Science in Context*, ed. B. Lightman. Chicago and London: University of Chicago Press.

Sloan, Philip R., ed. 2000. *Controlling Our Destinies: Historical, Philosophical, Ethical, and Theological Perspectives on the Human Genome Project.* Notre Dame, Ind.: University of Notre Dame Press.

Smith, Charles H., ed. 1991. *Alfred Russel Wallace: An Anthology of His Shorter Writings.* Oxford: Oxford University Press.

———. 1992. *Alfred Russel Wallace on Spiritualism, Man, and Evolution: An Analytical Essay.* Torrington, Conn.: Charles H. Smith.

Smith, Crosbie. 1998. *The Science of Energy: A Cultural History of Energy Physics in Victorian Britain.* Chicago: University of Chicago Press.

Smith, Roger. 1981. "Evolutionism in Mind and Society." In *Dictionary of the History of Science*, ed. W. F. Bynum, E. J. Browne, and Roy Porter. Princeton, N.J.: Princeton University Press.

Spencer, Frank. 1990. *Piltdown: A Scientific Forgery.* Oxford: Oxford University Press.

Spencer, Herbert. 1887. *The Factors of Organic Evolution.* London: Williams and Norgate.

Stafford, Barbara Maria. 1984. *Voyage into Substance: Art, Science, Nature and the Illustrated Travel Account, 1760–1840.* Cambridge, Mass.: MIT Press.

Stafford, Robert. 1989. *Scientist of Empire: Sir Roderick Murchison, Scientific Exploration, and Victorian Imperialism.* Cambridge: Cambridge University Press.

Stepan, Nancy Leys. 1982. *The Idea of Race in Science: Great Britain, 1800–1960.* Hamden, Conn.: Archon Books.

Stepan, Nancy Leys, and Sander Gilman. 1993. "Appropriating the Idioms of Science: The Rejection of Scientific Racism." In *The "Racial" Economy of Science: Toward a Democratic Future,* ed. Sandra Harding. Bloomington: Indiana University Press.

Stocking, George W. 1987. *Victorian Anthropology.* New York: Free Press.

Strahler, Arthur N. 1999. *Science and Earth History: The Evolution/Creation Controversy.* Amherst, N.Y.: Prometheus Books.

Sulloway, Frank. 1982. "Darwin's Conversion: The *Beagle* Voyage and Its Aftermath." *Journal of the History of Biology* 15: 327–398.

Suvin, Darko. 1983. *Victorian Science Fiction in the U.K.: The Discourses of Knowledge and Power.* Boston: G. K. Hall.

Taylor, Eugene. 1990. "William James on Darwin: An Evolutionary Theory of Consciousness." *Annals of the New York Academy of Science* 602: 7–33.

Tompkins, Jeremy R., ed. 1965. *D-Days at Dayton: Reflections on the Scopes Trial.* Baton Rouge: Louisiana State University Press.

Turner, Frank M. 1974. *Between Science and Religion: The Reaction to Scientific Naturalism in Late Victorian England.* New Haven, Conn.: Yale University Press.

———. 1993. *Contesting Cultural Authority: Essays in Victorian Intellectual Life.* Cambridge: Cambridge University Press.

Tyndall, John. 1874. *Address Delivered before the British Association, Assembled at Belfast.* London: Longmans, Green.

Van Riper, A. Bowdoin. 1993. *Men Among the Mammoths: Victorian Science and the Discovery of Human Prehistory.* Chicago: University of Chicago Press.

Viguerie, Richard. 1981. *The New Right: We're Ready to Lead.* Ottowa, Ill.: Caroline House.

Walker, Samuel. 1990. *In Defense of American Liberties: A History of the ACLU.* New York: Oxford University Press.

Wallace, Alfred Russel. 1853. *A Narrative of Travels on the Amazon and Rio Negro, with an Account of the Native Tribes, and Observations on the Climate, Geology, and Natural History of the Amazon Valley.* London: Reeve & Co. Reprint, New York: Dover Publications, 1972.

———. 1864. "The Origin of Human Races and the Antiquity of Man Deduced

from the Theory of 'Natural Selection.'" *Journal of the Anthropological Society of London* 2: clviii–clxx.
———. 1869a. "Sir Charles Lyell on Geological Climates and the Origin of Species." *London Quarterly Review* 126: 359–394.
———. 1869b. *The Malay Archipelago*. London: Macmillan and Company. Reprint of 10th ed. (1890), New York: Dover Publications, 1962.
———. 1870. *Contributions to the Theory of Natural Selection*. London: Macmillan & Co.
———. 1890. "Human Selection." *Fortnightly Review* 48: 325–337.
———. 1891. *Natural Selection and Tropical Nature: Essays on Descriptive and Theoretical Biology*. London and New York: Macmillan & Co. Reprint, Westmead, England: Gregg International, 1969.
———. 1892. "Human Progress: Past and Future." *Arena* 5: 145–159.
———. 1905. *My Life: A Record of Events and Opinions*. 2 vols. London: Chapman and Hall.
Whitcomb, John C., and Henry M. Morris. 1961. *The Genesis Flood: The Biblical Record and Its Scientific Implications*. Philadelphia: Presbyterian and Reformed Publishing Co.
Wilson, David Sloan, and Lee A. Dugatkin. 1992. "Altruism: Contemporary Debates." In *Keywords in Evolutionary Biology*, ed. E. F. Keller and E. A. Lloyd. Cambridge, Mass.: Harvard University Press.
Wilson, Leonard G. 1972. *Charles Lyell, the Years to 1841: The Revolution in Geology*. New Haven: Yale University Press.
Winch, Donald. 1987. *Malthus*. Oxford and New York: Oxford University Press.
Winsor, Mary P. 1991. *Reading the Shape of Nature*. Chicago: University of Chicago Press.
Winter, Alison. 1997. "The Construction of Orthodoxies and Heterodoxies in the Early Victorian Life Sciences." In *Victorian Science in Context*, ed. Bernard Lightman. Chicago: University of Chicago Press.
———. 1998. *Mesmerized: Powers of Mind in Victorian Britain*. Chicago and London: University of Chicago Press.
Worthen, Thomas D. 1991. *The Myth of Replacement: Stars, God, and Order in the Universe*. Tucson: University of Arizona Press.
Wright, Margaret R. 1997. "Marcella O'Grady Boveri (1863–1950): Her Three Careers in Biology." *Isis* 88: 631–635.
Yeo, Richard. 1993. *Defining Science: William Whewell, Natural Knowledge, and Public Debate in Early Victorian Britain*. Cambridge: Cambridge University Press.

Young, David. 1993. *The Discovery of Evolution.* Cambridge: Cambridge University Press.

Young, Robert M. 1985. *Darwin's Metaphor: Nature's Place in Victorian Culture*. Cambridge: Cambridge University Press.

INDEX

INDEX

Villa Borghese & Northern Rome
Encompasses the glorious park of Villa Borghese and the city's cultural hub. (Map p256)
Museo e Galleria Borghese
Tridente, Trevi & the Quirinale
Glamorous, debonair and touristy – this is Rome's designer-clad soul. (Map p255)
Museo Nazionale Romano: Palazzo Massimo alle Terme
Stazione Termini
Monti, Esquilino & San Lorenzo
Busy areas, speckled with glittering churches and some cool restaurants. (Map p255)
Basilica di Santa Maria Maggiore
Colosseum
Basilica di San Clemente
Ancient Rome
No other city has such an evocative ancient heart. (Map p252)
Basilica di San Giovanni in Laterano
Terme di Caracalla
Ciampino Airport (12km)
Southern Rome
Encompassing the beautiful, historic Via Appia Antica and ancient catacombs.
Via Appia Antica

Welcome to Rome

An ancient superpower turned modern metropolis, Rome is a mesmerising mix of haunting ruins, epic monuments and awe-inspiring art. Its romantic streets are made for leisurely exploring and its elegant piazzas provide the perfect backdrop for memorable al fresco nights.

Rome was founded more than 3000 years ago and over the centuries it has acquired a historical and artistic heritage that few cities can rival. Ancient icons such as the Colosseum, Roman Forum and Pantheon recall its golden age as the *caput mundi* (capital of the world), while Michelangelo's frescoes and Caravaggio's canvases testify to its starring role in the Renaissance and baroque eras. Priceless treasures adorn world-class museums, and celebrated masterpieces dazzle in the city's great art-rich churches, culminating in the overpowering spectacle that is St Peter's Basilica. Walk around the centre of town and without even trying you'll come across works by the big-name giants of Western art.

But a trip to Rome is as much about lapping up the dolce vita lifestyle as gorging on art and culture. Eating in boisterous, neighbourhood trattorias, whiling away hours at streetside cafes, people-watching on theatrical piazzas – these are all an integral part of the Roman experience.

haunting ruins, epic monuments and awe-inspiring art

Piazza Navona (p64)

ROME
0 500 m
0 0.25 miles
Chiesa di Santa Maria del Popolo
Vatican Museums
St Peter's Basilica
Tiber
Piazza Navona
Pantheon
Chiesa del Gesù
Basilica di Santa Maria in Trastevere
Isola Tiberina
Vatican City, Borgo & Prati
Home to a stunning wealth of artistic treasures.
Centro Storico
The city's tangled historic centre is packed with incredible sights.
(Map p250)
Ostia Antica (22km)
Leonardo da Vinci (Fiumicino) International Airport (26km)
Trastevere & Gianicolo
Enchantingly pretty, with tangled lanes, ochre palazzi (palaces) and a boho vibe.

Plan Your Trip
This Year in Rome

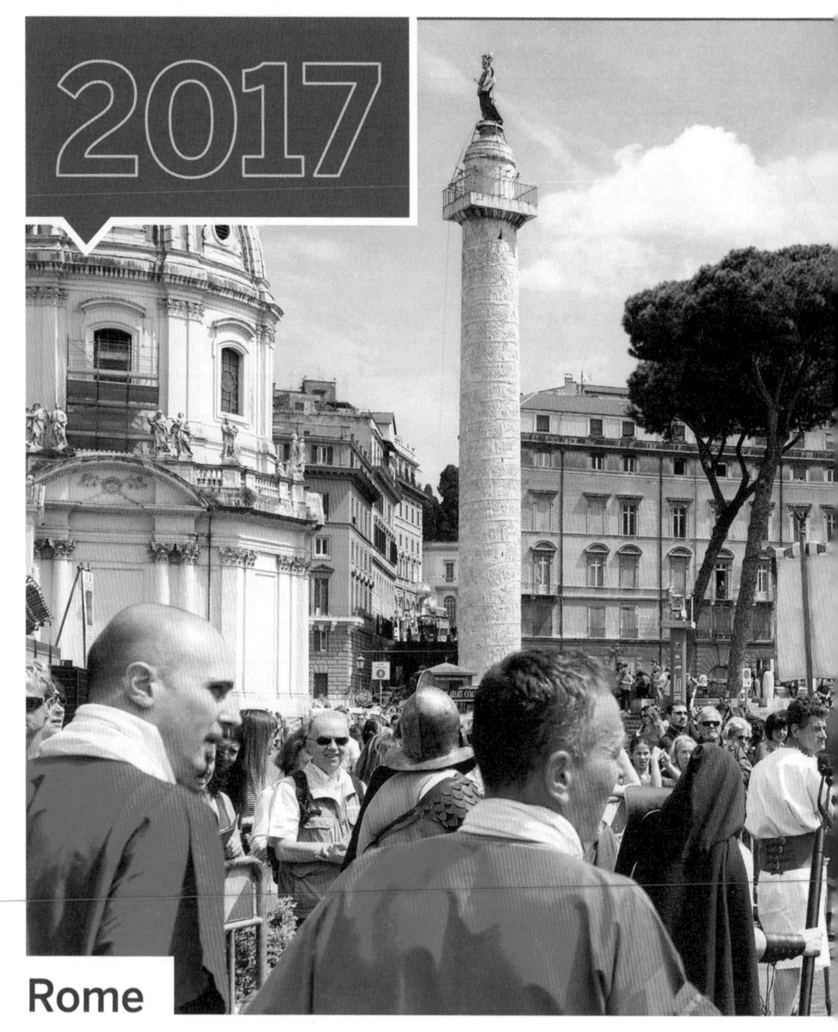

Rome's calendar bursts with events, ranging from traditional neighbourhood shindigs and saints' days to shopping bonanzas, catwalk parades and major cultural festivals.

Above: Natale di Roma (p9)

Top Festivals & Events

Carnevale – 26 Feb (p7)
Natale di Roma – 21 Apr (p9)
Lungo il Tevere – Summer (p11)
Romaeuropa – Late Sep to Dec (p14)

Plan Your Trip
This Year in Rome

January

As New Year celebrations fade, the winter cold digs in. It's a quiet time of year but the winter sales are a welcome diversion.

6 January

Epiphany

A witch known as La Befana delivers gifts to Italian kids for Epiphany, the last day of the Christmas holidays. To mark the occasion, a costumed procession makes its way down Via Concilizione to St Peter's Square.

17 January

Festa di Sant'Antonio Abate

Animal-lovers take their pets to be blessed at the Chiesa di Sant'Eusabio on Piazza Vittorio Emanuele in honour of the patron saint of animals.

Late January

Alta Roma

Fashionistas descend on town for the winter outing of Rome's top fashion event. Catwalk shows, held in venues across town, provide sneak previews of seasonal collections by local and international designers.

7 January

Shopping Sales

The first Saturday of the month sees bargain hunters descend on the city's shops for the winter sales. Until mid-February stores offer savings of between 20% and 50%.

February

Rome's winter quiet is shattered by high-spirited carnival celebrations, which signal that spring is on the way.

5 February

☆ Six Nations Rugby

The Azzurri kick off the Six Nations rugby tournament against Wales at the Stadio Olimpico. Further home matches follow on 11 February (against Ireland) and 11 March (against France).

14 February

☆ Valentine's Day

Romance breaks out across town. Some museums and galleries offer two-for-one ticket discounts while restaurants prepare special St Valentine Day's menus. Book early.

Mid-February

☆ Equilibrio Festival della Danza Nuova

The Auditorium Parco della Musica stages emerging talents and affirmed international choreographers at this festival of contemporary dance.

26 February

Carnevale

In the days leading up to Carnevale, Rome really goes to town with leaping horse shows on Piazza del Popolo, costumed parades down Via del Corso, street performers on Piazza Navona, and crowds of kids in fancy dress.

LUIGI DE POMPEIS / ALAMY STOCK PHOTO ©

Carnevale parade

Plan Your Trip
This Year in Rome

March

The onset of spring brings blooming flowers, rising temperatures and unpredictable rainfall. Unless Easter falls in late March, the city is fairly subdued and low-season prices still apply.

15 March

Re-Enactment of Caesar's Death
Romans remember the assassination of Caesar on the Ides of March, 44 BC. To get everyone in the mood, costumed performers re-enact the stabbing on the very spot it happened, in the ruins of Largo di Torre Argentina.

19 March

Festa di San Giuseppe
The Feast of St Joseph is celebrated in the Trionfale neighbourhood. Little stalls are set up to serve *fritelle* (fried pastries) and there's usually a market near the church of San Giuseppe.

Around 18 & 19 March

Giornate FAI di Primavera
Palazzi, churches and archaeological sites that are generally closed to the public open their doors for a weekend of special openings, courtesy of Italy's main conservation body, the Fondo Ambiente Italiano.

Late March

Maratona di Roma
Sightseeing becomes sport at Rome's annual marathon (www.maratonadiroma.it). The 42km route starts and finishes near the Colosseum, taking in many of the city's big attractions.

Maratona di Roma

April

April is a great month, with lovely, sunny weather, fervent Easter celebrations, azaleas on the Spanish Steps and Rome's birthday festivities. Expect high-season prices.

14–17 April

Easter

On Good Friday the pope leads a candlelit procession around the Colosseum, and there are other smaller parades around the city. At noon on Easter Sunday (16 April) the Pope blesses the crowds in St Peter's Square.

Mid-April

Mostra delle Azalee

From mid-April to early May the Spanish Steps are decorated with 600 vases of blooming, brightly coloured azaleas.

21 April

Natale di Roma

Rome celebrates its birthday with music, historical re-creations and fireworks. Events are staged throughout the city but the focus is Campidoglio and Circo Massimo.

25 April

Festa della Liberazione

Schools, shops and offices shut as Rome commemorates the liberation of Italy by Allied troops and resistance forces in 1945.

From left: Azaleas on the Spanish Steps; Natale di Roma
STEFANO PATERNA/ROBERT HARDING ©; EMIPRESS/SHUTTERSTOCK ©

Plan Your Trip
This Year in Rome

May

May is a busy, high-season month. The weather's perfect – usually warm enough to eat outside – and the city is looking gorgeous with blue skies and spring flowers.

1 May

☆ Primo Maggio
Thousands of fans troop to Piazza di San Giovanni in Laterano for Rome's free May Day rock concert. It's a mostly Italian affair with big-name local performers, but you might catch the occasional foreign guest star.

15 May

☆ Italian Open Tennis Tournament
The world's top tennis stars bash it out on the clay courts of the Foro Italico at the Internazionali BNL d'Italia, one of Europe's major tournaments.

Late May

☆ Coppa Italia Final
Football fans fill the Stadio Olimpico for the final of Italy's main football cup. Neither of Rome's two sides have won since Lazio raised the trophy in 2013.

Late May

☆ Show Jumping
Villa Borghese sets the attractive stage for Rome's annual horse-jumping event, known officially as the Concorso Ippico Internazionale di Piazza di Siena.

From left: Primo Maggio; Show jumping

June

Summer has arrived and with it hot weather and the Italian school holidays.

2 June

⁂ Festa della Repubblica

A big military parade along Via dei Fori Imperiali is the highlight of ceremonial events held to commemorate the birth of the Italian Republic in 1946. Presiding is the President of the Republic and other assorted worthies.

Mid-June

☆ Isola del Cinema

The Isola Tiberina provides the picturesque backdrop for this open-air film festival (www.isoladelcinema.com), which screens a range of Italian and international films with a focus on independent productions.

23 June

⁂ Festa di San Giovanni

The feast day of St John the Baptist is commemorated around the Basilica di San Giovanni in Laterano. Traditionally, stewed snails and *porchetta* (herbed suckling pig) are served.

29 June

⁂ Festa dei Santi Pietro e Paolo

Rome celebrates its two patron saints, Peter and Paul, with a mass at St Peter's Basilica and a street fair on Via Ostiense near the Basilica di San Paolo Fuori-le-Mura.

Late June

☆ Roma Incontro Il Mondo

Villa Ada (www.villaada.org) is transformed into a colourful multiethnic village for this popular annual event. There's a laid-back party vibe and an excellent program of concerts ranging from Roman rap to jazz and world music.

GARI WYN WILLIAMS / ALAMY STOCK PHOTO ©/ARTWORK BY ALBA GONZALES

Summer

⁂ Lungo il Tevere

Nightly crowds converge on the river Tiber for this popular summer-long event. Stalls, clubs, bars, restaurants and dance floors line the river bank as Rome's nightlife goes al fresco.

Plan Your Trip
This Year in Rome

July

Hot summer temperatures make sightseeing a physical endeavour, but come the cool of evening, the city's streets burst into life as locals come out to enjoy summer festivities.

June–July

☆ Rock in Roma
Dust down the denims for Rome's big rock fest. Headline acts from recent editions have included Iron Maiden, Bruce Springsteen and Primal Scream.

July–Early August

☆ Opera at the Terme di Caracalla
The hulking ruins of a vast 2nd-century baths complex provide the spectacular setting for the Teatro dell'Opera di Roma's summer season. Also ballet and the occasional rock concert.

July

☆ Luglio Suona Bene
From Sting to Santana, music legends take to the outdoor stage at the Auditorium Parco della Musica for a month-long series of concerts held as part of the Luglio Suona Bene (July Sounds Good) festival.

Mid-July

☆ Invito alla Danza
Rome's oldest dance festival provides a showcase for traditional and experimental dance. Top international performers take to the stage at the Teatro Villa Pamphilj.

Late July

Festa de' Noantri
Trastevere celebrates its roots with a raucous street party in the last two weeks of the month. Centred on Piazza Santa Maria, events kick off with a religious procession and continue with much eating, drinking, dancing and praying.

June–October

Estate Romana
Rome's big summer festival (www.estateromana.comune.roma.it) involves everything from concerts and dance performances to book fairs, puppet shows and late-night museum openings.

August

Rome melts in the heat as locals flee the city for their summer holidays. Many businesses shut down but hoteliers offer discounts and there are loads of summer events to enjoy.

June–September

Gay Village

The big annual event in Rome's gay calendar is held between June and September, usually at the Parco del Ninfeo in EUR. Expect huge crowds, DJs, dance music, film screenings, cabarets and theatrical performances.

5 August

Festa della Madonna della Neve

On 5 August rose petals are showered on celebrants in the Basilica di Santa Maria Maggiore to commemorate a miraculous 4th-century snowfall.

15 August

Ferragosto

The Festival of the Assumption is celebrated with almost total shutdown, as what seems like Rome's entire population decamps to the sea.

20 August

☆ Football Season Starts

While the rest of Italy basks in the summer sun, the nation's footballers return to work. The Serie A season kicks off around 20 August and the city's ardent fans can finally get their weekly fix.

LONELY PLANET/GETTY IMAGES ©

A S Roma fans watching football at the Stadio Olimpico (p198)

Plan Your Trip
This Year in Rome

September

Life returns to the city after the August torpor. The kids go back to school and locals return to work but there's still a relaxed summer vibe and the weather's perfect.

June–September

Night Visits

A lot of Rome's headline sights offer night visits over the summer and through September. In past years, these have included the Colosseum and Vatican Museums.

Mid-September

Taste of Rome

Foodies flock to the Auditorium Parco della Musica to revel in world food. Join Rome's top chefs for tastings, performances and three days of food-related events. Check www.tasteofroma.it.

Late September

Clubbing

Late September is a good time for partygoers as the city's main clubs return to town after their summer exodus. Curtain-raiser events are a guarantee of big nights and sweaty dance floors.

Late September–Early December

Romaeuropa

Established international performers join emerging stars at Rome's premier dance and drama festival (www.romaeuropa.net). Events range from avant-garde dance performances to installations and readings.

From left: Roman cuisine; Orchestra dell'Accademia Nazionale di Santa Cecilia, Auditorium Parco della Musica (p192)

October

Autumn is a good time to visit – the warm weather is holding, Romaeuropa ensures plenty of cultural action and, with the schools back, there are far fewer tourists around.

Early October

☆ Start of Santa Cecilia Symphony Season

Rome's premier orchestra, the Orchestra dell'Accademia Nazionale di Santa Cecilia, returns to its home stage at the Auditorium Parco della Musica for the start of the symphonic season. For concert details see www.santacecilia.it.

Early October–January

Fotografia Festival

Photographs covering a range of styles and subjects are juxtaposed against the contemporary architecture of the MACRO modern art museum at Rome's premier photographic festival (www.fotografia festival.it).

Late October–Mid-December

☆ Chamber Music Concerts

Designed by baroque genius Francesco Borromini, the Chiesa di Sant'Agnese in Agone on Piazza Navona hosts a series of chamber music concerts.

B O'KANE / ALAMY STOCK PHOTO ©

Mid-October

Festa del Cinema di Roma

Held at the Auditorium Parco della Musica in mid-October, Rome's film festival rolls out the red carpet for Hollywood hotshots and bigwigs from Italian cinema. Consult the program at www.romacinemafest.it.

This Year in Rome

November

Although the wettest month, November has its compensations – low-season prices, excellent jazz concerts and no queues outside the big sights. Autumn is also great for foodies.

1 November

Ognissanti

Celebrated as a national holiday, All Saints' Day commemorates the Saint Martyrs, while All Souls' Day, on 2 November, is set aside to honour the deceased. Many Romans leave flowers on tombs at the Cimitero di Campo Verano.

Mid-November

☆ Festival Internazionale di Musica e Arte Sacra

Over several days in mid-November, the Vienna Philharmonic Orchestra and other top ensembles perform a series of classical concerts in Rome's four papal basilicas and other churches. Check the program at www.festivalmusicaeartesacra.net.

Late November

☆ Start of Opera Season

Towards the end of November the opera season gets underway at the Teatro dell'Opera di Roma, the city's opera house. The theatre is also home to Rome's principal ballet corps with the dance season starting in December.

VINZO/ GETTY IMAGES ©

Mid-November

☆ Rome Jazz Festival

For jazz fans, the Auditorium Parco della Musica is the place to be in mid-November, as performers from around the world play to appreciative audiences during the three-week Roma Jazz Festival (www.romajazzfestival.it).

December

The build-up to Christmas feels festive, as the city twinkles in anticipation. Every church has a presepe *(nativity scene) displayed, from intricate small tableaux to life-sized extravaganzas.*

6 December

Piazza Navona Christmas Fair

Rome's most beautiful baroque square becomes a big, brash marketplace as brightly lit market stalls set up shop, selling everything from nativity scenes to stuffed toys and teeth-cracking *torrone* (nougat).

8 December

Immaculate Conception

Tradition dictates that the Pope, in his capacity as the Bishop of Rome, celebrates the Immaculate Conception in Piazza di Spagna. Earlier in the day Rome's fire brigade place a garland of flowers atop the Colonna dell'Immaculata in adjacent Piazza Mignanelli.

Early December

Christmas Lights

Crowds fill Via del Corso for the annual switching on of the Christmas lights early in December. Over the river in the Vatican, a huge Christmas tree and life-sized nativity scene adorn St Peter's Square.

31 December

Capodanno

Rome is a noisy place to be on New Year's Eve as big firework displays usher in the new year and outdoor concerts are held across town, most notably on Piazzas del Quirinale and del Popolo.

Piazza Navona Christmas Fair

Plan Your Trip

Need to Know

Daily Costs

Budget
Less than €100

- Dorm bed: €15–€30
- Double room in a budget hotel: €50–€110
- Pizza or pasta: €6–€12

Midrange
€100–250

- Double room in a hotel: €110–€200
- Lunch and dinner in local restaurants: €25–€45
- Museum admission: €4–€15
- Roma Pass, a three-day card covering museum entry and public transport: €36

Top End
More than €250

- Double room in a four- or five-star hotel: €200–€450
- Top restaurant dinner: over €45
- Opera ticket: €40–€200
- City-centre taxi ride: €10–€15
- Auditorium concert ticket: €25–€90

Advance Planning

- **Two months before** Book high-season accommodation.
- **Three to four weeks before** Check for concerts at www.auditorium.com and www.operaroma.it.
- **One to two weeks before** Reserve tables at A-list restaurants. Sort out tickets to the pope's weekly audience at St Peter's.
- **Few days before** Book for the Museo e Galleria Borghese (compulsory) and for the Vatican Museums (advisable to avoid queues).

Useful Websites

- **Lonely Planet** (www.lonelyplanet.com/rome) Destination low-down, hotel bookings and traveller forum.
- **060608** (www.060608.it) Rome's official tourist website.
- **Coopculture** (www.coopculture.it) Information and ticket booking for Rome's monuments.
- **Vatican Museums** (www.vatican.va) Book tickets and avoid the queues.
- **Auditorium** (www.auditorium.com) Check concert listings.

Currency

Euro (€)

Language

Italian

Visas

Not required by EU citizens. Not required by nationals of Australia, Canada, New Zealand and the USA for stays of up to 90 days.

Money

ATMs are widespread. Major credit cards are widely accepted but some smaller shops, trattorias and *pensioni* (small hotels or guesthouses) might not take them.

Mobile Phones

Local SIM cards can be used in European, Australian and unlocked US phones. Other phones must be set to roaming.

Time

Western European Time (GMT/UTC plus one hour)

For more, see the **Survival Guide** (p233)

When to Go

In spring and early autumn there's good weather and many festivals and outdoor events. It's also busy and peak rates apply.

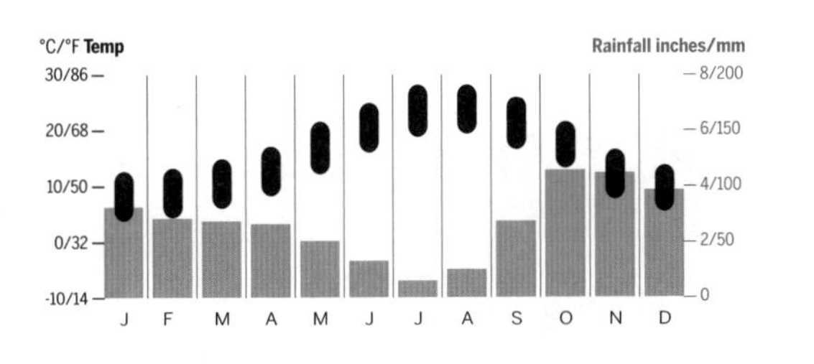

Arriving in Rome

Leonardo da Vinci (Fiumicino) Airport
Direct trains to Stazione Termini run from 6.23am to 11.23pm, €14; slower trains run to Trastevere, Ostiense and Tiburtina stations from 5.57am to 10.42pm, €8; buses to Stazione Termini from 5.35am to midnight, €4 to €9; private transfers cost from €13 per person; taxis cost €48 (fixed fare to within the Aurelian walls).

Ciampino Airport Buses to Stazione Termini run from 7.45am to 11.59pm, €4; private transfers cost from €13 per person; taxis cost €30 (fixed fare to within the Aurelian walls).

Stazione Termini Airport buses and trains as well as international trains arrive at Stazione Termini. From here, continue by bus, metro or taxi.

Getting Around

Rome's main public-transport hub is Stazione Termini, the only point at which the city's two main metro lines cross. The metro is quicker than surface transport but the network is limited and the bus is often a better bet. Children under 10 travel free.

Metro The main lines are: A (orange; 5.30am to 9.30pm Thursday to Sunday, replacement bus MA1–MA2 to 11.30pm, to 1.30am Saturday) and B (blue; 5.30am to 11.30pm Monday to Thursday, to 1.30am Friday and Saturday).

Buses Most routes pass through Stazione Termini. Buses run 5.30am to midnight, with limited services throughout the night.

Sleeping

Rome is expensive and with the city busy year-round, you'll want to book as far ahead as you can to secure the best deal and the place you want.

Accommodation options range from palatial five-star hotels to hostels, B&Bs, convents, *pensioni* (small hotels or guesthouses) and a good range of Airbnb options. Hostels are the cheapest, offering dorm beds and private rooms. B&Bs range from simple homestyle set-ups to chic boutique outfits with prices to match, while religious institutions provide basic, value-for-money accommodation but may insist on a curfew. Hotels are plentiful and there are many budget, family-run *pensioni* in the Termini area.

Plan Your Trip
Top Days in Rome

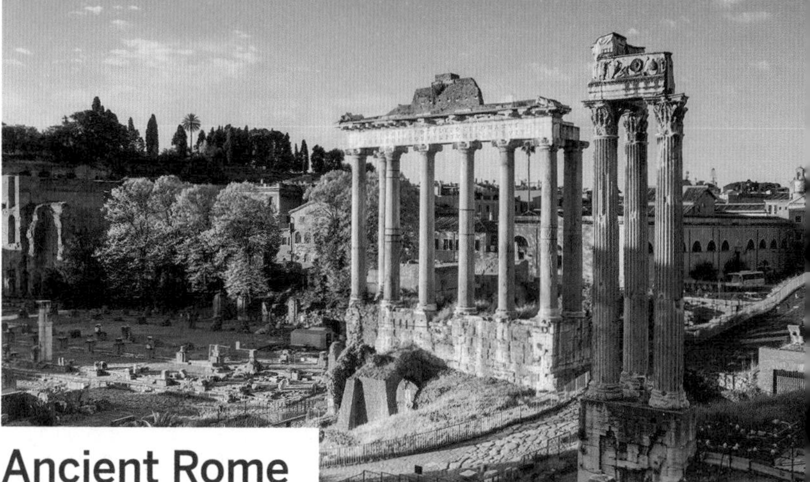

JULIAN ELLIOTT PHOTOGRAPHY / GETTY IMAGES ©

Ancient Rome

The Colosseum is an appropriate high on which to start your odyssey in Rome. Next, head to the nearby crumbling scenic ruins of the Palatino, followed by the Roman Forum. After lunch enjoy 360-degree views from Il Vittoriano and classical art at the Capitoline Museums.

Day 01

❶ Colosseum (p36)

More than any other monument, it's the Colosseum that symbolises the power and glory of ancient Rome. Visit its broken interior and imagine the roar of the 50,000-strong crowd as the gladiators fought for their entertainment.

➔ Colosseum to Palatino

Walk south down the Via di San Gregorio to the Palatino.

❷ Palatino (p60)

The gardens and ruins of the Palatino (included with the Colosseum ticket) are an atmospheric place to explore, with great views across Circo Massimo and the Roman Forum. The Palatino was the most exclusive part of ancient Rome, home of the imperial palace, and is still today a hauntingly beautiful site.

➔ Palatino to Roman Forum

Still in the Palatino, follow the path down past the Vigna Barberino to enter the Roman Forum near the Arco di Tito.

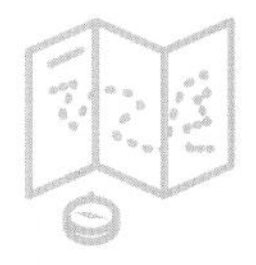

MAREMAGNUM/GETTY IMAGES ©

❸ Roman Forum (p80)

Sprawled beneath the Palatino, the Forum was the empire's nerve centre, a teeming hive of law courts, temples, piazzas and shops. See where the vestal virgins lived and the Curia, where senators debated matters of state.

➡ Roman Forum to Terre e Domus

Exit the Forum onto Via dei Fori Imperiali and head up Via Alessandrina through the Imperial Forums to Terre e Domus near Trajan's Column.

❹ Lunch at Terre e Domus (p124)

Lunch on earthy Lazio food and wine at this bright modern restaurant just off Piazza Venezia.

➡ Terre e Domus to Il Vittoriano

Return to Piazza Venezia and follow up to the mountainous monument Il Vittoriano.

❺ Il Vittoriano (p75)

Il Vittoriano is an ostentatious, overpowering mountain of white marble. Love it or hate it, it's an impressive sight, but for an even more mind-blowing view, take the glass lift to the top and you'll be rewarded with 360-degree views across the whole of Rome.

➡ Il Vittoriano to Campidoglio & Capitoline Museums

Descend from Il Vittoriano and head left to the sweeping staircase, La Cordonata. Climb the stairs to reach Piazza del Campidoglio and the Capitoline Museums.

❻ Campidoglio & Capitoline Museums (p72)

With wonderful views over the Forum, the piazza atop the Capitoline hill (Campidoglio) was designed by Michelangelo and is flanked by the world's oldest national museums. The Capitoline Museums harbour some of Rome's most spectacular ancient art, including the iconic depiction of Romulus and Remus sat under a wolf, the *Lupa capitolina*.

From left: Tempio di Saturno (p82), Roman Forum; Il Vittoriano (p75)

Plan Your Trip
Top Days in Rome

Vatican City & Centro Storico

On day two, hit the Vatican. Blow your mind at the Sistine Chapel and Vatican Museums, then complete your tour at St Peter's Basilica. Dedicate the afternoon to sniffing around the historic centre, including Piazza Navona and the Pantheon.

❶ Vatican Museums (p40)

With more than 7km of exhibits, it'd be hard to see it all in a morning, but make a beeline for the Pinacoteca, the Museo Pio-Clementino, Galleria delle Carte Geografiche, Stanze di Raffaello (Raphael Rooms) and the Sistine Chapel.

➡ Vatican Museums to Fa-Bìo

🚶 From the Vatican Museums entrance, turn downhill and follow the walls towards Piazza del Risorgimento. Take a left down Via Vespasiano and then the first right to Via Germanico and Fa-Bìo.

❷ Lunch at Fa-Bìo (p132)

Grab a light lunch bite at this tiny takeaway. It's very popular so you'll need to squeeze through to the counter to order your *panino* (sandwich), salad or smoothie, all made with quality organic products.

➡ Fa-Bìo to St Peter's Basilica

🚶 From the takeaway, double back to Piazza del Risorgimento, then follow the crowds to reach St Peter's Basilica.

ARTIE PHOTOGRAPHY (ARTIE NG)/GETTY IMAGES ©

❸ St Peter's Basilica (p46)

Approaching St Peter's Square from the side, you'll see it as Bernini intended: a surprise. Visit this beautiful public square and the church itself, home to Michelangelo's *Pietà* and a breathtaking dome – it's worth climbing the latter for astounding views.

➔ St Peter's Basilica to Castel Sant'Angelo

From near St Peter's Square, walk along the Borgo Sant'Angelo to reach Castel Sant'Angelo.

❹ Castel Sant'Angelo (p49)

If you're not feeling overwhelmed by sightseeing, visit the interior of this ancient Roman tomb that became a fortress.

➔ Castel Sant'Angelo to Piazza Navona

Cross the river via the pedestrianised Ponte Sant'Angelo, then follow the river eastwards for around 300m before turning right inland at the next bridge, following Via G Zanardelli to reach Piazza Navona.

❺ Piazza Navona (p64)

This vast baroque square is a showpiece of the *centro storico,* and full of vibrant life. The lozenge-shaped space is an echo of its ancient origins as the site of a stadium.

➔ Piazza Navona to Pantheon

It's a short walk eastwards from Piazza Navona to Piazza della Rotonda, where you'll find the Pantheon.

❻ Pantheon (p50)

This 2000-year-old temple, now a church, is an extraordinary building, the innovative design of which has served to inspire generations of architects and engineers.

➔ Pantheon to Armando al Pantheon

There are plenty of excellent restaurants around the Pantheon, and Armando al Pantheon, within sight of the temple, is one of Rome's best local restaurants.

❼ Dinner at Armando al Pantheon (p127)

Go for a taste of authentic Roman cuisine at long-time favourite Armando al Pantheon.

From left: St Peter's Square (p49); Castel Sant'Angelo (p49)

Top Days in Rome

Villa Borghese, Tridente & Trevi

Start your day at the brilliant Museo e Galleria Borghese, before rambling around the shady avenues of the surrounding park of Villa Borghese. Next, explore the Tridente neighbourhood, including the Spanish Steps and Via dei Condotti, before heading to the Trevi Fountain.

❶ Museo e Galleria Borghese (p54)

Book ahead and start your day at the Museo e Galleria Borghese, one of Rome's best art museums. The highlight is a series of astonishing sculptures by baroque genius Gian Lorenzo Bernini.

➲ Museo e Galleria Borghese to Villa Borghese

Work your way through the leafy paths of Villa Borghese towards the Pincio.

❷ Villa Borghese (p57)

Meander through the lovely, rambling park of Villa Borghese, formerly the playground of the mighty Borghese family. En route you'll pass the Piazza di Siena and walk along tree-shaded lanes to reach the Pincio, a panoramic terrace offering great views across Rome.

➲ Villa Borghese to the Spanish Steps

From the Pincio, exit along Viale Gabriele D'Annunzio and follow on to the top of the Spanish Steps.

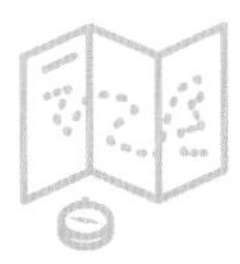

RICHARD CUMMINS/GETTY IMAGES ©

❸ Spanish Steps (p100)

The Spanish Steps are a glorious flight of ornamental rococo steps, with views over the glittering, designer-store-packed streets of the Tridente district and Piazza di Spagna.

➡ The Spanish Steps to Via dei Condotti

Walk down the Spanish Steps to Via dei Condotti.

❹ Via dei Condotti (p151)

Via dei Condotti is Rome's most exclusive shopping street, lined by big-name designers and jewellers such as Prada, Bulgari, Fendi and Salvatore Ferragamo. Even if you haven't got cash to splash, it's well worth a wander to window-shop and people-watch.

➡ Via dei Condotti to Enoteca Regionale Palatium

From Via Condotti, turn south along Via Belsiana to reach the Enoteca Regionale Palatium.

❺ Lunch at Enoteca Regionale Palatium (p130)

Specialising in produce from Lazio, this sleek wine bar is a great place to try local wines and delicacies, such as Frascati white wine and *porchetta* (pork roasted with herbs).

➡ Enoteca Regionale Palatium to Piazza Del Popolo

Returning along Via Belsiana, turn right at Via Vittoria, then left at Via del Babuino, to reach Piazza del Popolo.

❻ Piazza del Popolo (p89)

The huge, oval Piazza del Popolo dates from the 16th century and is overlooked by Chiesa di Santa Maria del Popolo, which contains a remarkable array of masterpieces.

➡ Piazza del Popolo to Trevi Fountain

Walk back along Via del Corso, then turn left up Via Sabini to reach the Trevi Fountain.

❼ Trevi Fountain (p86)

End your day at this foaming, fantastical baroque fountain designed by Nicola Salvi, where you can toss in a coin to ensure a return visit to Rome.

From left: View from Pincio Hill Gardens (p58); Spanish Steps (p100) and Chiesa della Trinità dei Monti (p101)

Plan Your Trip
Top Days in Rome

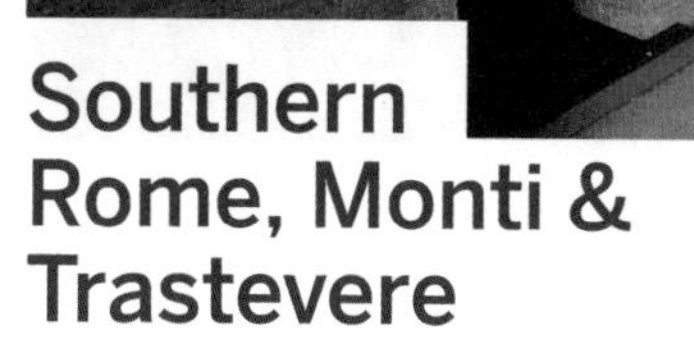

Southern Rome, Monti & Trastevere

On your fourth day, venture out to Via Appia Antica and the catacombs. Start the afternoon by visiting the Museo Nazionale Romano: Palazzo Massimo alle Terme, then drop by the Basilica di Santa Maria Maggiore. Finish with an evening in Trastevere.

Day 04

1 Catacombe di San Sebastiano (p78)

Start your day underground, by taking a tour of one of the three networks of catacombs that are open to the public. It's a fascinating and chilling experience to see the tunnels where early Christians buried their dead.

Catacombe di San Sebastiano to Villa di Massenzio

Walk around 100m south along Via Appia and you'll see Villa di Massenzio on your left.

2 Villa di Massenzio (p79)

The best preserved part of Maxentius' 4th-century ruined palace is the Circo di Massenzio, which was once a racetrack with the capacity for 10,000 people.

Villa di Massenzio to Mausoleo di Cecilia Metella

Walk 50m or so onwards along Via Appia to the Mausoleo di Cecilia Metella.

3 Mausoleo di Cecilia Metella (p79)

With travertine walls and an interior deco-rated with a sculpted frieze bearing Gaelic

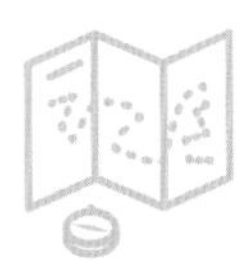

BORIS-B/SHUTTERSTOCK ©

shields, ox skulls and festoons, this great, rotund tomb is an imposing sight.

➡ Mausoleo di Cecilia to Qui Non se More Mai

🚶 From the tomb, continue up the road for a few metres to Qui Non se More Mai.

❹ Lunch at Qui Non se More Mai (p143)

Fortify yourself for the afternoon ahead with a lunch of hearty Roman pasta and expertly grilled meat at this rustic restaurant (closed Sunday and Monday).

➡ Qui Non se More Mai to Palazzo Massimo alle Terme

🚌 After lunch, hop on a bus to Termini station to visit the Palazzo Massimo alle Terme.

❺ Museo Nazionale Romano: Palazzo Massimo alle Terme (p68)

This light-filled museum holds part of the Museo Nazionale Romano collection, with a splendid array of classical carving and an unparalleled selection of ancient Roman frescoes.

➡ Palazzo Massimo alle Terme to Basilica di Santa Maria Maggiore

🚶 Walk about 200m southwest along Via Massimo d'Azeglio to reach Basilica di Santa Maria Maggiore.

❻ Basilica di Santa Maria Maggiore (p104)

One of Rome's four patriarchal basilicas, this monumental church stands on the summit of the Esquilino hill, on the spot where snow is said to have miraculously fallen in the summer of AD 358.

➡ Basilica di Santa Maria Maggiore to Trastevere

🚌 From Termini station, which is a short walk from the basilica, you can take a bus to Trastevere.

❼ Trastevere (p138)

Spend the evening wandering the charismatic streets of Trastevere. This district is as popular with locals as it is with tourists, and is a beguiling place for an evening stroll before settling on a place for dinner.

From left: Villa di Massenzio (p79); people dining on a cobbled street

Plan Your Trip

Hotspots For...

👁 **Trevi Fountain** Throw a coin into the fountain where Anita Ekberg cavorted in *La dolce vita*. (p86)

👁 **Spanish Steps** (pictured above) Grab a perch and enjoy the parade of people on the piazza below. (p100)

🍷 **Stravinskij Bar** The swank Hotel de Russie provides the setting for stylish cocktails. (p172)

🍷 **Salotto 42** Join the beautiful people for an *aperitivo* at this hip bar. (p172)

🛍 **Fendi** Top up your wardrobe at the flagship store of the Fendi fashion house. (p157)

👁 **Museo e Galleria Borghese** Houses the best baroque sculpture in town and some seriously good Old Masters. (p54)

☆ **Auditorium Parco della Musica** (pictured below) Rome's contemporary arts hub occupies a striking Renzo Piano–designed complex. (p192)

✕ **Glass Hostaria** Creative modern food in the heart of atmospheric Trastevere. (p139)

☆ **Alexanderplatz** Get into the swing at Rome's most celebrated jazz club. (p189)

🍷 **Il Tiaso** Modish Pigneto bar with indie art, a chilled vibe and live music. (p173)

SPIRITUAL

St Peter's Basilica A monument to architectural genius and papal ambition, this is the greatest church in the Catholic world. (p46)

Vatican Museums Go face to face with Michelangelo's celebrated frescoes in the Sistine Chapel. (p40)

Stadio Olimpico For football fans a trip to Rome's Olympic stadium is a spiritual experience. (p198)

Caffè Sant'Eustachio The coffee at this busy bar is nothing short of heavenly. (p170)

Aroma Colosseum views and Michelin-starred food are the perfect marriage. (p141)

UNDERGROUND

Basilica di San Clemente (pictured above) Duck under this multi-layered basilica to discover a pagan temple and 1st-century house. (p96)

Via Appia Antica The Appian Way is home to Rome's most famous catacombs. (p76)

Flavio al Velavevodetto A classic Roman trattoria set in a hill made out of smashed vases. (p140)

Lanificio 159 A great factory venue for gigs, club nights, markets and exhibitions. (p182)

Locanda Atlantide Get down and dirty at this basement club. (p176)

HISTORY BUFFS

Colosseum Rome's breathtaking arena encapsulates all the drama of ancient Rome. (p36)

Palatino According to legend, this is where Romulus and Remus founded Rome. (p60)

Armando al Pantheon A family-run restaurant near the Pantheon, famed for its authentic Roman food. (p127)

Chiostro del Bramante Caffè Enjoy a coffee in the historic confines of a Renaissance cloister. (p125)

Terme di Caracalla (pictured above) Towering Roman ruins set the stage for summer opera. (p98)

Plan Your Trip
What's New

Co.So

Join hipsters, mixologists and aficionados at Co.So (p174), one of the city's new breed of cocktail bars in the bohemian Pigneto district.

Enoteca La Torre

After years of success in Viterbo, chef Danilo Ciavattino has transferred his fine-dining restaurant Enoteca La Torre (p135) to the romantic riverside environs of Villa Laetitia.

Pasticceria De Bellis

Pastry making becomes fine art at Pasticceria De Bellis (p125), a designer pastry shop selling a range of edible mini-masterpieces in the historic centre.

Temakinho

The vibrant Monti neighbourhood sets the stage for sushi, sake and cocktails at the popular Brazilian-Japanese Temakinho (p136).

La Ciambella

A laid-back eatery serving everything from breakfast to pizzas and cocktails, La Ciambella (p126) sits over an ancient baths complex near the Pantheon.

Spot

Refined mid-century furnishings, glassware and objets d'art take centre stage at Spot (p159), a fascinating shop in Monti.

Plan Your Trip

For Free

Need to Know

- **Transport** Holders of the Roma Pass are entitled to free public transport.
- **Wi-Fi** Free wi-fi is available in many hostels, hotels, bars and cafes.
- **Tours** To take a free tour check out www.newromefreetour.com.

Art

Feast on fine art in Rome's churches. They're all free and many contain priceless treasures by big-name artists such as Michelangelo, Raphael, Bernini and Caravaggio. Major art churches include St Peter's Basilica (p46), the Basilica di San Pietro in Vincoli (p105), Chiesa di San Luigi dei Francesi (p67), and Chiesa di Santa Maria del Popolo (p88).

All state-run museums are free on the first Sunday of the month, including the Museo Carlo Bilotti (p59). The Vatican Museums (p40) are free on the last Sunday of the month.

Monuments

Some of Rome's best known sites are free.

A pagan temple turned church, the Pantheon (p50) is a staggering work of architecture with its record-breaking dome and echoing interior.

You don't have to spend a penny to admire the Trevi Fountain (p86), although most people throw a coin in to ensure their return to Rome.

According to legend, if you tell a lie with your hand in the Bocca della Verità (Mouth of Truth; p75), it'll bite your hand off.

Piazzas & Parks

People-watching on Rome's piazzas is a signature city experience. Top spots include Piazza Navona (p64), Campo de' Fiori (p67) and Piazza di Spagna (p101).

It doesn't cost a thing to enjoy Rome's most famous park, Villa Borghese (p57). Greenery can also be found at Villa Celimontana (p99) and Gianicolo.

Above: Piazza Navona (p64)

Plan Your Trip
Family Travel

DOUG OGDEN/DESIGN PICS/GETTY IMAGES ©

Despite a reputation as a highbrow cultural destination, Rome has a lot to offer kids. Child-specific sights might be thin on the ground but if you know where to go there's plenty to keep the little ones occupied and parents happy.

Need to Know

- **Getting Around** Cobbled streets make getting around with a pram difficult.
- **Eating Out** In a restaurant ask for a *mezza porzione* (child's portion) and *seggiolone* (highchair).
- **Supplies** Buy baby formula and sterilising solutions at pharmacies. Disposable nappies (diapers; *pannolini*) are available from supermarkets and pharmacies.
- **Transport** Under 10s travel free on all public transport.

History for Kids

Everyone wants to see the Colosseum (p36) and it doesn't disappoint, especially if accompanied by tales of bloodthirsty gladiators and hungry lions. For maximum effect prep your kids beforehand with a Rome-based film.

Spook your teens with a trip to the catacombs on Via Appia Antica (p76). These pitch-black tunnels, full of tombs and ancient burial chambers, are fascinating, but not suitable for children under about seven.

Hands-On Activities

Kids love throwing things, so they'll enjoy flinging a coin into the Trevi Fountain (p86). And if they ask, you can tell them that about €3000 is thrown in on an average day.

Another favourite is putting your hand in the Bocca della Verità (p75), the Mouth of Truth. Just don't tell a fib, otherwise the mouth will bite it off.

ILPO MUSTO / ALAMY STOCK PHOTO ©

Food for Kids

Pizza al taglio (by the slice) is a godsend for parents. It's cheap (about €1 buys two slices of *pizza bianca* – with rosemary, salt and olive oil), easy to get hold of (there are hundreds of takeaways around town), and works wonders on flagging spirits.

Ice cream is another manna from heaven, served in *coppette* (tubs) or *coni* (cones). Child-friendly flavours include *fragola* (strawberry), *cioccolato* (chocolate) and *bacio* (with hazelnuts).

Run in the Park

When the time comes to let the kids off the leash, head to Villa Borghese (p57), the most central of Rome's main parks. There's plenty of space to run around in – though it's not absolutely car-free – and you can hire family bikes.

A Family Day Out

Many of Rome's ancient ruins can be boring for children – they just look like piles of old stones – but Ostia Antica (p94) is different. Here your kids can run along the ancient town's streets, among shops, and up the tiers of its impressive amphitheatre. A trip to Ostia also means a quick ride on a train. However, note that there's little shade on the site so bring water and hats, and take all the usual precautions.

Best Food Stops

Forno Roscioli (p124)
Fatamorgana (p138)
Gelateria del Teatro (p125)
Forno di Campo de' Fiori (p125)
Trapizzino (p140)

From left: Colosseum (p36); Forno Roscioli (p124)

TOP EXPERIENCES

The very best to see and do

Colosseum

A monument to raw, merciless power, the Colosseum (Colosseo) is the most thrilling of Rome's ancient sights. It was here that gladiators met in mortal combat and condemned prisoners fought off wild beasts in front of baying, bloodthirsty crowds. Two thousand years on and it's Italy's top tourist attraction, drawing more than five million visitors a year.

Great For...

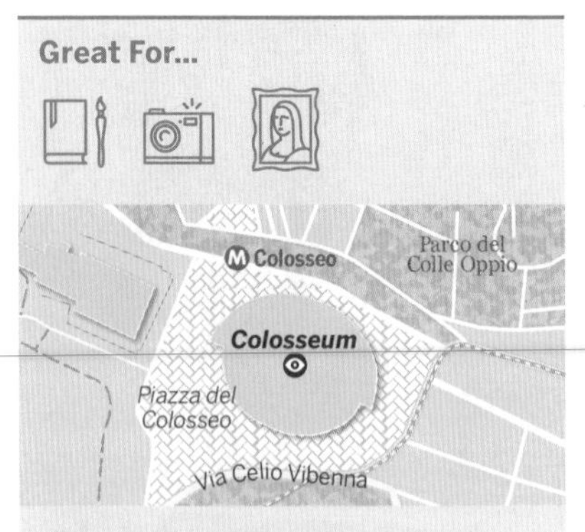

Need to Know

Map p252; ☎06 3996 7700; www.coopculture.it; Piazza del Colosseo; adult/reduced incl Roman Forum & Palatino €12/7.50; ⏱8.30am-1hr before sunset; Ⓜ Colosseo

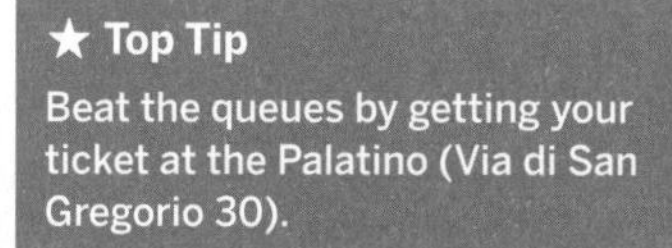

★ Top Tip

Beat the queues by getting your ticket at the Palatino (Via di San Gregorio 30).

EVAN REINHEIMER/GETTY IMAGES ©

Built by Vespasian (r AD 69–79) in the grounds of Nero's vast Domus Aurea complex, it was inaugurated in AD 80, eight years after it had been commissioned. To mark the occasion, Vespasian's son and successor Titus (r 79–81) staged games that lasted 100 days and nights, during which 5000 animals were slaughtered. Trajan (r 98–117) later topped this, holding a marathon 117-day killing spree involving 9000 gladiators and 10,000 animals.

The 50,000-seat arena was originally known as the Flavian Amphitheatre, and although it was Rome's most fearsome arena it wasn't the biggest – the Circo Massimo could hold up to 250,000 people. The name Colosseum, when introduced in medieval times, was not a reference to its size but to the Colosso di Nerone, a giant statue of Nero that stood nearby.

With the fall of the Roman Empire in the 5th century, the Colosseum was abandoned and gradually became overgrown. In the Middle Ages it served as a fortress for two of the city's warrior families, the Frangipani and the Annibaldi. Later, during the Renaissance and baroque periods, it was plundered of its precious travertine, and marble stripped from it was used to make huge palaces such as Palazzo Venezia, Palazzo Barberini and Palazzo Cancelleria.

More recently, pollution and vibrations caused by traffic and the metro have taken their toll. It has recently undergone a €25-million clean-up, the first in its 2000-year history.

Exterior

The outer walls have three levels of arches, framed by Ionic, Doric and Corinthian

Interior of the Colosseum

columns. These were originally covered in travertine, and marble statues filled the niches on the 2nd and 3rd storeys. The upper level, punctuated with windows and slender Corinthian pilasters, had supports for 240 masts that held up a huge canvas awning over the arena, shielding the spectators from sun and rain. The 80 entrance arches, known as vomitoria, allowed the spectators to enter and be seated in a matter of minutes.

☑ Don't Miss

The hypogeum, a network of dank tunnels that extended beneath the main arena. Visits require advance booking and cost €9 on top of the normal Colosseum ticket.

DEVASAHAYAM CHANDRA DHAS/GETTY IMAGES ©

Arena

The arena originally had a wooden floor covered in sand to prevent the combatants from slipping and to soak up the blood. It could also be flooded for mock sea battles. Trapdoors led down to the hypogeum, a subterranean complex of corridors, cages and lifts beneath the arena floor.

Stands

The *cavea,* for spectator seating, was divided into three tiers: magistrates and senior officials sat in the lowest tier, wealthy citizens in the middle, and the plebs in the highest tier. Women (except for vestal virgins) were relegated to the cheapest sections at the top. And as in modern stadiums, tickets were numbered and spectators were assigned a precise seat in a specific sector – in 2015 restorers uncovered traces of red numerals on the arches, indicating how the sectors were numbered. The podium, a broad terrace in front of the tiers of seats, was reserved for the emperor, senators and VIPs.

Hypogeum

The hypogeum served as the stadium's backstage area. Sets for the various battle scenes were prepared here and hoisted up to the arena by a complicated system of pulleys. Caged animals were kept here and gladiators would gather here before show time, having come in through an underground corridor from the nearby Ludus Magnus (Gladiator School).

The hypogeum, and top tier, are open to the public by guided tour only.

Take a Break

Head up to Cavour 313 (p170) for a postarena break. The highlight here is its wine offerings, but you can also snack on cheese and cured-meat platters.

Vatican Museums hallway

Vatican Museums

Founded in the 16th century, the Vatican Museums boast one of the world's greatest art collections. Highlights include spectacular classical statuary, rooms frescoed by Raphael, and the Michelangelo-decorated Sistine Chapel.

Housing the museums are the lavishly decorated halls and galleries of the Palazzo Apostolico Vaticano. This vast 5.5-hectare complex consists of two palaces – the Vatican palace (nearer to St Peter's) and the Belvedere Palace – joined by two long galleries. Inside are three courtyards: the Cortile della Pigna, the Cortile della Biblioteca and, to the south, the Cortile del Belvedere. You'll never cover it all in one day, so it pays to be selective.

Great For...

☑ Don't Miss

The *Laocoön* and other magnificent sculptures in the Museo Pio-Clementino.

Pinacoteca

Often overlooked by visitors, the papal picture gallery contains Raphael's last work, *La Trasfigurazione* (Transfiguration; 1517–20), and paintings by Giotto, Fra Angelico, Filippo Lippi, Perugino, Titian, Guido Reni, Guercino, Pietro da Cortona, Caravaggio and Leonardo da Vinci, whose haunting *San*

Spiral staircase

WIBOWO RUSLI/GETTY IMAGES ©

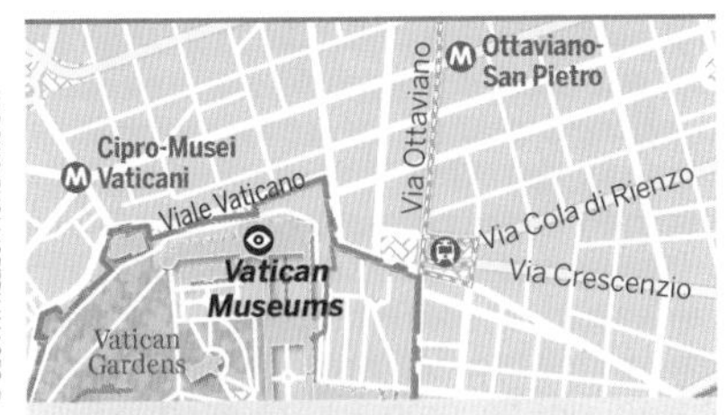

Need to Know

Musei Vaticani; 06 6988 4676; http://mv.vatican.va; Viale Vaticano; adult/reduced €16/8, last Sun of month free; 9am-4pm Mon-Sat, 9am-12.30pm last Sun of month; M Ottaviano-San Pietro

Take a Break

Search out Pizzarium (p132) for some of Rome's best sliced pizza.

★ Top Tip

Avoid queues by booking tickets online (http://biglietteriamusei.vatican.va/musei/tickets/do); the booking fee costs €4.

Gerolamo (St Jerome; c 1480) was never finished.

Museo Chiaramonti & Braccio Nuovo

The Museo Chiaramonti is effectively the long corridor that runs down the eastern side of the Belvedere Palace. Its walls are lined with thousands of statues and busts representing everything from immortal gods to playful cherubs and ugly Roman patricians. Near the end of the hall, off to the right, is the Braccio Nuovo (New Wing; currently closed for restoration), which contains a famous statue of the Nile as a reclining god covered by 16 babies.

Museo Pio-Clementino

This stunning museum contains some of the Vatican Museums' finest classical statuary, including the peerless *Apollo Belvedere* and the 1st-century *Laocoön*, both in the **Cortile Ottagono** (Octagonal Courtyard). Before you go into the courtyard take a moment to admire the 1st-century *Apoxyomenos*, one of the earliest-known sculptures to depict a figure with a raised arm.

To the left as you enter the courtyard, the **Apollo Belvedere** is a 2nd-century Roman copy of a 4th-century-BC Greek bronze. A beautifully proportioned representation of the sun god Apollo, it's considered one of the great masterpieces of classical sculpture. Nearby, the **Laocoön** depicts a muscular Trojan priest and his two sons in mortal struggle with two sea serpents.

Back inside, the **Sala degli Animali** is filled with sculpted creatures and some magnificent 4th-century mosaics. Continuing on, you come to the **Sala delle Muse**, centred on the *Torso Belvedere*, another of the museum's must-sees. A fragment of a muscular 1st-century-BC Greek sculpture, this was found in Campo de' Fiori and used by Michelangelo as a model for his *ignudi* (male nudes) in the Sistine Chapel. It's currently undergoing restoration.

The next room, the **Sala Rotonda**, contains a number of colossal statues, including a gilded-bronze *Ercole* (Hercules) and an exquisite floor mosaic. The enormous basin in the centre of the room was found at Nero's Domus Aurea and is made out of a single piece of red porphyry stone.

Museo Gregoriano Egizio

Founded by Gregory XVI in 1839, this museum contains pieces taken from Egypt in Roman times. The collection is small, but there are fascinating exhibits, including the *Trono di Ramses II* (part of a statue of the seated king), vividly painted sarcophagi dating from around 1000 BC, and some macabre mummies.

Museo Gregoriano Etrusco

At the top of the 18th-century Simonetti staircase, the Museo Gregoriano Etrusco contains artefacts unearthed in the Etruscan tombs of northern Lazio, as well as a superb collection of vases and Roman antiquities. Of particular interest is the *Marte di Todi* (Mars of Todi), a black bronze of a warrior dating to the late 5th century BC.

Galleria delle Carte Geografiche & Sala Sobieski

The last of three galleries – the other two are the **Galleria dei Candelabri** (Gallery of the Candelabra) and the **Galleria degli Arazzi** (Tapestry Gallery) – this 120m-long corridor is hung with 40 huge topograph-

Ceiling in the Galleria della Carte Geographiche

ical maps. These were created between 1580 and 1583 for Pope Gregory XIII based on drafts by Ignazio Danti, one of the leading cartographers of his day.

Beyond the gallery, the **Sala Sobieski** is named after an enormous 19th-century painting depicting the victory of the Polish King John III Sobieski over the Turks in 1683.

Stanze di Raffaello

These four frescoed chambers, currently undergoing partial restoration, were part of Pope Julius II's private apartments. Raphael himself painted the Stanza della Segnatura (1508–11) and the Stanza d'Eliodoro (1512–14), while the Stanza dell'Incendio di Borgo (1514–17) and Sala di Costantino (1517–24) were decorated by students following his designs.

The first room you come to is the **Sala di Costantino**, which features a huge fresco depicting Constantine's defeat of Maxentius at the battle of Milvian Bridge.

The **Stanza d'Eliodoro**, which was used for private audiences, takes its name from the *Cacciata d'Eliodoro* (Expulsion of Heliodorus from the Temple), an allegorical work reflecting Pope Julius II's policy of forcing foreign powers off Church lands. To its right, the *Messa di Bolsena* (Mass of Bolsena) shows Julius paying homage to the relic of a 13th-century miracle at the lakeside town of Bolsena. Next is the *Incontro di Leone Magno con Attila* (Encounter of Leo the Great with Attila) by Raphael and his school, and, on the fourth wall, the *Liberazione di San Pietro* (Liberation of St Peter), a brilliant work illustrating Raphael's masterful ability to depict light.

The **Stanza della Segnatura**, Julius' study and library, was the first room that Raphael painted, and it's here that you'll find his great masterpiece, *La Scuola di Atene* (The School of Athens), featuring philosophers and scholars gathered around Plato and Aristotle. The seated figure in front of the steps is believed to be Michelangelo, while the figure of Plato is said to be a portrait of Leonardo da Vinci, and Euclide (the bald man bending over) is Bramante. Raphael also included a self-portrait in the lower right corner – he's the second figure from the right.

The most famous work in the **Stanza dell'Incendio di Borgo** is the *Incendio di*

☑ Best Time to Visit

Tuesday and Thursday are quietest; Wednesday mornings are good while everyone is at the pope's weekly audience; and generally afternoon is better than the morning. Avoid Monday when many other museums are shut.

☑ Audioguides

On the whole, exhibits are not well labelled, so consider hiring an audioguide (€7) or buying the excellent *Guide to the Vatican Museums and City* (€14).

Borgo (Fire in the Borgo), which depicts Pope Leo IV extinguishing a fire by making the sign of the cross. The ceiling was painted by Raphael's master, Perugino.

Sistine Chapel

The jewel in the Vatican's crown, the Sistine Chapel (Cappella Sistina) is home to two of the world's most famous works of art: Michelangelo's ceiling frescoes and his *Giudizio Universale* (Last Judgment).

The chapel was originally built for Pope Sixtus IV, after whom it's named, and consecrated on 15 August 1483. However, apart from the wall frescoes and floor, little remains of the original decor, which was sacrificed to make way for Michelangelo's two masterpieces. The first, the ceiling, was commissioned by Pope Julius II and painted between 1508 and 1512; the second, the spectacular *Giudizio Universale,* was painted between 1535 and 1541.

Michelangelo's ceiling design, which is best viewed from the chapel's main entrance in the far east wall, covers the entire 800-sq-metre surface. With painted architectural features and a cast of colourful biblical characters, it's centred on nine panels depicting scenes from the Creation, the story of Adam and Eve, the Fall, and the plight of Noah.

As you look up from the east wall, the first panel is the *Drunkenness of Noah,* followed by *The Flood* and the *Sacrifice of Noah.* Next, *Original Sin and Banishment from the Garden of Eden* famously depicts Adam and Eve being sent packing after accepting the forbidden fruit from Satan, represented by a snake with the body of a woman coiled around a tree. The *Creation of Eve* is then followed by the *Creation of Adam.* This, one of the most famous images in Western art, shows a bearded God pointing his finger at Adam, thus bringing him to life. Completing the sequence are the *Separation of Land from Sea;* the *Creation of the Sun, Moon and Plants;* and the *Separation of Light from Darkness,* featuring a fearsome God reaching out to touch the sun. Set around the central panels are 20 athletic male nudes, known as *ignudi.*

Opposite, on the west wall, is Michelangelo's mesmeric *Giudizio Universale,* showing Christ – in the centre near the top – passing sentence over the souls of the dead as they are torn from their graves to face him. The saved get to stay up in heaven (in the upper right), the damned are sent down to face the demons in hell (in the bottom right).

Near the bottom, on the right, you'll see a man with donkey ears and a snake wrapped around him. This is Biagio de Cesena, the papal master of ceremonies, who was a fierce critic of Michelangelo's composition. Another famous figure is St Bartholomew, just beneath Christ, holding his own flayed skin. The face in the skin is

God the Father with Angels on the ceiling of the Stanza dell'Incendio di Borgo (p43)

said to be a self-portrait of Michelangelo, its anguished look reflecting the artist's tormented faith.

The chapel's walls also boast superb frescoes. Painted between 1481 and 82 by a crack team of Renaissance artists, including Botticelli, Ghirlandaio, Pinturicchio, Perugino and Luca Signorelli, they represent events in the lives of Moses (to the left looking at the *Giudizio Universale*) and Christ (to the right). Highlights include Botticelli's *Temptations of Christ* and Perugino's *Handing over of the Keys*.

As well as providing a showcase for priceless art, the Sistine Chapel also serves an important religious function as the place where the conclave meets to elect a new pope.

★ Sistine Chapel Myth One

It's often said Michelangelo worked alone. He didn't. Throughout the job, he employed a steady stream of assistants to help with the plasterwork.

★ Sistine Chapel Myth Two

A popular myth is that Michelangelo painted the ceilings lying down. In fact, he designed a curved scaffolding that allowed him to work standing up.

St Peter's Basilica and St Peter's Square

HANS-PETER MERTEN/GETTY IMAGES ©

St Peter's Basilica

In this city of outstanding churches, none can hold a candle to St Peter's Basilica, Italy's largest, richest and most spectacular basilica.

Great For...

☑ Don't Miss

The *Pietà*, Michelangelo's hauntingly sad depiction of a youthful Mary cradling the body of Jesus.

The original church was commissioned by the emperor Constantine and built around 349 on the site where St Peter is said to have been buried between AD 64 and 67. But like many medieval churches, it eventually fell into disrepair and it wasn't until the mid-15th century that efforts were made to restore it, first by Pope Nicholas V and then, rather more successfully, by Julius II.

In 1506 construction began on a design by Bramante, but when the architect died in 1514, building ground to a halt. In 1547 Michelangelo took the project on. He simplified Bramante's plans and drew up designs for what was to become his greatest architectural achievement, the dome. He never lived to see it built, though, and it was left to Giacomo della Porta, Domenico Fontana and Carlo Maderno to complete

Interior of St Peter's Basilica

WIBOWO RUSLI/GETTY IMAGES ©

the basilica, which was finally consecrated in 1626.

Facade

Built between 1608 and 1612, Maderno's immense facade is 48m high and 118.6m wide. Eight 27m-high columns support the upper attic, on which 13 statues stand representing Christ the Redeemer, St John the Baptist and the 11 apostles. The central balcony, the **Loggia della Benedizione**, is where the pope stands to deliver his Urbi et Orbi blessing at Christmas and Easter.

Interior

At the beginning of the right aisle is Michelangelo's hauntingly beautiful **Pietà**. Sculpted when the artist was 25 (in 1499), it's the only work he ever signed – his

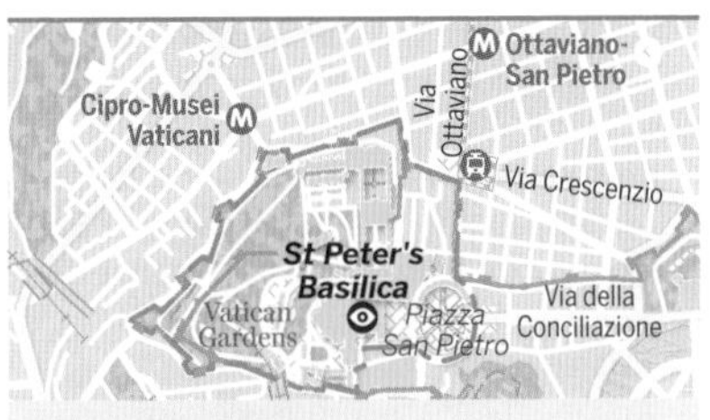

Need to Know

Basilica di San Pietro; www.vatican.va; St Peter's Sq; 7am-7pm summer, to 6.30pm winter; M Ottaviano-San Pietro; FREE

Take a Break

Search out Fa-Bìo (p132) for a freshly made sandwich or a healthy lunchtime salad.

★ Top Tip

Strict dress codes are enforced, so no shorts, miniskirts or bare shoulders.

signature is etched into the sash across the Madonna's breast.

On a pillar just beyond the *Pietà*, Carlo Fontana's gilt and bronze **monument to Queen Christina of Sweden** commemorates the far-from-holy Swedish monarch who converted to Catholicism in 1655.

Moving on, you'll come to the **Cappella di San Sebastiano**, home of Pope John Paul II's tomb, and the **Cappella del Santissimo Sacramento**, a sumptuously decorated baroque chapel.

Dominating the centre of the basilica is Bernini's 29m-high **baldachin**. Supported by four spiral columns and made with bronze taken from the Pantheon, it stands over the **high altar**, which itself sits on the site of St Peter's grave.

Above the baldachin, Michelangelo's **dome** soars to a height of 119m. Based on Brunelleschi's cupola in Florence, it's supported by four massive stone **piers** named after the saints whose statues adorn the Bernini-designed niches – Longinus, Helena, Veronica and Andrew.

At the base of the **Pier of St Longinus** is Arnolfo di Cambio's much-loved

13th-century bronze **statue of St Peter**, whose right foot has been worn down by centuries of caresses.

Dominating the tribune behind the altar is Bernini's extraordinary **Cattedra di San Pietro**, centred on a wooden seat that was once thought to have been St Peter's but in fact dates to the 9th century.

To the right of the throne, Bernini's **monument to Urban VIII** depicts the pope flanked by the figures of Charity and Justice.

Near the head of the left aisle are the so-called **Stuart monuments**. On the right is the monument to Clementina Sobieska, wife of James Stuart, by Filippo Barigioni, and on the left is Canova's vaguely erotic monument to the last three members of the Stuart clan, the pretenders to the English throne who died in exile in Rome.

Dome

From the **dome** (with/without lift €7/5; ⏲8am-5.45pm summer, to 4.45pm winter; Ⓜ Ottaviano-San Pietro) entrance on the right of the basilica's main portico, you can walk the 551 steps to the top or take a small lift halfway and then follow on foot for the last 320 steps. Either way, it's a long, steep climb. But make it to the top, and you're rewarded with stunning views.

Museo Storico Artistico

Accessed from the left nave, the **Museo Storico Artistico** (Tesoro; adult/reduced €7/5; ⏲9am-6.15pm summer, to 5.15pm winter; Ⓜ Ottaviano-San Pietro) sparkles with sacred relics. Highlights include a tabernacle by Donatello and the 6th-century *Crux Vaticana* (Vatican Cross).

St Peter's Square

Vatican Grottoes

Extending beneath the basilica, the **Vatican Grottoes** (9am-6pm summer, to 5pm winter; M Ottaviano-San Pietro) FREE contain the tombs and sarcophagi of numerous popes, as well as several columns from the original 4th-century basilica. The entrance is in the Pier of St Andrew.

St Peter's Tomb

Excavations beneath the basilica have uncovered part of the original church and what archaeologists believe is the **Tomb of St Peter** (06 6988 5318; admission €13, over 15s only; M Ottaviano-San Pietro).

The excavations can only be visited by guided tour. To book a spot, email the Ufficio Scavi (scavi@fsp.va) as early as possible.

★ The World's Largest Church

Contrary to popular opinion, St Peter's Basilica is not the world's largest church – the Basilica of Our Lady of Peace in Yamoussoukro on the Ivory Coast is bigger.

USABIN/GETTY IMAGES ©

What's Nearby?

St Peter's Square Piazza

(Piazza San Pietro; M Ottaviano-San Pietro)

Overlooked by St Peter's Basilica, the Vatican's central square was laid out between 1656 and 1667 to a design by Gian Lorenzo Bernini. Seen from above, it resembles a giant keyhole with two semicircular colonnades, each consisting of four rows of Doric columns, encircling a giant ellipse that straightens out to funnel believers into the basilica. The effect was deliberate – Bernini described the colonnades as representing 'the motherly arms of the church'.

Castel Sant'Angelo Museum, Castle

(06 681 91 11; www.castelsantangelo.beniculturali.it; Lungotevere Castello 50; adult/reduced €7/3.50; 9am-7.30pm Tue-Sun; Piazza Pia)

With its chunky round keep, this castle is an instantly recognisable landmark. Built as a mausoleum for the emperor Hadrian, it was converted into a papal fortress in the 6th century and named after an angelic vision that Pope Gregory the Great had in 590. Nowadays, it houses the **Museo Nazionale di Castel Sant'Angelo** and its eclectic collection of paintings, sculpture, military memorabilia and medieval firearms.

★ Crowning Glory

Near the main entrance, a red floor disk marks the spot where Charlemagne and later Holy Roman Emperors were crowned by the pope.

Pantheon

A striking 2000-year-old temple, now a church, the Pantheon is Rome's best-preserved ancient monument and one of the most influential buildings in the Western world. Its greying, pock-marked exterior might look its age, but inside it's a different story, and it's a unique and exhilarating experience to pass through its vast bronze doors and gaze up at the largest unreinforced concrete dome ever built.

Great For...

Need to Know

Map p250; www.pantheonroma.com; Piazza della Rotonda; 8.30am-7.30pm Mon-Sat, 9am-6pm Sun; Largo di Torre Argentina) FREE

★ Top Tip

Mass is celebrated at the Pantheon at 5pm on Saturday and 10.30am on Sunday.

In its current form the Pantheon dates to around AD 125. The original temple, built by Marcus Agrippa in 27 BC, burnt down in AD 80, and although it was rebuilt by Domitian, it was struck by lightning and destroyed for a second time in AD 110. The emperor Hadrian had it reconstructed between AD 118 and 125, and it's this version that you see today.

Hadrian's temple was dedicated to the classical gods – hence the name Pantheon, a derivation of the Greek words *pan* (all) and *theos* (god) – but in 608 it was consecrated as a Christian church and it's now officially known as the Basilica di Santa Maria ad Martyres.

Thanks to this consecration, it was spared the worst of the medieval plundering that reduced many of Rome's ancient buildings to near dereliction. But it didn't escape entirely unscathed – its gilded-bronze roof tiles were removed and bronze from the portico was used by Bernini for the baldachino at St Peter's Basilica.

Exterior

The dark-grey pitted exterior faces onto busy, cafe-lined Piazza della Rotonda. And while its facade is somewhat the worse for wear, it's still an imposing sight. The monumental entrance **portico** consists of 16 Corinthian columns, each 13m high and each made of Egyptian granite, supporting a triangular **pediment**. Behind the columns, two 20-tonne **bronze doors** – 16th-century restorations of the original portal – give onto the central rotunda.

Rivets and holes in the brickwork indicate where marble-veneer panels were originally placed.

Interior of the Pantheon

The Inscription

For centuries the inscription under the pediment – 'M:AGRIPPA.L.F.COS.TERTIUM. FECIT' (Marcus Agrippa, son of Lucius, consul for the third time built this) – led scholars to think that the current building was Agrippa's original temple. However, 19th-century excavations revealed traces of an earlier temple and historians realised that Hadrian had simply kept Agrippa's original inscription.

Interior

Although impressive from outside, it's only when you get inside that you can really appreciate the Pantheon's full size. With light streaming in through the **oculus** (the 8.7m-diameter hole in the centre of the dome), the cylindrical marble-clad interior seems vast.

Opposite the entrance is the church's main **altar**, over which hangs a 7th-century icon of the *Madonna col Bambino* (Madonna and Child). To the left are the tombs of the artist Raphael, King Umberto I and Margherita of Savoy. Over on the opposite side of the rotunda is the tomb of King Vittorio Emanuele II.

☑ Don't Miss

The entrance doors – these 7m-high bronze portals provide a suitably grand entrance.

ITSTUDIO/SHUTTERSTOCK ©

Dome

The Pantheon's dome, considered the Romans' most important architectural achievement, was the largest dome in the world until Brunelleschi beat it with his Florentine cupola. Its harmonious appearance is due to a precisely calibrated symmetry – its diameter is exactly equal to the building's interior height of 43.3m. At its centre, the oculus, which symbolically connected the temple with the gods, plays a vital structural role by absorbing and redistributing the dome's huge tensile forces.

What's Nearby?

Basilica di Santa Maria Sopra Minerva Basilica

(Map p250; www.santamariasopraminerva.it; Piazza della Minerva 42; ⏲6.45am-7pm Mon-Fri, 6.45am-12.30pm & 3.30-7pm Sat, 8am-12.30pm & 3.30-7pm Sun; 🚌Largo di Torre Argentina) Built on the site of three pagan temples, including one to the goddess Minerva, the Dominican Basilica di Santa Maria Sopra Minerva is Rome's only Gothic church. However, little remains of the original 13th-century structure and these days the main drawcard is a minor Michelangelo sculpture and the colourful, art-rich interior.

✕ Take a Break

For an uplifting drink head to La Casa del Caffè Tazza d'Oro (p171), a busy cafe serving some of the best coffee in town.

Museo e Galleria Borghese

Housing what's often referred to as the 'queen of all private art collections', this spectacular gallery boasts some of the city's finest art treasures, including a series of sensational sculptures by Gian Lorenzo Bernini and important paintings by the likes of Caravaggio, Titian, Raphael and Rubens.

Great For...

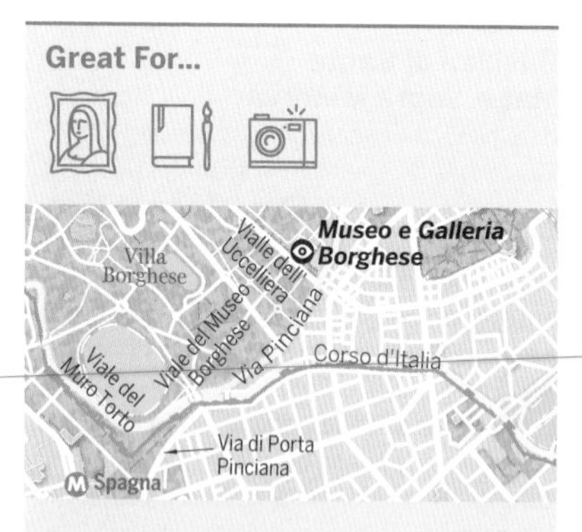

Need to Know

Map p256; ☎06 3 28 10; www.galleriaborghese.it; Piazzale del Museo Borghese 5; adult/reduced €11/6.50; 9am-7pm Tue-Sun; Via Pinciana

Sale IV (p57) with the *Ratto di Proserpina*

★ **Top Tip**
Remember to prebook your ticket, and take ID when you pick it up.

The museum's collection was formed by Cardinal Scipione Borghese (1579–1633), the most knowledgeable and ruthless art collector of his day. It was originally housed in his residence near St Peter's, but in the 1620s he had it transferred to his new villa just outside Porta Pinciana. And it's here, in the villa's central building, the Casino Borghese, that you'll see it today.

Over the centuries the villa has undergone several overhauls, most notably in the late 1700s when Prince Marcantonio Borghese added much of the lavish neoclassical decor.

The villa is divided into two parts: the ground-floor museum and the upstairs picture gallery.

Entrance & Ground Floor

The **entrance hall** features 4th-century floor mosaics of fighting gladiators and a 2nd-century *Satiro Combattente* (Fighting Satyr). High on the wall is a gravity-defying bas-relief of a horse and rider falling into the void by Pietro Bernini (Gian Lorenzo's father).

Sala I is centred on Antonio Canova's daring depiction of Napoleon's sister, Paolina Bonaparte Borghese, reclining topless as **Venere Vincitrice** (Venus Victrix; 1805–08). Its suggestive pose and technical virtuosity is typical of Canova's elegant, mildly erotic neoclassical style.

But it's Gian Lorenzo Bernini's spectacular sculptures – flamboyant depictions of pagan myths – that really steal the show. Just look at Daphne's hands morphing into

Villa Borghese

leaves in the swirling *Apollo e Dafne* (1622–25) in **Sala III**, or Pluto's hand pressing into the seemingly soft flesh of Persephone's thigh in the *Ratto di Proserpina* (Rape of Proserpina; 1621–22) in **Sala IV**.

Caravaggio, one of Cardinal Scipione's favourite artists, dominates **Sala VIII**. You'll see a dissipated *Bacchino malato* (Young Sick Bacchus; 1593–94), the strangely beautiful *La Madonna dei Palafrenieri* (Madonna of the Palafrenieri; 1605–06) and *San Giovanni Battista* (St John the Baptist; 1609–10), probably his last work. There's also the much-loved *Ragazzo col Canestro di Frutta* (Boy with a Basket of Fruit; 1593–95) and dramatic *Davide con la Testa di Golia* (David with the Head of Goliath; 1609–10) – Goliath's head is said to be a self-portrait.

☑ **Don't Miss**

Canova's *Venere Vincitrice*, his sensual portrayal of Paolina Bonaparte Borghese.

PHANT/GETTY IMAGES ©

Picture Gallery

With works representing the best of the Tuscan, Venetian, Umbrian and northern European schools, the upstairs picture gallery offers a wonderful snapshot of Renaissance art.

In **Sala IX** don't miss Raphael's extraordinary *La Deposizione di Cristo* (The Deposition; 1507) and his charming *Dama con Liocorno* (Lady with a Unicorn; 1506). In the same room you'll find Fra Bartolomeo's superb *Adorazione del Bambino* (Adoration of the Christ Child; 1499) and Perugino's *Madonna col Bambino* (Madonna and Child; early 16th century).

Next door, Correggio's *Danäe* (1530–31) shares wall space with a willowy Venus, as portrayed by Cranach in his *Venere e Amore che Reca Il Favo do Miele* (Venus and Cupid with Honeycomb; 1531).

Moving on, **Sala XIV** boasts two self-portraits by Bernini, and **Sala XVIII** contains two significant works by Rubens: *Compianto su Cristo morto* (The Deposition; 1603) and *Susanna e I Vecchioni* (Susanna and the Elders; 1605–07).

To finish off, Titian's early masterpiece *Amor Sacro e Amor Profano* (Sacred and Profane Love; 1514) in **Sala XX** is one of the collection's most prized works.

What's Nearby?

Villa Borghese Park

(Map p256; entrances at Piazzale San Paolo del Brasile, Piazzale Flaminio, Via Pinciana, Via Raimondo, Largo Pablo Picasso; dawn-dusk; Porta Pinciana) Locals, lovers, tourists, joggers – no one can resist the lure of Rome's most

Take a Break

There's a bar in the basement entrance area, but for a more memorable meal head across the park to the Caffè delle Arti (p142).

celebrated park. Originally the 17th-century estate of Cardinal Scipione Borghese, it covers about 80 hectares of wooded glades, gardens and grassy banks. Among its attractions are several excellent museums, the landscaped **Giardino del Lago** (Map p256; boat hire per person €3 for 20 minutes; 7am-9pm), and **Piazza di Siena** (Map p256; Porta Pinciana), a dusty arena used for Rome's top equestrian event in May.

Museo Nazionale Etrusco di Villa Giulia — Museum

(Map p256; www.villagiulia.beniculturali.it; Piazzale di Villa Giulia; adult/reduced €8/4; 8.30am-7.30pm Tue-Sun; Via delle Belle Arti) Pope Julius III's 16th-century villa provides the charming setting for Italy's finest collection of Etruscan and pre-Roman treasures. Exhibits, many of which came from burial tombs in the surrounding Lazio region, range from bronze figurines and black *bucchero* tableware to temple decorations, terracotta vases and a dazzling display of sophisticated jewellery.

Must-sees include a polychrome terracotta statue of Apollo, the 6th-century-BC *Sarcofago degli Sposi* (Sarcophagus of the Betrothed) and the *Euphronios Krater*, a celebrated Greek vase.

Further finds relating to the Umbri and Latin peoples are on show in the nearby **Villa Poniatowski** (Map p256; 06 321 96 98; www.villagiulia.beniculturali.it; Piazzale di Villa Giulia; incl Museo Nazional Etrusco di Villa Giulia adult/reduced €8/4; 9am-1.30pm Tue-Sat, booking necessary; Via delle Belle Arti). You'll need to book to enter here, with Sunday visits restricted to guided tours run by the Coop Arteingioco. Call 06 4423 9949 for details.

Galleria Nazionale d'Arte Moderna e Contemporanea — Gallery

(Map p256; 06 3229 8221; www.gnam.beniculturali.it; Viale delle Belle Arti 131, disabled entrance Via Gramsci 73; adult/reduced €8/4; 8.30am-7.30pm Tue-Sun; Piazza Thorvaldsen) Housed in a vast belle époque palace, this oft-overlooked gallery is an unsung gem. Its superlative collection runs the gamut from neoclassical sculpture to abstract expressionism, with works by many of the most important exponents of 19th- and 20th-century art.

There are canvases by the *macchiaioli* (Italian Impressionists) and futurists Boccioni and Balla, as well as sculptures by Canova and major works by Modigliani, De Chirico and Guttuso. International artists represented include Van Gogh, Cézanne, Monet, Klimt, and Alberto Giacometti, whose trademark stick-figures share a room with a Jackson Pollock canvas, a curvaceous Henry Moore sculpture and a hanging mobile by Alexander Calder.

Pincio Hill Gardens — Gardens

(Map p256; MFlaminio) Overlooking Piazza del Popolo, the 19th-century Pincio Hill is

Classical fountain, Villa Borghese

named after the Pinci family, who owned this part of Rome in the 4th century. It's quite a climb up from the piazza, but at the top you're rewarded with lovely views over to St Peter's and the Gianicolo Hill. Alternatively, you can approach from the top of the Spanish Steps. From the gardens you can strike out to explore Villa Borghese, the Villa dei Medici, or the Chiesa della Trinità dei Monti at the top of the Spanish Steps.

Museo Carlo Bilotti Gallery

(Map p256; ☎06 06 08; www.museocarlobilotti.it; Viale Fiorello La Guardia; ⊙10am-4pm Tue-Fri winter, 1-7pm Tue-Fri summer, 10am-7pm Sat & Sun year-round; 🚌Porta Pinciana) FREE The Orangery of Villa Borghese provides the handsome setting for the art collection of billionaire cosmetics magnate Carlo Bilotti. The main focus are 18 works by Giorgio de Chirico (1888–1978), one of Italy's foremost modern artists, but also of note is a Warhol portrait of Bilotti's wife and daughter.

☑ Best Views

For unforgettable views over Rome's rooftops and domes, make your way to the Pincio Hill Gardens in the southwest of Villa Borghese.

★ Top Tip

Monday is not a good time for exploring Villa Borghese. Sure, you can walk in the park, but its museums and galleries are all shut – they are only open Tuesday through Sunday.

BWZENITH/GETTY IMAGES ©

Palatino

Rising above the Roman Forum, the Palatino (Palatine Hill) is an atmospheric area of towering pine trees, majestic ruins and memorable views. According to legend, this is where Romulus and Remus were saved by a wolf and where Romulus founded Rome in 753 BC. Archaeological evidence can't prove the myth, but it has dated human habitation here to the 8th century BC.

Great For...

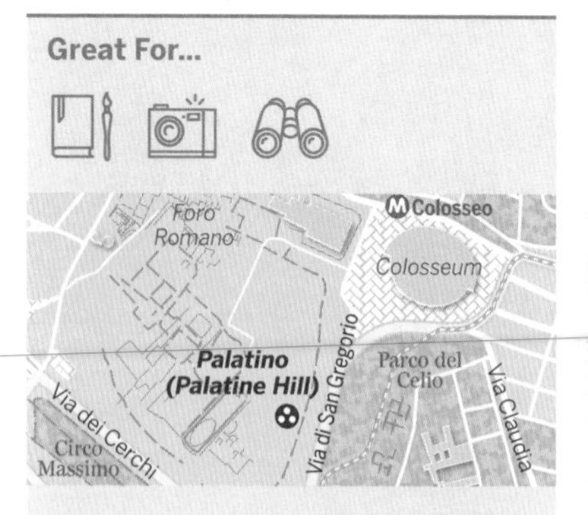

Need to Know

Palatine Hill; Map p252; ☎06 3996 7700; www.coopculture.it; Via di San Gregorio 30 & Via Sacra; adult/reduced incl Colosseum & Roman Forum €12/7.50; ⊙8.30am-1hr before sunset; Ⓜ Colosseo

Stadio (p62)

★ Top Tip

The best spot for a picnic is the grassy Vigna Barberini near the Orti Farnesiani.

The Palatino was ancient Rome's most exclusive neighbourhood. The emperor Augustus lived here all his life and successive emperors built increasingly opulent palaces. But after Rome's fall, it fell into disrepair and in the Middle Ages churches and castles were built over the ruins. Later, wealthy Renaissance families established gardens on the hill.

Most of the Palatino as it appears today is covered by the ruins of the emperor Domitian's 1st-century complex, which served as the main imperial palace for 300 years.

Stadio

On entering the Palatino from Via di San Gregorio, head uphill until you come to the first recognisable construction, the **stadio**. This sunken area, which was part of the main imperial palace, was used by the emperor for private games. To the southeast of the stadium are the remains of a complex built by Septimius Severus, comprising baths (**Terme di Settimio Severo**) and a palace (**Domus Severiana**) where, if they're open, you can visit the **Arcate Severiane** (Severian Arches; Map p252; ☎06 3996 7700; www.coopculture.it; admission incl in Palatino ticket; ⌚8.30am-1hr before sunset Tue & Fri; Ⓜ Colosseo), a series of arches built to facilitate further development.

Domus Augustana & Domu Flavia

Next to the *stadio* are the ruins of the **Domus Augustana** (Emperor's Residence), the emperor's private quarters in the imperial palace. Over two levels, rooms lead off a *peristilio* (porticoed courtyard) on each floor. You can't get to the lower level,

Domus Augustana

but from above you can see the basin of a fountain.

Over on the other side of the Museo Palatino is the **Domus Flavia**, the public part of the palace. The Domus was centred on a grand columned peristyle – the grassy area with the base of an octagonal fountain – off which the main halls led.

Museo Palatino

The **Museo Palatino** (admission incl in Palatino ticket; 8.30am-1hr before sunset; M Colosseo) houses a small collection of finds from the Palatino. The downstairs section illustrates the history of the hill from its origins to the Republican age, while upstairs you'll find artefacts from the Imperial age, including a beautiful 1st-century bronze, the *Erma di Canefora*.

☑ Don't Miss

The sight of the Roman Forum laid out beneath you from the viewing balcony on the Orti Farnesiani.

MARCPO/GETTY IMAGES ©

Casa di Livia & Casa di Augusto

Among the best-preserved buildings on the Palatino is the **Casa di Livia** (Map p252; 06 3996 7700; www.coopculture.it; incl Casa di Augusto €4; guided tour 1pm daily, prebooking necessary; M Colosseo), northwest of the Domus Flavia. Home to Augustus' wife Livia, it was built around an atrium leading onto what were once frescoed reception rooms. Nearby, the **Casa di Augusto** (Map p252; 06 3996 7700; www.coopculture.it; incl Casa di Livia €4; guided tour 1pm daily, prebooking necessary; M Colosseo), Augustus' separate residence, contains superb frescoes in vivid reds, yellows and blues.

Criptoportico

Reached from near the Orti Farnesiani, the **criptoportico** is a 128m tunnel where Caligula is said to have been murdered, and which Nero used to connect his Domus Aurea with the Palatino. It is now used for temporary exhibitions.

Orti Farnesiani

Covering the Domus Tiberiana (Tiberius' Palace) in the northwest of the Palatino, the **Orti Farnesiani** is one of Europe's earliest botanical gardens. Named after Cardinal Alessandro Farnese, who had it laid out in the mid-16th century, it commands breathtaking views over the Roman Forum.

✕ Take a Break

There are no great options in the immediate vicinity so hotfoot it to Terre e Domus (p124) for some rousing regional fare.

Fontana del Moro (p66)

Piazza Navona

With its ornate fountains, exuberant baroque palazzi *(mansions) and pavement cafes, Piazza Navona is central Rome's elegant showcase square. Long a hub of local life, it hosted Rome's main market for close on 300 years, and today attracts a colourful daily circus of street performers, hawkers, artists, tourists, fortune-tellers and pigeons.*

Great For...

Need to Know

Map p250; Corso del Rinascimento

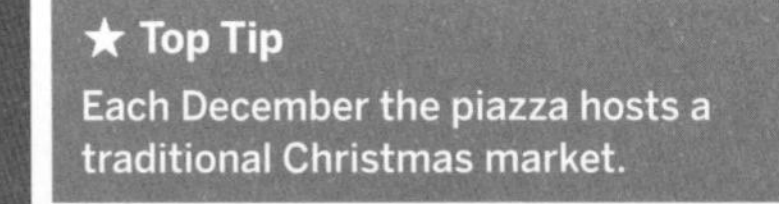
★ Top Tip
Each December the piazza hosts a traditional Christmas market.

SLOW IMAGES/GETTY IMAGES ©

Stadio di Domiziano

Like many of Rome's landmarks, the piazza sits over an ancient monument. The 30,000-seat **Stadio di Domiziano** (Domitian's Stadium; Map p250; 06 4568 6100; www.stadiodomiziano.com; Via di Tor Sanguigna 3; adult/reduced €8/6; 10am-7pm Sun-Fri, to 8pm Sat; Corso del Rinascimento), the subterranean remains of which can be accessed from Via di Tor Sanguigna, used to host athletic meets – hence the name Navona, a corruption of the Greek word *agon,* meaning 'public games'.

Fountains

The piazza's grand centrepiece is Bernini's **Fontana dei Quattro Fiumi** (Fountain of the Four Rivers; Map p250), a showy fountain featuring four muscular personifications of the rivers Nile, Ganges, Danube and Plate.

The **Fontana del Moro** (Map p250), at the southern end of the square, was designed by Giacomo della Porta in 1576.

At the northern end of the piazza, the 19th-century **Fontana del Nettuno** (Map p250) depicts Neptune fighting with a sea monster, surrounded by sea nymphs.

Main Buildings

Overlooking Bernini's Fontana dei Quattro Fiumi is the **Chiesa di Sant'Agnese in Agone** (Map p250; www.santagneseinagone.org; concerts €13; 9.30am-12.30pm & 3.30-7pm Mon-Sat, 10am-1pm & 4-8pm Sun), an elaborate baroque church designed by Francesco Borromini.

Further down, the 17th-century **Palazzo Pamphilj** (Map p250) was built for Pope Innocent X and now houses the Brazilian Embassy.

Piazza Navona

What's Nearby?

Chiesa di San Luigi dei Francesi — Church

(Map p250; Piazza di San Luigi dei Francesi 5; 10am-12.30pm & 3-7pm, closed Thu afternoon; Corso del Rinascimento) Church to Rome's French community since 1589, this opulent baroque *chiesa* (church) is home to a celebrated trio of Caravaggio paintings: the *Vocazione di San Matteo* (The Calling of Saint Matthew), the *Martirio di San Matteo* (The Martyrdom of Saint Matthew) and *San Matteo e l'angelo* (Saint Matthew and the Angel), known collectively as the St Matthew cycle.

☑ Don't Miss

Bernini's Fontana dei Quattro Fiumi, the piazza's high-camp central fountain.

CHANCLOS/SHUTTERSTOCK ©

Museo Nazionale Romano: Palazzo Altemps — Museum

(Map p250; 06 3996 7700; www.coopculture.it; Piazza Sant'Apollinare 44; adult/reduced €7/3.50; 9am-7.45pm Tue-Sun; Corso del Rinascimento) Just north of Piazza Navona, Palazzo Altemps is a beautiful late-15th-century *palazzo*, housing the best of the Museo Nazionale Romano's formidable collection of classical sculpture. Many pieces come from the celebrated Ludovisi collection, amassed by Cardinal Ludovico Ludovisi in the 17th century.

Campo de' Fiori — Piazza

(Map p250; Corso Vittorio Emanuele II) Noisy, colourful 'Il Campo' is a major focus of Roman life: by day it hosts one of Rome's best-known markets, while at night it morphs into a raucous open-air pub. For centuries the square was the site of public executions, and it was here that the philosopher Giordano Bruno was burned at the stake for heresy in 1600. The spot is marked by a sinister statue of the hooded monk, created by Ettore Ferrari and unveiled in 1889.

Palazzo Farnese — Historic Building

(Map p250; www.inventerrome.com; Piazza Farnese; admission €5; guided tours 3pm, 4pm & 5pm Mon, Wed & Fri; Corso Vittorio Emanuele II) Home of the French embassy, this formidable Renaissance *palazzo*, one of Rome's finest, was started in 1514 by Antonio da Sangallo the Younger, continued by Michelangelo and finished by Giacomo della Porta. Inside, it boasts a series of frescoes by Annibale Carracci that are said by some to rival Michelangelo's in the Sistine Chapel. The highlight, painted between 1597 and 1608, is the monumental ceiling fresco *Amori degli Dei* (The Loves of the Gods) in the recently restored Galleria dei Carracci.

Take a Break

Grab a bite in a Renaissance cloister at the Chiostro del Bramante Caffè (p125).

Museo Nazionale Romano: Palazzo Massimo alle Terme

Every day, thousands of tourists, commuters and passers-by hurry past this towering neo-Renaissance palazzo without giving it a second glance. They don't know what they're missing. This is one of Rome's great museums, an oft-overlooked treasure trove of classical art. The sculpture is truly impressive but what really takes the breath away is the collection of vibrantly coloured frescoes and mosaics.

Great For...

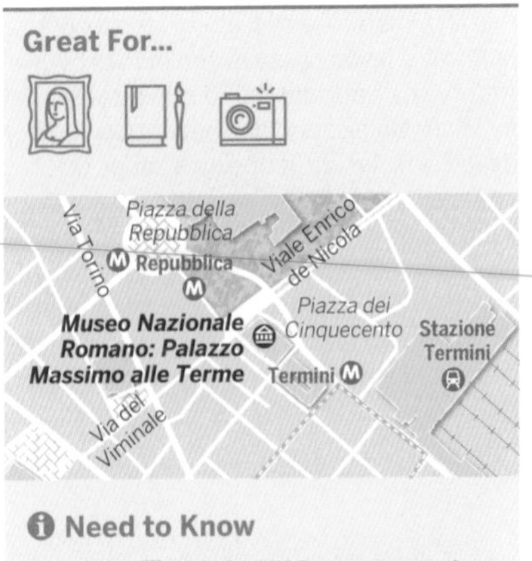

Need to Know

Map p255; ☎06 3996 7700; www.coopculture.it; Largo di Villa Peretti 1; adult/reduced €7/3.50; 9am-7.45pm Tue-Sun; Ⓜ Termini

Ancient Roman mosaic (p70)

★ **Top Tip**

Your ticket also gives admission to the other three sites of the Museo Nazionale Romano.

CHRISTIAN HANDL/ROBERT HARDING ©

Sculpture

The ground and 1st floors are devoted to sculpture, examining imperial portraiture as propaganda and including some breathtaking works of art.

Ground-floor showstoppers include the 5th-century-BC *Niobide morente* (Dying Niobid) and two 2nd-century-BC Greek bronzes – the *Pugile* (Boxer) and the *Principe ellenistico* (Hellenistic Prince). Upstairs, look out for *Il discobolo* (Discus Thrower), a muscular 2nd-century-AD copy of an ancient Greek work. Another admirable body belongs to the graceful *Ermafrodite dormiente* (Sleeping Hermaphrodite).

Also fascinating are the elaborate bronze fittings that belonged to Caligula's ceremonial ships.

Frescoes & Mosaics

On the 2nd floor you'll find the museum's thrilling exhibition of ancient mosaics and frescoes. These vibrantly coloured panels were originally used as interior decor and provide a more complete picture of the inside of a grand ancient Roman villa than you'll see anywhere else in the world. There are intimate *cubicula* (bedroom) frescoes focusing on nature, mythology, domestic and sensual life, and delicate landscape paintings from the winter *triclinium* (dining room).

The museum's crowning glory is a room of frescoes from Villa Livia, one of the homes of Augustus' wife Livia Drusilla. The frescoes depict a paradisiacal garden full of a wild tangle of roses, violets, pomegranates, irises and camomile under a deep-blue sky. These decorated a summer triclinium, a large living and dining area

Third-century marble sarcophagus, Museo Nazionale Romano: Terme di Diocleziano

built half underground to provide protection from the heat. The lighting mimics the modulation of daylight and highlights the richness of the millennia-old colours.

Basement

The basement contains a coin collection that's far more absorbing than you might expect, tracing the Roman Empire's use of coins for propaganda purposes. There's also jewellery dating back several millennia that looks as good as new, and the disturbing remains of a mummified eight-year-old girl, the only known example of mummification dating from the Roman Empire.

☑ Don't Miss

Il discobolo; the athletic pose of the discuss thrower is a homage to the male physique.

GONZALO AZUMENDI/GETTY IMAGES ©

What's Nearby?

Museo Nazionale Romano: Terme di Diocleziano — Museum

(Map p255; ☎06 3996 7700; www.coopculture.it; Viale Enrico de Nicola 78; adult/reduced €7/3.50; ⏰9am-7.30pm Tue-Sun; Ⓜ Termini) The Terme di Diocleziano was ancient Rome's largest bath complex, covering about 13 hectares and with a capacity for 3000 people. This branch of the Museo Nazionale Romano supplies a fascinating insight into Roman life through memorial inscriptions and other artefacts. Outside, the vast, elegant cloister was constructed from drawings by Michelangelo.

Chiesa di Santa Maria della Vittoria — Church

(Map p255; Via XX Settembre 17; ⏰8.30am-noon & 3.30-6pm; Ⓜ Repubblica) This modest church is an unlikely setting for an extraordinary work of art – Bernini's extravagant and sexually charged *Santa Teresa trafitta dall'amore di Dio* (Ecstasy of St Teresa). This daring sculpture depicts Teresa, engulfed in the folds of a flowing cloak, floating in ecstasy on a cloud while a teasing angel pierces her repeatedly with a golden arrow.

Galleria Nazionale d'Arte Antica: Palazzo Barberini — Gallery

(Map p255; ☎06 3 28 10; www.galleriabarberini.benicultural.it; Via delle Quattro Fontane 13; adult/reduced €7/3.50, incl Palazzo Corsini, valid 3 days €9/4.50; ⏰8.30am-7pm Tue-Sun; Ⓜ Barberini) Commissioned to celebrate the Barberini family's rise to papal power, Palazzo Barberini is a sumptuous baroque palace that impresses even before you go inside and start on the breathtaking art. Many high-profile architects worked on it, including rivals Bernini and Borromini: the former contributed a large squared staircase, the latter a helicoidal one.

✕ Take a Break

Drop into Panella L'Arte del Pane (p135) for a coffee and gourmet snack.

Palazzo dei Conservatori (p74)

Capitoline Museums

Dating to 1471, the Capitoline Museums are the world's oldest public museums. Their collection of classical sculpture is one of Italy's finest, including crowd-pleasers such as the iconic Lupa capitolina (Capitoline Wolf), but there's also a formidable picture gallery with masterpieces by the likes of Titian, Tintoretto, Rubens and Caravaggio.

Great For...

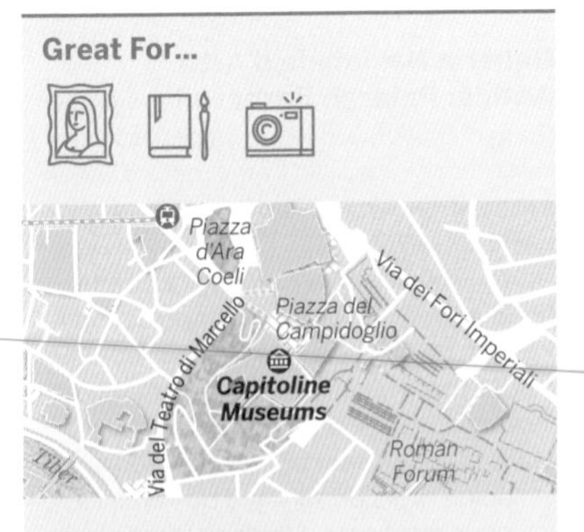

Need to Know

Musei Capitolini; Map p252; 06 06 08; www.museicapitolini.org; Piazza del Campidoglio 1; adult/reduced €11.50/9.50; 9.30am-7.30pm, last admission 6.30pm; Piazza Venezia

★ Top Tip

In a tunnel between the two *palazzi*, the Tabularium commands inspiring views over the Roman Forum.

MAREMAGNUM/GETTY IMAGES ©

The museums occupy two stately palazzi on **Piazza del Campidoglio** (Map p252). The entrance is in **Palazzo dei Conservatori** (Map p252), where you'll find the original core of the sculptural collection and the Pinacoteca (picture gallery).

Palazzo dei Conservatori: 1st Floor

Before you start on the sculpture collection proper, check out the marble body parts littered around the ground-floor **courtyard**. The mammoth head, hand and feet all belonged to a 12m-high statue of Constantine that stood in the Basilica di Massenzio in the Roman Forum.

Of the sculpture on the 1st floor, the Etruscan *Lupa capitolina* (Capitoline Wolf) is the most famous. Dating to the 5th century BC, the bronze wolf stands over her suckling wards, Romulus and Remus, who were added in 1471.

Other crowd-pleasers include the *Spinario*, a delicate 1st-century-BC bronze of a boy removing a thorn from his foot, and Gian Lorenzo Bernini's *Medusa* bust.

Also on this floor, in the modern **Esedra di Marco Aurelio**, is the original of the equestrian statue that stands outside in Piazza del Campidoglio.

Palazzo dei Conservatori: 2nd Floor

The 2nd floor is given over to the **Pinacoteca**, the museum's picture gallery.

Each room harbours masterpieces but two stand out: the **Sala Pietro da Cortona**, which features Pietro da Cortona's famous depiction of the *Ratto delle sabine* (Rape of the Sabine Women; 1630), and the **Sala di**

Sculptures in the Palazzo Nuovo

Santa Petronilla, named after Guercino's huge canvas *Seppellimento di Santa Petronilla* (The Burial of St Petronilla; 1621–23). This airy hall also boasts two works by Caravaggio: *La Buona Ventura* (The Fortune Teller; 1595) and *San Giovanni Battista* (John the Baptist; 1602).

Tabularium

A tunnel links Palazzo dei Conservatori to Palazzo Nuovo via the **Tabularium**, ancient Rome's central archive.

Palazzo Nuovo

Palazzo Nuovo (Map p252) contains some unforgettable show-stoppers. Chief among them is the **Galata Morente** (Dying Gaul), a Roman copy of a 3rd-century-BC Greek original that movingly depicts the anguish of a dying Gaul warrior.

Another superb figurative piece is the the *Venere Capitolina* (Capitoline Venus), a sensual yet demure portrayal of the nude goddess.

☑ Don't Miss

The *Galata Morente* (Dying Gaul) in the Sala del Gladiatore in Palazzo Nuovo.

VIACHESLAV LOPATIN/SHUTTERSTOCK ©

What's Nearby?

Il Vittoriano — Monument

(Map p252; Piazza Venezia; 9.30am-5.30pm summer, to 4.30pm winter; Piazza Venezia) FREE Love it or loathe it, as most locals do, you can't ignore Il Vittoriano (aka the Altare della Patria; Altar of the Fatherland), the massive mountain of white marble that towers over Piazza Venezia. Begun in 1885 to honour Italy's first king, Victor Emmanuel II, it incorporates the **Museo Centrale del Risorgimento** (Map p252; www.risorgimento.it; adult/reduced €5/2.50; 9.30am-6.30pm, closed 1st Mon of month), a small museum documenting Italian unification, and the **Tomb of the Unknown Soldier**.

For Rome's best 360-degree views, take the **Roma dal Cielo** (Map p252; adult/reduced €7/3.50; 9.30am-6.30pm Mon-Thu, to 7.30pm Fri-Sun) lift to the top.

Bocca della Verità — Monument

(Mouth of Truth; Piazza Bocca della Verità 18; donation €0.50; 9.30am-5.50pm summer, to 4.50pm winter; Piazza Bocca della Verità) A bearded face carved into a giant marble disc, the *Bocca della Verità* is one of Rome's most popular curiosities. Legend has it that if you put your hand in the mouth and tell a lie, the Bocca will slam shut and bite your hand off.

The mouth, which was originally part of a fountain, or possibly an ancient manhole cover, now lives in the portico of the **Chiesa di Santa Maria in Cosmedin**, a handsome medieval church.

Take a Break

Head up to the 2nd floor of Palazzo dei Conservatori for a bite at the panoramic Caffè Capitolino (p124).

Via Appia Antica

Ancient Rome's regina viarum *(queen of roads) is now one of Rome's most exclusive addresses, a beautiful cobbled thoroughfare flanked by grassy fields, ancient ruins and towering pine trees. But it has a dark history – it was here that Spartacus and 6000 of his slave rebels were crucified, and the early Christians buried their dead in the underground catacombs.*

Great For...

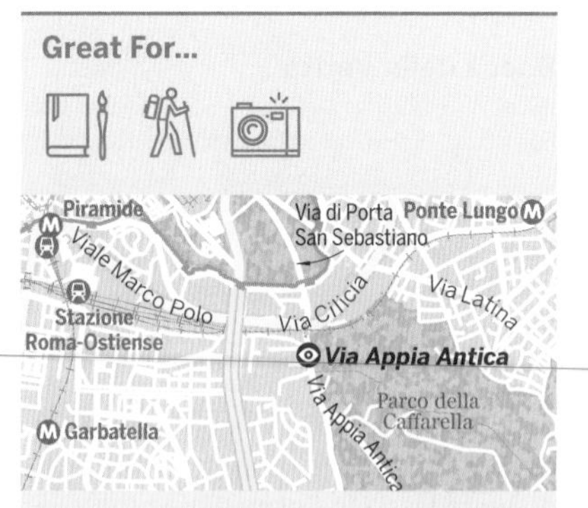

Need to Know

Appian Way; 06 513 53 16; www.parcoappiaantica.it; bike hire hour/day €3/15; Info Point 9.30am-1pm & 2-5.30pm Mon-Fri, 9.30am-6.30pm Sat & Sun, to 5pm winter; Via Appia Antica

★ Top Tip

The stretch near the Basilica di San Sebastiano is traffic-free on Sunday.

JANNIS WERNER/GETTY IMAGES ©

Heading southeast from Porta San Sebastiano, Via Appia Antica was named after Appius Claudius Caecus, who laid the first 90km section in 312 BC. It was later extended in 190 BC to reach Brindisi, some 540km away on the southern Adriatic coast.

Catacombe di San Sebastiano

The **Catacombe di San Sebastiano** (☎06 785 03 50; www.catacombe.org; Via Appia Antica 136; adult/reduced €8/5; ⏰10am-5pm Mon-Sat, closed Dec; 🚌Via Appia Antica) were the first burial chambers to be called catacombs, the name deriving from the Greek *kata* (near) and *kymbas* (cavity), because they were located near a cave. During the persecutory reign of Vespasian from AD 258, they are said to have provided a safe haven for the remains of Saints Peter and Paul.

The 1st level is now almost completely destroyed, but frescoes, stucco work and epigraphs can be seen on the 2nd level. There are also three perfectly preserved mausoleums and a plastered wall with hundreds of invocations to Peter and Paul, engraved by worshippers in the 3rd and 4th centuries.

Above the catacombs, the **Basilica di San Sebastiano** preserves one of the arrows allegedly used to kill St Sebastian, and the column to which he was tied.

Catacombe di San Callisto

Founded at the end of the 2nd century and named after Pope Calixtus I, the **Catacombe di San Callisto** (☎06 513 01 51; www.catacombe.roma.it; Via Appia Antica 110 & 126; adult/reduced €8/5; ⏰9am-noon & 2-5pm, closed Wed & Feb; 🚌Via Appia Antica)

Ruins along Via Appia Antica

became the official cemetery of the newly established Roman Church. In the 20km of tunnels explored to date, archaeologists have found the tombs of 500,000 people and seven popes who were martyred in the 3rd century.

The patron saint of music, St Cecilia, was also buried here, though her body was later removed to the Basilica di Santa Cecilia in Trastevere. When her body was exhumed in 1599, it was apparently perfectly preserved.

Catacombe di Santa Domitilla

Among Rome's largest and oldest, the **Catacombe di Santa Domitilla** (☎06 511 03 42; www.domitilla.info; Via delle Sette Chiese 283; adult/reduced €8/5; ⏲9am-noon & 2-5pm Wed-Mon, closed Jan; 🚌Via Appia Antica) were established on the private burial ground of Flavia Domitilla, niece of the emperor Domitian. They contain Christian wall paintings and the haunting underground Chiesa di SS Nereus e Achilleus, a 4th-century church dedicated to two Roman soldiers martyred by Diocletian.

☑ **Don't Miss**

The ruins of Villa di Massenzio, littering the green fields by the side of the cobbled Via.

DAVID SOANES PHOTOGRAPHY/GETTY IMAGES ©

What's Nearby?

Mausoleo di Cecilia Metella Ruin

(☎06 3996 7700; www.coopculture.it; Via Appia Antica 161; adult/reduced incl Terme di Caracalla & Villa dei Quintili €7/4; ⏲9am-1hr before sunset Tue-Sun; 🚌Via Appia Antica) Dating to the 1st century BC, this great drum of a mausoleum encloses a burial chamber, now roofless. In the 14th century it was converted into a fort by the Caetani family, who were related to Pope Boniface VIII, and used to frighten passing traffic into paying a toll.

Villa di Massenzio Ruin

(☎06 780 13 24; www.villadimassenzio.it; Via Appia Antica 153; ⏲9am-1pm Tue-Sat; 🚌Via Appia Antica) The outstanding feature of Maxentius' enormous 4th-century palace complex is the **Circo di Massenzio** (Via Appia Antica 153; 🚌Via Appia Antica), Rome's best-preserved ancient racetrack – you can still make out the starting stalls used for chariot races. The 10,000-seat arena was built by Maxentius around 309, but he died before ever seeing a race here.

✕ **Take a Break**

Just south of the Mausoleo di Cecilia Metellia, Qui Non se More Mai (p143) is good for grilled meats and authentic Roman pastas.

Tempio di Saturno (p82)

Roman Forum

The Roman Forum was ancient Rome's showpiece centre, a grandiose district of temples, basilicas and vibrant public spaces. Nowadays, it's a collection of impressive, if badly labelled, ruins that can leave you drained and confused. But if you can get your imagination going, there's something wonderfully compelling about walking in the footsteps of Julius Caesar and other legendary figures of Roman history.

Great For...

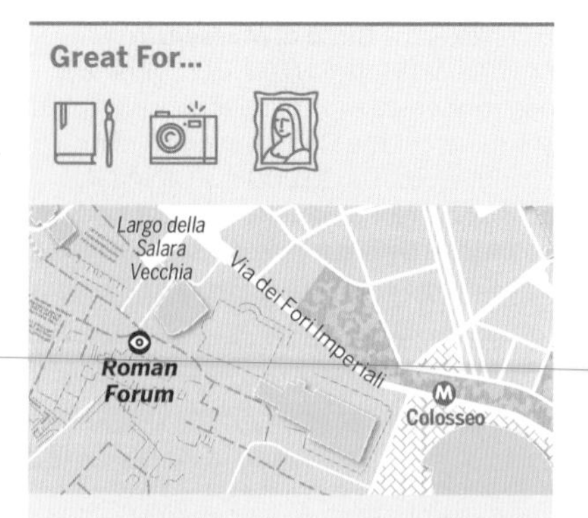

Need to Know

Foro Romano; Map p252; ☎06 3996 7700; www.coopculture.it; Largo della Salara Vecchia & Via Sacra; adult/reduced incl Colosseum & Palatino €12/7.50; ⏱8.30am-1hr before sunset; 🚌Via dei Fori Imperiali

★ **Top Tip**

Exit near the Arco di Settimuio Severo to continue up to Campidoglio and the Capitoline Museums.

AFRIANDI/GETTY IMAGES ©

Originally an Etruscan burial ground, the Forum was first developed in the 7th century BC, growing over time to become the social, political and commercial hub of the Roman Empire. In the Middle Ages it was reduced to pasture land and extensively plundered for its marble. The area was systematically excavated in the 18th and 19th centuries and work continues to this day.

Via Sacra to Campidoglio

Entering the Forum from Largo della Salara Vecchia, you'll see the **Tempio di Antonino e Faustina** (Map p252) ahead to your left. Erected in AD 141, this was transformed into a church in the 8th century, the **Chiesa di San Lorenzo in Miranda** (Map p252). To your right is the 179 BC **Basilica Fulvia Aemilia** (Map p252).

At the end of the path, you'll come to **Via Sacra** (Map p252), the Forum's main thoroughfare, and the **Tempio di Giulio Cesare** (Tempio del Divo Giulio; Map p252), which stands on the spot where Julius Caesar was cremated.

Heading right brings you to the **Curia** (Map p252), the original seat of the Roman Senate, though what you see today is a reconstruction of how it looked in the reign of Diocletian (r 284–305).

At the end of Via Sacra, the **Arco di Settimio Severo** (Arch of Septimius Severus; Map p252) is dedicated to the eponymous emperor and his sons, Caracalla and Geta. Close by, the **Colonna di Foca** (Column of Phocus; Map p252) rises above what was once the Forum's main square, **Piazza del Foro**.

The eight granite columns that rise behind the Colonna are all that survive of the

Tempio di Antonino e Faustina

Tempio di Saturno (Temple of Saturn; Map p252), an important temple that doubled as the state treasury.

Tempio di Castore e Polluce & Casa delle Vestali

From the path that runs parallel to Via Sacra, you'll pass the stubby ruins of the **Basilica Giulia** (Map p252). At the end of the basilica, three columns remain from the 5th-century-BC **Tempio di Castore e Polluce** (Temple of Castor and Pollux; Map p252). Nearby, the 6th-century **Chiesa di Santa Maria Antiqua** (Map p252) is the oldest Christian church in the Forum.

☑ Don't Miss

The Basilca di Massenzio, to get some idea of the scale of ancient Rome's mammoth buildings.

TREVOR COLLENS/ALAMY STOCK PHOTO ©

Back towards Via Sacra is the **Casa delle Vestali** (House of the Vestal Virgins; Map p252), home of the virgins who tended the flame in the adjoining **Tempio di Vesta** (Map p252).

Via Sacra towards the Colosseum

Heading up Via Sacra past the **Tempio di Romolo** (Temple of Romulus; Map p252), you'll come to the **Basilica di Massenzio** (Basilica di Costantino; Map p252), the largest building in the Forum.

Beyond the basilica, the **Arco di Tito** (Arch of Titus; Map p252) was built in AD 81 to celebrate Vespasian and Titus' victories against rebels in Jerusalem.

What's Nearby?

Imperial Forums Archaeological Site

(Fori Imperiali; Map p252; Via dei Fori Imperiali; 🚌Via dei Fori Imperiali) The forums of Trajan, Augustus, Nerva and Caesar are known collectively as the Imperial Forums. These were largely buried when Mussolini bulldozed Via dei Fori Imperiali through the area in 1933, but excavations have since unearthed much of them. The standout sights are the Mercati di Traiano (Trajan's Markets), accessible through the Museo dei Fori Imperiali, and the landmark **Colonna di Traiano** (Trajan's Column; Map p252).

Mercati di Traiano Museo dei Fori Imperiali Museum

(Map p252; ☎06 06 08; www.mercatiditraiano.it; Via IV Novembre 94; adult/reduced €11.50/9.50; ⌚9.30am-7.30pm, last admission 6.30pm; 🚌Via IV Novembre) This striking museum brings to life the **Mercati di Traiano**, emperor Trajan's great 2nd-century market complex, while also providing a fascinating introduction to the Imperial Forums with multimedia displays, explanatory panels and a smattering of archaeological artefacts.

✕ Take a Break

Continue up to the Capitoline Museums to enjoy inspiring views and coffee at the Caffè Capitolino (p124).

Roman Forum

In ancient times, a forum was a market place, civic centre and religious complex all rolled into one, and the greatest of all was the Roman Forum (Foro Romano). Situated between the Palatino (Palatine Hill), ancient Rome's most exclusive neighbourhood, and the Campidoglio (Capitoline Hill), it was the city's busy, bustling centre. On any given day it teemed with activity. Senators debated affairs of state in the **Curia** ❶, shoppers thronged the squares and traffic-free streets and crowds gathered under the **Colonna di Foca** ❷ to listen to politicians holding forth from the **Rostrum** ❷. Elsewhere, lawyers worked the courts in basilicas including the **Basilica di Massenzio** ❸, while the Vestal Virgins quietly went about their business in the **Casa delle Vestali** ❹.

Special occasions were also celebrated in the Forum: religious holidays were marked with ceremonies at temples such as **Tempio di Saturno** ❺ and **Tempio di Castore e Polluce** ❻, and military victories were honoured with dramatic processions up Via Sacra and the building of monumental arches like **Arco di Settimio Severo** ❼ and **Arco di Tito** ❽.

The ruins you see today are impressive but they can be confusing without a clear picture of what the Forum once looked like. This spread shows the Forum in its heyday, complete with temples, civic buildings and towering monuments to heroes of the Roman Empire.

TOP TIPS

» Get grandstand views of the Forum from the Palatino and Campidoglio.

» Visit first thing in the morning or late afternoon; crowds are worst between 11am and 2pm.

» In summer it gets hot in the Forum and there's little shade, so take a hat and plenty of water.

Colonna di Foca & Rostrum
The free-standing, 13.5m-high Column of Phocus is the Forum's youngest monument, dating to AD 608. Behind it, the Rostrum provided a suitably grandiose platform for pontificating public speakers.

Campidoglio (Capitoline Hill)

ADMISSION

Although valid for two days, admission tickets only allow for one entry into the Forum, Colosseum and Palatino.

Tempio di Saturno
Ancient Rome's Fort Knox, the Temple of Saturn was the city treasury. In Caesar's day it housed 13 tonnes of gold, 114 tonnes of silver and 30 million sestertii worth of silver coins.

JONATHAN SMITH/GETTY IMAGES ©

LONELY PLANET/GETTY IMAGES ©

Tempio di Castore e Polluce
Only three columns of the Temple of Castor and Pollux remain. The temple was dedicated to the Heavenly Twins after they supposedly led the Romans to victory over the Latin League in 496 BC.

Arco di Settimio Severo

One of the Forum's signature monuments, this imposing triumphal arch commemorates the military victories of Septimius Severus. Relief panels depict his campaigns against the Parthians.

Curia

This big barn-like building was the official seat of the Roman Senate. Most of what you see is a reconstruction, but the interior marble floor dates to the 3rd-century reign of Diocletian.

Basilica di Massenzio

Marvel at the scale of this vast 4th-century basilica. In its original form the central hall was divided into enormous naves; now only part of the northern nave survives.

JULIUS CAESAR

Julius Caesar was cremated on the site where the Tempio di Giulio Cesare now stands.

1 7 2 3 6 4 8

Tempio di Giulio Cesare

Casa delle Vestali

White statues line the grassy atrium of what was once the luxurious 50-room home of the Vestal Virgins. The virgins played an important role in Roman religion, serving the goddess Vesta.

Arco di Tito

Said to be the inspiration for the Arc de Triomphe in Paris, the well-preserved Arch of Titus was built by the emperor Domitian to honour his elder brother Titus.

Trevi Fountain

The recently restored Fontana di Trevi is Rome's largest and most celebrated fountain. A foaming ensemble of mythical figures, wild horses and cascading rock falls, it takes up the entire side of the 17th-century Palazzo Poli.

Great For...

☑ **Don't Miss**

The contrasting seahorses, or moods of the sea.

Immortalised by Anita Ekberg's dip in Federico Fellini's film *La dolce vita* – apparently she wore waders under her iconic black ballgown – the Trevi Fountain is one of Rome's great must-see sights. It was completed in 1762 and named Trevi in reference to the *tre vie* (three roads) that converge on it.

The water still comes from the Aqua Virgo, an underground aqueduct that was built by General Agrippa during the reign of Augustus some 2000 years ago. Then, as now, the water flows in from the Salone springs around 19km away.

The Design

The fountain's design, conceived by Nicola Salvi in 1732, depicts Neptune, the god of the sea, in a shell-shaped chariot being led by Tritons and two seahorses, one wild,

one docile, representing the moods of the sea. In the niche to the left of Neptune, a statue represents Abundance; to the right is Salubrity.

On the eastern side is a strange conical urn. Known as the *Assso di coppe* (Ace of Cups), this was supposedly placed by Nicola Salvi to block the view of a busybody barber who had been a vocal critc of Salvi's design during the fountain's construction.

Throw Your Money In

The famous tradition (since the 1954 film *Three Coins in the Fountain*) is to toss a coin into the water and thus ensure you'll one day return to Rome. About €3000 is thrown in on an average day. For years much of this was scooped up by local thieves but in 2012 the city authorities clamped down, making it illegal to remove

Need to Know

Fontana di Trevi; Map p255; Piazza di Trevi; (M)Barberini)

Take a Break

For a hearty meal make for Colline Emiliane (p130), up towards Piazza Barberini.

★ Top Tip

The fountain's dazzling white stone photographs best in the soft late-afternoon light.

coins from the water. The money is now collected daily and handed over to the Catholic charity Caritas.

What's Nearby?

Palazzo del Quirinale Palace

(Map p255; ☎06 4 69 91; www.quirinale.it; Piazza del Quirinale; admission €10, ½hr tour €1.50, 2½hr tour €10; ⏲9.30am-4pm Tue, Wed & Fri-Sun, closed Aug; (M)Barberini) Overlooking Piazza del Quirinale, this immense palace is the official residence of Italy's head of state, the Presidente della Repubblica. For almost three centuries it was the pope's summer residence, but in 1870 Pope Pius IX begrudgingly handed the keys over to Italy's new king. Later, in 1948, it was given to the Italian state.

Chiesa di Santa Maria del Popolo

A magnificent repository of art, this is one of Rome's earliest and richest Renaissance churches, with lavish chapels decorated by artists such as Caravaggio, Bernini, Raphael and Pinturicchio.

Great For...

☑ Don't Miss

Pinturicchio's wonderful frescoes.

The first chapel was built here in 1099 to exorcise the ghost of Nero, who was secretly buried on this spot and whose ghost was thought to haunt the area. It had since been overhauled, but the church's most important makeover came when Bramante renovated the presbytery and choir in the early 16th century and Pinturicchio added a series of frescoes. Bernini further reworked the church in the 17th century.

Cerasi Chapel

The church's dazzling highlight is the Cappella Cerasi with its two works by Caravaggio: the *Conversion of Saul* and the *Crucifixion of St Peter*, dramatically spotlit via the artist's use of light and shade. The former is the second version, as the first was rejected by the patron.

The central altarpiece painting is the *Assumption* by Annibale Carracci.

Interior of Chiesa di Santa Maria del Popolo

SONNET SYLVAIN/GETTY IMAGES ©

Chigi Chapel

Raphael designed the Cappella Chigi, dedicated to his patron Agostino Chigi, but never lived to see it completed. Bernini finished the job more than 100 years later, contributing statues of Daniel and Habakkuk to the altarpiece. Only the floor mosaics were retained from Raphael's original design, including that of a kneeling skeleton, placed there to remind the living of the inevitable.

Delle Rovere Chapel

The chapel, built in the late 15th century, features works by Pinturicchio and his school. Frescoes in the lunettes depict episodes from the life of St Jerome, while the main altarpiece shows the Nativity with St Jerome.

Need to Know

Map p256; Piazza del Popolo; 7am-noon & 4-7pm Mon-Sat, 7.30am-1.30pm & 4.30-7.30pm Sun; Ⓜ Flaminio

Take a Break

Treat yourself to a sublime ice cream from the magnificent Fatamorgana (p129).

★ Top Tip

Look out for the oldest stained-glass windows in Rome.

What's Nearby?

Piazza del Popolo Piazza

(Map p256; Ⓜ Flaminio) This dazzling piazza was laid out in 1538 to provide a grandiose entrance to what was then Rome's main northern gateway. It has since been remodelled several times, most recently by Giuseppe Valadier in 1823.

Guarding its southern approach are Carlo Rainaldi's twin 17th-century churches, **Chiesa di Santa Maria dei Miracoli** (Map p256) and **Chiesa di Santa Maria in Montesanto** (Map p256). In the centre, the 36m-high **obelisk** (Map p256) was brought by Augustus from ancient Egypt and originally stood in Circo Massimo.

Porta del Popolo Landmark

(Map p256; Ⓜ Flaminio) On the northern flank of Piazza del Popolo, the Porta del Popolo was created by Bernini in 1655 to celebrate Queen Christina of Sweden's defection to Catholicism,

Interior of Basilica di San Giovanni in Laterano

Basilica di San Giovanni in Laterano

For a thousand years this landmark cathedral was the most important church in Christendom. It was the first Christian basilica built in the city, and until the early 14th century, was the pope's main place of worship. It's still Rome's official cathedral and the pope's seat as the bishop of Rome.

Great For...

Need to Know

Piazza di San Giovanni in Laterano 4; basilica/cloister free/€5; 7am-6.30pm, cloister 9am-6pm; M San Giovanni

★ Top Tip

Look down as well as up – the basilica has a beautiful inlaid marble floor.

GRAFALEX/SHUTTERSTOCK ©

The oldest of Rome's four papal basilicas, it was commissioned by the Emperor Constantine and consecrated by Pope Sylvester I in 324. From then until 1309, when the papacy moved to Avignon, it was the principal pontifical church, and the adjacent Palazzo Laterano was the pope's official residence. Both buildings fell into disrepair during the papacy's French interlude, and when Pope Gregory XI returned to Rome in 1377 he preferred to decamp to the fortified Vatican rather than stay in the official papal digs.

Over the centuries the basilica has been revamped several times, most notably by Borromini in the 17th century, and by Alessandro Galilei, who added the immense white facade in 1735.

The Facade

Surmounted by 15 7m-high statues – Christ with St John the Baptist, John the Evangelist and the 12 Apostles – Galilei's huge facade is an imposing work of late-baroque classicism. Behind the colossal columns there are five sets of doors in the portico. The **central bronze doors** were moved here from the Curia in the Roman Forum, while, on the far right, the **Holy Door** is only opened in Jubilee years.

The Interior

The enormous marble-clad interior owes much of its present look to Francesco Borromini, who was called in by Pope Innocent X to decorate it for the 1650 Jubilee. Divided into a central nave and four minor aisles, it's a breathtaking sight, measuring 130m

Portico in Basilica di San Giovanni in Laterano

(length) by 55.6m (width) by 30m (height). Up above, the spectacular **gilt ceiling** was created at different times, but the central section, which is set around Pope Pius IV's carved coat of arms, dates to the 1560s. Beneath your feet, the beautiful inlaid **mosaic floor** was laid down by Pope Martin V in 1425.

The **central nave** is lined with 18th-century sculptures of the apostles, each 4.6m high and each set in a heroic pose in its own dramatic niche. At the head of the nave, an elaborate Gothic **baldachin** stands over the papal altar. Dating to the 14th century, this towering ensemble is said to contain the relics of the heads of Saints Peter and Paul. In front, a double staircase leads down to the **confessio** and the Renaissance tomb of Pope Martin V.

Behind the altar, the massive **apse** is decorated with sparkling mosaics. Parts of these date to the 4th century, but most were added in the 1800s.

At the other end of the basilica, on the first pillar in the right-hand nave, is an incomplete Giotto fresco. While admiring this, cock your ear towards the next column, where a monument to Pope Sylvester II is said to creak when the death of a pope is imminent.

☑ Don't Miss

The serene cloister; well worth the small entrance fee as it's a lovely spot to collect your thoughts.

The Cloister

To the left of the altar, the basilica's 13th-century cloister is a lovely, peaceful place with graceful twisted columns set around a central garden. Lining the ambulatories are marble fragments from the original church, including the remains of a 5th-century papal throne and inscriptions of two papal bulls.

On the cloister's western side, four columns support a slab of marble that medieval Christians believed represented the height of Jesus.

Take a Break

There are few good eateries right by the basilica, so head towards the Colosseum for some tasty cafe fare at Cafè Cafè (p139).

Teatro, Ostia Antica

Day Trip: Ostia Antica

Rome's answer to Pompeii, the Scavi Archeologici di Ostia Antica is one of Italy's most underappreciated archaeological sites. The amazingly preserved ruins of Rome's main seaport provide a thrilling glimpse into the workings of an ancient town.

Great For...

☑ Don't Miss

The views over the site from atop the Terme di Nettuno.

Founded in the 4th century BC, Ostia (the name means the mouth, or *ostium*, of the Tiber) grew to become a great port and commercial centre with a population of around 50,000.

Decline set in after the fall of the Roman Empire, and by the 9th century the city had largely been abandoned, its citizens driven off by barbarian raids and outbreaks of malaria. Over subsequent centuries, it was plundered of marble and building materials, and its ruins were gradually buried in river silt, hence their survival.

To get to the site from Rome, take the Ostia Lido train from Stazione Porta San Paolo (next to Piramide metro station) and get off at Ostia Antica.

Roman column, Ostia Antica

LEADINGLIGHTS/GETTY IMAGES ©

Need to Know

06 5635 0215; www.ostiaantica.beniculturali.it; Viale dei Romagnoli 717; adult/reduced €8/4, plus possible exhibition supplement €3; 8.30am-6.15pm Tue-Sun summer, earlier closing winter

Take a Break

Try the **Ristorante Monumento** (06 565 00 21; www.ristorantemonumento.it; Piazza Umberto I 8; fixed-price lunch menu €14, meals €25-30; 12.30-3.30pm & 8-11pm Tue-Sun) in the picturesque borgo near the entrance to the Scavi.

★ Top Tip

Come on a weekday when the site is much quieter.

The Ruins

Near the entrance, **Porta Romana** gives onto the **Decumanus Maximus**, the site's central strip, which runs over 1km to Porta Marina, the city's original sea-facing gate.

On the Decumanus, the **Terme di Nettuno** is a must-see. This baths complex, one of 20 that originally stood in town, dates to the 2nd century and boasts some superb mosaics, including one of Neptune driving his sea-horse chariot. In the centre of the complex are the remains of an arcaded **palestra** (gym).

Next to the Terme is the **Teatro**, an amphitheatre built by Agrippa and later enlarged to hold 4000 people.

The grassy area behind the amphitheatre is the **Piazzale delle Corporazioni** (Forum of the Corporations), home to the offices of Ostia's merchant guilds. The mosaics that line the perimeter – ships, dolphins, a lighthouse, an elephant – are thought to represent the businesses housed on the square: ships and dolphins indicated shipping agencies, while the elephant probably referred to a business involved in the ivory trade.

The Forum, Ostia's main square, is overlooked by what remains of the **Capitolium**, a temple built by Hadrian and dedicated to Jupiter, Juno and Minerva.

Nearby is another highlight: the **Thermopolium**, an ancient cafe. Check out the bar, frescoed menu, kitchen and small courtyard where customers would have relaxed next to a fountain.

Across the road are the remains of the 2nd-century-AD **Terme del Foro**, originally the city's largest baths complex. Here, in the *forica* (public toilet), you can see 20 well-preserved latrines set sociably in a long stone bench.

Basilica superiore, Basilica di San Clemente

Basilica di San Clemente

Nowhere better illustrates the various stages of Rome's turbulent past than this fascinating, multi-layered church in the shadow of the Colosseum.

Great For...

☑ **Don't Miss**

The temple to Mithras, deep in the bowels of the basilica.

Basilica Superiore

The ground-floor basilica superiore contains some glorious works of medieval art. These include a golden 12th-century apse mosaic, the *Trionfo della Croce* (Triumph of the Cross), showing the Madonna and St John the Baptist standing by a cross on which Christ is represented by 12 white doves. Also impressive are Masolino's 15th-century frescoes in the **Cappella di Santa Caterina**, depicting a crucifixion scene and episodes from the life of St Catherine.

Basilica Inferiore

Steps lead down to the 4th-century basilica inferiore, mostly destroyed by Norman invaders in 1084, but with some faded 11th-century frescoes illustrating the life of San Clemente.

Basilica di San Clemente

EURASIA/ROBERTHARDING/GETTY IMAGES ©

Need to Know

www.basilicasanclemente.com; Via di San Giovanni in Laterano; excavations adult/reduced €10/5; 9am-12.30pm & 3-6pm Mon-Sat, 12.15-6pm Sun; Via Labicana

Take a Break

Treat yourself to a slap-up trattoria meal at Il Bocconcino (p141).

★ Top Tip

Bring a sweater – the temperatures drops as you descend underground.

Follow down another level and you'll find yourself walking an ancient lane leading to a 1st-century Roman house and a dark, 2nd-century **temple to Mithras**, with an altar showing the god slaying a bull. Beneath it all, you can hear the eerie sound of a subterranean river flowing through a Republic-era drain.

What's Nearby?

Basilica di SS Quattro Coronati — Basilica

(Via dei Santissimi Quattro Coronati 20; 10-11.45am & 4-5.45pm Mon-Sat, 4-5.45pm Sun; Via di San Giovanni in Laterano) This brooding fortified church harbours some lovely 13th-century frescoes and a delightful hidden cloister. The frescoes, in the **Oratorio di San Silvestro**, depict the story of the Donation of Constantine, a notorious forged document with which the emperor Constantine ceded control of Rome and the Western Roman Empire to the papacy.

To access the Oratorio, ring the bell in the entrance courtyard. You might also have to ring for the cloister, which is situated off the northern aisle.

Chiesa di Santo Stefano Rotondo — Church

(www.santo-stefano-rotondo.it; Via di Santo Stefano Rotondo 7; 10am-1pm & 2-5pm winter, 10am-1pm & 3-6pm summer; Via della Navicella) Set in its own secluded grounds, this haunting church boasts a porticoed facade and a round, columned interior. But what really gets the heart racing is the graphic wall decor – a cycle of 16th-century frescoes depicting the tortures suffered by many early Christian martyrs.

Describing them in 1846, Charles Dickens wrote: 'Such a panorama of horror and butchery no man could imagine in his sleep, though he were to eat a whole pig, raw, for supper.'

Terme di Caracalla

ALESSANDROCALZOLARO/GETTY IMAGES ©

Terme di Caracalla

The remains of the Terme di Caracalla, the emperor Caracalla's vast baths complex, are among Rome's most awe-inspiring ruins. The original 10-hectare complex comprised baths, gyms, libraries, shops and gardens.

Great For...

☑ Don't Miss

A white marble slab used in an ancient board game.

Inaugurated in AD 216, the baths remained in continuous use until 537, when the invading Visigoths cut off Rome's water supply. Excavations in the 16th and 17th centuries unearthed a number of important sculptures on the site, many of which found their way into the Farnese family's art collection.

In its heyday, the complex attracted between 6000 and 8000 people every day, while, underground, hundreds of slaves sweated in 9.5km of tunnels, tending to the intricate plumbing systems.

The Ruins

Most of the ruins are what's left of the central bathhouse. This was a huge rectangular edifice bookended by two *palestre* (gyms) and centred on a *frigidarium* (cold room), where bathers would stop after spells in the warmer *tepidarium* and dome-capped *caldaria* (hot room).

Inside the baths complex

MAC99/GETTY IMAGES ©

Need to Know

06 3996 7700; www.coopculture.it; Viale delle Terme di Caracalla 52; adult/reduced €6/3; 9am-1hr before sunset Tue-Sun, 9am-2pm Mon; Viale delle Terme di Caracalla

Take a Break

Head over to the Colosseum area for a light meal at Cafè Cafè (p139).

Top Tip

Opera fans should check for summer performances at the Terme.

As you traverse the ruins towards the *palestra orientale,* look out for a slab of white, pockmarked marble on your right. This is a board from an ancient game called '*tropa*' (the hole game).

Underground, archaeologists have discovered a temple dedicated to the Persian god Mithras.

In summer the ruins are used to stage opera and ballet performances.

What's Nearby?

Villa Celimontana Park

(7am-sunset; Via della Navicella) With its grassy banks and colourful flower beds, this leafy park is a wonderful place to escape the crowds and enjoy a summer picnic. At its centre is a 16th-century villa housing the Italian Geographical Society.

Basilica di Santa Sabina Basilica

(06 5 79 41; Piazza Pietro d'Illiria 1; 6.30am-12.45pm & 3-8pm; Lungotevere Aventino) This solemn basilica, one of Rome's most beautiful medieval churches, was founded by Peter of Illyria in around AD 422. It was enlarged in the 9th century and again in 1216, just before it was given to the newly founded Dominican order – note the tombstone of Muñoz de Zamora, one of the order's founding fathers, in the nave floor. A 20th-century restoration returned it to its original look.

Priorato dei Cavalieri di Malta Historic Building

(Piazza dei Cavalieri di Malta; closed to the public; Lungotevere Aventino) Fronting an ornate cypress-shaded piazza, the Roman headquarters of the Cavalieri di Malta (Knights of Malta) boast one of Rome's most celebrated views. It's not immediately apparent, but look through the keyhole in the Priorato's green door and you'll see the dome of St Peter's Basilica perfectly aligned at the end of a hedge-lined avenue.

Spanish Steps

Rising above Piazza di Spagna, the Spanish Steps provide a perfect people-watching perch and you'll almost certainly find yourself taking stock here at some point.

The Spanish Steps area has long been a magnet for foreigners. In the late 1700s it was much loved by English travellers on the Grand Tour, and was known locally as *'er ghetto de l'inglesi'* (the English ghetto). Keats lived for a short time in some rooms overlooking the Spanish Steps, and died here of tuberculosis at the age of 25. Later, in the 19th century, Charles Dickens visited, noting how artists' models would hang around in the hope of being hired to sit for a painting.

Great For...

☑ Don't Miss

The sweeping rooftop views from the top of the Steps.

The Steps

Although Piazza di Spagna was named after the nearby Spanish embassy to the Holy See, the monumental 135-step staircase (known in Italian as the Scalinata della Trinità dei Monti) was designed by an Italian,

Barcaccia and the Spanish Steps

Need to Know

Map p255; Spagna

Take a Break

Treat yourself to some regional cuisine at nearby Enoteca Regionale Palatium (p130).

★ Top Tip

Visit in late April or early May to see the Steps ablaze with brightly coloured azaleas.

Francesco de Sanctis, and built in 1725 with money bequeathed by a French diplomat.

Chiesa della Trinità dei Monti

The landmark **Chiesa della Trinità dei Monti** (Map p255; Piazza Trinità dei Monti; 6.30am-8pm Tue-Sun; Spagna) was commissioned by King Louis XII of France and consecrated in 1585. Apart from the great rooftop views from outside, it boasts some wonderful frescoes by Daniele da Volterra. His *Deposizione* (Deposition), in the second chapel on the left, is regarded as a masterpiece of mannerist painting.

Piazza di Spagna

At the foot of the steps, the fountain of a sinking boat, the **Barcaccia** (1627), is believed to be by Pietro Bernini, father of the more famous Giani Lorenzo. The bees and suns that decorate the structure, which was sunken to compensate for the low pressure of the feeder aqueduct, represent the Barbarini family who commissioned the fountain.

Opposite, **Via dei Condotti** is Rome's most exclusive shopping strip, while to the southeast, **Piazza Mignanelli** is dominated by the Colonna dell'Immacolata, built in 1857 to celebrate Pope Pius IX's declaration of the Immaculate Conception.

What's Nearby?

Keats–Shelley House Museum

(Map p255; 06 678 42 35; www.keats-shelley-house.org; Piazza di Spagna 26; adult/reduced €5/4, ticket gives discount for Casa di Goethe; 10am-1pm & 2-6pm Mon-Fri, 11am-2pm & 3-6pm Sat; Spagna) The Keats-Shelley House is where Romantic poet John Keats died of TB at the age of 25, in February 1821. A year later, fellow poet Percy Bysshe Shelley drowned off the coast of Tuscany. The small apartment evokes the impoverished lives of the poets, and is now a small museum crammed with memorabilia, from faded letters to death masks.

Auditorium Parco della Musica

ALEXANDRE ZVEIGER/SHUTTERSTOCK ©

Modern Architecture

Rome is best known for its classical architecture, but the city also boasts a string of striking modern buildings, many created and designed by the 21st century's top starchitects.

Great For...

☑ **Don't Miss**

The outlying EUR district is home to some impressive rationalist architecture.

Auditorium Parco della Musica Cultural Centre

(☎06 8024 1281; www.auditorium.com; Viale Pietro de Coubertin 10; guided tours adult/reduced €9/7; ⏲11am-8pm Mon-Sat, 10am-6pm Sun; 🚊Viale Tiziano) Designed by archistar Renzo Piano and inaugurated in 2002, Rome's flagship cultural centre is an audacious work of architecture consisting of three grey pod-like concert halls set round a 3000-seat amphitheatre.

Excavations during its construction revealed remains of an ancient Roman villa, which are now on show in the Auditorium's small **Museo Archeologico** (⏲10am-8pm Mon-Sat summer, 11am-6pm Mon-Sat winter, 10am-6pm Sun year-round) FREE.

Guided tours (for a minimum of 10 people) depart hourly between 11.30am and 4.30pm Saturday and Sunday, and by arrangement from Monday to Friday.

Palazzo della Civiltà del Lavoro

TICHR/GETTY IMAGES ©

Museo Nazionale delle Arti del XXI Secolo (MAXXI) Gallery

(☎06 320 19 54; www.fondazionemaxxi.it; Via Guido Reni 4a; adult/reduced €11/8; ⌚11am-7pm Tue-Sun, to 10pm Sat; 🚋Viale Tiziano) As much as the exhibitions, the highlight of Rome's leading contemporary-art gallery is the Zaha Hadid–designed building it occupies. Formerly a barracks, the curved concrete structure is striking inside and out, with a multilayered geometric facade and a cavernous light-filled interior full of snaking walkways and suspended staircases.

The gallery has a small permanent collection but more interesting are the temporary exhibitions.

Museo d'Arte Contemporanea di Roma (MACRO) Gallery

(☎06 06 08; www.museomacro.org; Via Nizza 138, cnr Via Cagliari; adult/reduced €9.50/7.50; ⌚10.30am-7.30pm Tue-Sun; 🚌Via Nizza) Along with MAXXI, this is Rome's most important contemporary-art gallery. Occupying a converted Peroni brewery, it hosts temporary exhibitions and displays works from its permanent collection of post-1960s Italian art.

Vying with the exhibits for your attention is the museum's sleek black-and-red interior design. The work of French architect Odile Decq, it retains much of the building's original structure while also incorporating a sophisticated steel-and-glass finish.

Museo dell'Ara Pacis Museum

(☎06 06 08; www.arapacis.it; Lungotevere in Auga; adult/reduced €10.50/8.50, audioguide €4; ⌚9am-7pm, last admission 6pm; Ⓜ Flaminio) The first modern construction in Rome's historic centre since WWII, Richard Meier's controversial and widely detested glass-and-marble pavilion houses the *Ara Pacis Augustae* (Altar of Peace), Augustus' great monument to peace. One of the most important works of ancient Roman sculpture, the vast marble altar – measuring 11.6m by 10.6m by 3.6m – was completed in 13 BC.

Palazzo della Civiltà del Lavoro Historic Building

(Palace of the Workers; Quadrato della Concordia; Ⓜ EUR Magliana) Dubbed the Square Colosseum, the Palace of the Workers is EUR's architectural icon, a rationalist masterpiece clad in gleaming white travertine.

ℹ Need to Know

The Auditorium and MAXXI can be accessed by tram 2 from Piazzale Flaminio.

✕ Take a Break

Enjoy splendid trattoria fare at Al Gran Sasso (p130) near the Museo dell'Ara Pacis.

★ Top Tip

Take in a gig at the Auditorium to experience its perfect acoustics.

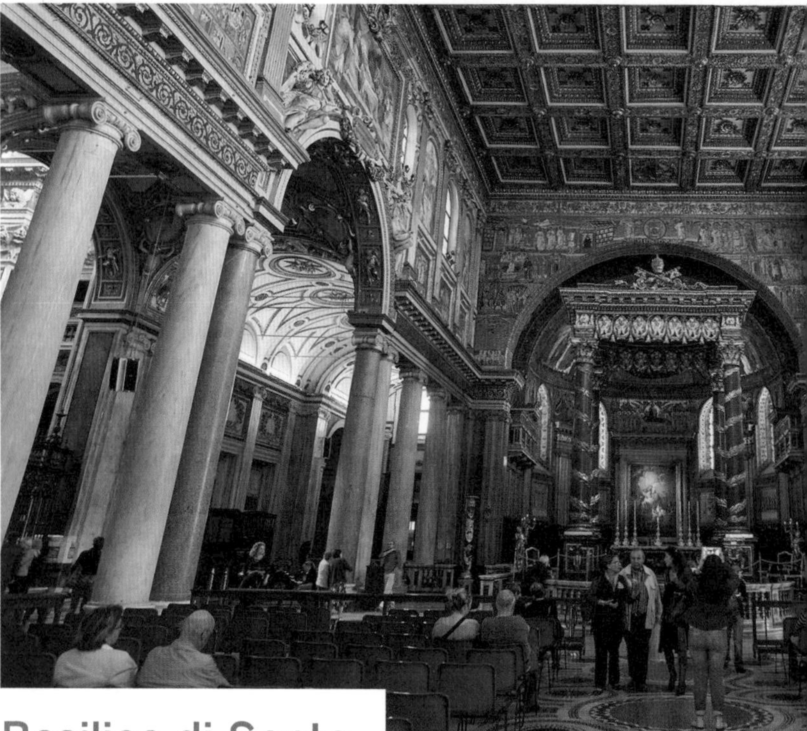

Basilica di Santa Maria Maggiore

One of Rome's four patriarchal basilicas, this monumental church stands on the summit of the Esquiline Hill, on the site of a miraculous snowfall in the summer of AD 358.

Great For...

☑ **Don't Miss**

The luminous 13th-century apse mosaics by Jacopo Torriti.

The basilica, much altered over the centuries, is something of an architectural hybrid, with a 14th-century Romanesque belfry, an 18th-century baroque facade, a largely baroque interior and a series of glorious 5th-century mosaics.

Exterior

Outside, the exterior is decorated with glimmering 13th-century mosaics, protected by Ferdinand Fuga's 1741 baroque loggia. Rising behind, the belfry, Rome's tallest, tops out at 75m.

On the piazza in front of the church, the 18.78m-high column originally stood in the Basilica of Massenzio in the Roman Forum.

Interior

The vast interior retains its original structure, despite the basilica's many overhauls.

Interior of Basilica di Santa Maria Maggiore

Need to Know

Map p255; Piazza Santa Maria Maggiore; basilica/museum/loggia/archaeological site free/€3/5/5; 7am-7pm, museum & loggia 9am-5.30pm; Piazza Santa Maria Maggiore

Take a Break

Relax over a filling meal at the highly rated Trattoria Monti (p135).

★ Top Tip

Come on 5 August to see the historic snowfall re-created with thousands of white petals.

Particularly spectacular are the **5th-century mosaics** in the **triumphal arch** and **nave**, depicting Old Testament scenes. The central image in the **apse**, signed by Jacopo Torriti, dates from the 13th century and represents the coronation of the Virgin Mary. Beneath your feet, the nave floor is a fine example of 12th-century Cosmati paving.

The **baldachin** over the high altar is heavy with gilt cherubs; the altar itself is a porphyry sarcophagus, which is said to contain the relics of St Matthew and other martyrs. The *Madonna col Bambino* (Madonna and Child) panel above the altar is believed to date from the 12th to 13th centuries.

A simple stone plaque embedded in the floor to the right of the altar marks the spot where Gian Lorenzo Bernini and his father Pietro are buried. Steps lead down to the **confessio** where a statue of Pope Pius IX kneels before a reliquary containing a fragment of Jesus' manger.

Through the souvenir shop on the right-hand side of the church is a **museum** with a glittering collection of religious artefacts. Most interesting, however, is the upper loggia, where you'll get a close look at the facade's iridescent 13th-century mosaics, created by Filippo Rusuti. You'll also see Bernini's helical staircase.

What's Nearby?

Basilica di San Pietro in Vincoli — Basilica

(Map p252; Piazza di San Pietro in Vincoli 4a; 8am-12.20pm & 3-7pm summer, to 6pm winter; MCavour) Pilgrims and art lovers flock to this 5th-century basilica for two reasons: to marvel at Michelangelo's colossal *Moses* (1505) sculpture and to see the chains that supposedly bound St Peter when he was imprisoned in the Carcere Mamertino (near the Roman Forum).

Access to the church is via a flight of steps through a low arch that leads up from Via Cavour.

Interior of Basilica di Santa Maria in Trastevere

Basilica di Santa Maria in Trastevere

The glittering Basilica di Santa Maria in Trastevere is said to be the oldest church dedicated to the Virgin Mary in Rome. Inside, its golden mosaics are a spectacle to behold.

Great For...

☑ **Don't Miss**

The Cappella Avila and its cleverly constructed 17th-century dome.

Dating to the early 3rd century, the basilica was commissioned by Pope Callixtus III on the site where, according to legend, a fountain of oil had miraculously sprung from the ground. The basilica has been much altered over the centuries and its current Romanesque form is the result of a 12th-century revamp. The portico came later, added by Carlo Fontana in 1702.

The Exterior

Rising above the four papal statues on Domenico Fontana's 18th-century **porch**, the basilica's restrained 12th-century **facade** is most notable for its beautiful medieval mosaic. This glittering gold banner depicts Mary feeding Jesus surrounded by 10 women bearing lamps.

Towering above the church is a 12th-century Romanesque **bell tower**, complete

Golden ceiling

JULIAN ELLIOTT PHOTOGRAPHY/GETTY IMAGES ©

with its very own mosaic – look in the small niche near the top.

Mosaics & Interior Design

The basilica's main drawcard is its golden 12th-century mosaics. In the **apse**, look out for the dazzling depiction of Christ and his mother flanked by various saints, and, on the far left, Pope Innocent II holding a model of the church. Beneath this is a series of six mosaics by Pietro Cavallini (c 1291) illustrating the life of the Virgin.

The interior boasts a typical 12th-century design with three naves divided by 21 **Roman columns**, some plundered from the Terme di Caracalla. On the right of the altar, near a spiralling Paschal candlestick, is an inscription, *Fons Olei*, which marks the spot where the miraculous oil fountain supposedly sprung. The spiralling **Cosmatesque floor** was relaid in the 1870s, a re-creation of the 13th-century original. Up above, the coffered **golden ceiling** was designed by Domenichino in 1617.

Also worth a look is the **Cappella Avila**, the last chapel on the left, with its stunning 17th-century dome. Antonio Gherardi's clever design depicts four angels holding the circular base of a large lantern whose columns rise to give the effect of a second cupola within a larger outer dome.

Piazza Santa Maria in Trastevere

Outside the basilica, Piazza Santa Maria in Trastevere is the neighbourhood's focal square. By day it's full of mums with strollers, chatting locals and guidebook-toting tourists; by night it's the domain of foreign students, young Romans and out-of-towners, all out for a good time. The fountain in the centre of the square is of Roman origin and was restored by Carlo Fontana in 1692.

Need to Know

Piazza Santa Maria in Trastevere; 7.30am-9pm; Viale di Trastevere, Viale di Trastevere

Take a Break

Enjoy a tasty meal at the hit foodie hotspot Pianostrada Laboratorio di Cucina (p138).

★ Top Tip

Take some coins to drop in the light box and illuminate the mosaics.

Ceiling fresco, Chiesa del Gesù

Chiesa del Gesù

An imposing example of Counter-Reformation architecture, Rome's most important Jesuit church is a treasure trove of glorious baroque art.

Great For...

☑ **Don't Miss**

Andrea Pozzo's lavish tomb of Jesuit founder St Ignatius Loyola.

The church, which was consecrated in 1584, is fronted by an impressive and much-copied facade by Giacomo della Porta. But more than the masonry, the star turn here is the lavish marble-clad interior. Of the art on display, the most astounding work is the hypnotic *Trionfo del Nome di Gesù* (Triumph of the Name of Jesus; 1679), the swirling ceiling fresco by Giovanni Battista Gaulli. The artist, better known as Il Baciccia, also created much of the stucco decoration and the cupola frescoes.

In the northern transept, the **Cappella di Sant'Ignazio** houses the tomb of Ignatius Loyola, the Spanish soldier who founded the Jesuits in 1540. Designed by baroque maestro Andrea Pozzo, the altar-tomb is an opulent marble-and-bronze affair with lapis lazuli–encrusted columns, and, on top, a lapis lazuli globe representing the Trinity.

Chiesa del Gesù facade

DALLAS STRIBLEY/GETTY IMAGES ©

Need to Know

Map p250; www.chiesadelgesu.org; Piazza del Gesù; 7am-12.30pm & 4-7.45pm, St Ignatius rooms 4-6pm Mon-Sat, 10am-noon Sun; Largo di Torre Argentina

Take a Break

Towards the Pantheon, La Ciambella (p126) is a great spot for a pasta lunch.

★ Top Tip

Have some coins to hand for the light machines in the dark interior.

On either side are sculptures whose titles neatly encapsulate the Jesuit ethos: to the left, *Fede che vince l'Idolatria* (Faith Defeats Idolatry); to the right, *Religione che flagella l'Eresia* (Religion Lashing Heresy).

The Spanish saint lived in the church from 1544 until his death in 1556. You can visit his **private rooms**, together with a corridor adorned with Andrea Pozzo frescoes, to the right of the main church.

What's Nearby?

Museo Nazionale Romano: Crypta Balbi — Museum

(Map p250; 06 3996 7700; www.coopculture.it; Via delle Botteghe Oscure 31; adult/reduced €7/3.50; 9am-7.45pm Tue-Sun; Via delle Botteghe Oscure) The least known of the Museo Nazionale Romano's four museums, the Crypta Balbi sits over the ruins of several medieval buildings, themselves set atop the Teatro di Balbo (13 BC). Archaeological finds illustrate the urban development of the surrounding area, while the museum's underground excavations, visitable by guided tour, provide an interesting insight into Rome's multilayered past.

Palazzo Venezia — Palace

(Map p250; Piazza Venezia; Piazza Venezia) Built between 1455 and 1464, this was the first of Rome's great Renaissance palaces. For centuries it served as the embassy of the Venetian Republic, but it's most readily associated with Mussolini, who installed his office here in 1929, and famously made speeches from the balcony. Nowadays, it's home to the tranquil **Museo Nazionale del Palazzo Venezia** (Map p250; 06 678 01 31; http://museopalazzovenezia.beniculturali.it; Via del Plebiscito 118; adult/reduced €5/2.50; 8.30am-7.30pm Tue-Sun; Piazza Venezia) and its eclectic collection of Byzantine and early Renaissance paintings, furniture, ceramics, bronze figures, weaponry and armour.

Galleria Doria Pamphilj

Hidden behind the grimy grey exterior of Palazzo Doria Pamphilj, this wonderful gallery boasts one of Rome's richest private art collections, with works by Raphael, Tintoretto, Brueghel, Titian, Caravaggio, Bernini and Velázquez.

Great For...

☑ Don't Miss

The *Ritratto di papa Innocenzo X*, generally considered the gallery's greatest masterpiece.

Palazzo Doria Pamphilj dates to the mid-15th century, but its current look was largely the work of the current owners, the Doria Pamphilj family, who acquired it in the 18th century. The Pamphilj's golden age, during which the family collection was started, came during the papacy of one of their own, Innocent X (r 1644–55).

The opulent picture galleries, decorated with frescoed ceilings and gilded mirrors, are hung with floor-to-ceiling paintings. Masterpieces abound, but look out for Titian's *Salomè con la testa del Battista* (Salome with the Head of John the Baptist) and two early Caravaggios: *Riposo durante la fuga in Egitto* (Rest During the Flight into Egypt) and *Maddalene Penitente* (Penitent Magdalen). The undisputed star, though, is Velázquez' **Ritratto di papa Innocenzo X**, his portrait of an implacable Pope Innocent X, who grumbled that the depiction was

Interior, Galleria Doria Pamphilj

'too real'. For a comparison, check out Gian Lorenzo Bernini's sculptural interpretation of the same subject.

The excellent free audioguide, narrated by Jonathan Pamphilj, brings the place alive with family anecdotes and background information.

What's Nearby?

Chiesa di Sant'Ignazio di Loyola — Church

(Map p250; Piazza di Sant'Ignazio; ⌚7.30am-7pm Mon-Sat, 9am-7pm Sun; 🚌Via del Corso) Flanking a delightful rococo piazza, this important Jesuit church boasts a Carlo Maderno facade and two celebrated *trompe l'œil* frescoes by Andrea Pozzo (1642–1709). One cleverly depicts a fake dome, whilst the other, on the nave ceiling, shows St Ignatius Loyola being welcomed into paradise by Christ and the Madonna.

Need to Know

Map p250; ☎06 679 73 23; www.dopart.it; Via del Corso 305; adult/reduced €11/7.50; ⌚9am-7pm, last admission 6pm; 🚌Via del Corso

Take a Break

A short walk north, Osteria dell'Ingegno (p128) offers seasonal pastas and a piazza-side setting.

★ Top Tip

Make sure to pick up the free audioguide.

Piazza Colonna — Piazza

(Map p250; 🚌Via del Corso) Together with the adjacent Piazza di Montecitorio, this stylish piazza is Rome's political nerve centre. On its northern flank, the 16th-century **Palazzo Chigi** (Map p250; www.governo.it; ⌚guided visits 9am-1pm Sat Oct-May, bookings required) FREE has been the official residence of Italy's prime minister since 1961. In the centre, the 30m-high **Colonna di Marco Aurelio** (Map p250) was completed in AD 193 to honour Marcus Aurelius' military victories.

Palazzo di Montecitorio — Historic Building

(Map p250; ☎800 012955; www.camera.it; Piazza di Montecitorio; ⌚guided visits noon-2.30pm 1st Sun of month; 🚌Via del Corso) FREE Home to Italy's Chamber of Deputies, this baroque *palazzo* was built by Bernini in 1653, expanded by Carlo Fontana in the late 17th century, and given an art nouveau facelift in 1918. Visits take in the mansion's lavish reception rooms and the main chamber where the 630 deputies debate beneath a beautiful Liberty-style skyline.

Sunset over the Teatro di Marcello

Life in the Ghetto

The Jewish Ghetto, centred on lively Via Portico d'Ottavia, is an atmospheric area studded with artisans' studios, vintage clothes shops, kosher bakeries and popular trattorias.

Rome's Jewish community harks back to the 2nd century BC, making it one of the oldest in Europe. Confinement in the Ghetto was first enforced in 1555 when Pope Paul IV ushered in a period of official intolerance that lasted, on and off, until the 20th century.

Great For...

☑ Don't Miss

Poking around the ruins beneath the landmark Teatro di Marcello.

Museo Ebraico di Roma Synagogue, Museum

(Jewish Museum of Rome; Map p250; ☎06 6840 0661; www.museoebraico.roma.it; Via Catalana; adult/reduced €11/8; ⏰10am-6.15pm Sun-Thu, 9am-3.15pm Fri summer, 10am-4.15pm Sun-Thu, 9am-1.15pm Fri winter; 🚌Lungotevere de' Cenci) The historical, cultural and artistic heritage of Rome's Jewish community is chronicled in this small but engrossing museum. Housed in the city's early 20th-century synagogue, Europe's second largest, it displays parchments, precious fabrics,

Fontana della Tartarughe

GORAN BOGICEVIC/SHUTTERSTOCK ©

marble carvings, and a collection of 17th- and 18th-century silverware. Documents and photos attest to life in the Ghetto and the hardships suffered by the city's Jewry during WWII.

Fontana delle Tartarughe Fountain

(Map p250; Piazza Mattei; Via Arenula) This playful, much-loved fountain features four boys gently hoisting tortoises up into a bowl of water. Created by Giacomo della Porta and Taddeo Landini in the late 16th century, it's the subject of a popular local legend, according to which it was created in a single night.

The story goes that the Duke of Mattei had it built to save his engagement and prove to his prospective father-in-law that despite gambling his fortune away he was still a good catch. And amazingly, it worked. The father was impressed and allowed Mattei to marry his daughter.

In reality, the fountain was no overnight sensation and took three years to craft (1581–1584). The tortoises, after whom it's named, were added by Bernini during a restoration in 1658.

Need to Know

Information on the area is available at the **Jewish Info Point** (06 9838 1030; www.jewishcommunityofrome.com; Via Santa Maria del Pianto 1; 8.30am-6.30pm Mon-Thu, 8.30am-2pm Fri, 1hr after end of Shabat-11pm Sat, 11am-6pm Sun; Via Arenula).

Take a Break

For a taste of traditional Roman-Jewish cooking, Piperno (p129) is a local institution.

★ Top Tip

A short hop away, the Isola Tiberina is worth a quick detour.

Area Archeologica del Teatro di Marcello e del Portico d'Ottavia Archaeological Site

(Map p250; entrances Via del Teatro di Marcello 44 & Via Portico d'Ottavia 29; 9am-7pm summer, 9am-6pm winter; Via del Teatro di Marcello) FREE To the east of the Jewish Ghetto, the **Teatro di Marcello** (Theatre of Marcellus; Map p250; Via del Teatro di Marcello) is the star turn of this dusty archaeological area. This 20,000-seat mini-Colosseum was planned by Julius Caesar and completed in 11 BC by Augustus, who named it after a favourite nephew, Marcellus. In the 16th century a *palazzo*, which now contains several exclusive apartments, was built on top of the original structure.

Iconic Treasures & Ruins

Follow in the footsteps of an ancient Roman on this whistle-stop tour of the city's iconic treasures. The route transports you back to Rome's glorious golden age.

Start: Ⓜ Colosseo
Distance: 1.5km
Duration: 3 hours plus

6 It's worth stopping off at **Il Vittoriano** (p75), the massive mountain of white marble, for its unforgettable 360-degree views over the city.

RICHARD I'ANSON/SHUTTERSTOCK ©

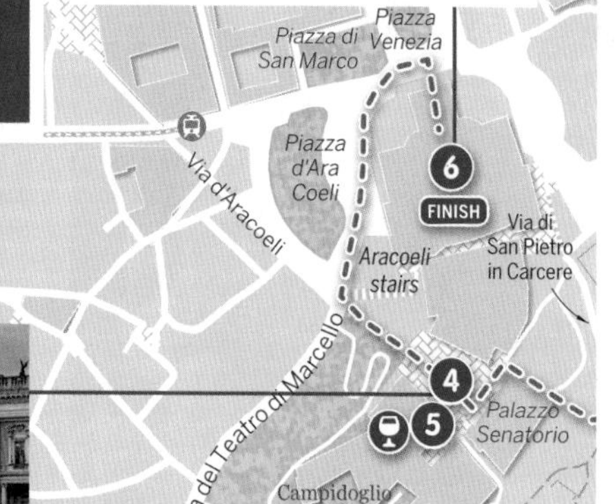

4 The Michelangelo-designed **Piazza del Campidoglio** (p74) sits atop the Capitoline hill, one of the seven hills on which Rome was founded.

ANSHAR/SHUTTERSTOCK ©

5 One of Rome's top museum complexes, the **Capitoline Museums** (p72) are home to the city's finest ancient sculpture.

Take a Break

The Caffè Capitolino (p124) is a refined spot for an uplifting coffee.

2 The **Palatino** (p60) was the site of the emperor's palace and home to the cream of imperial society.

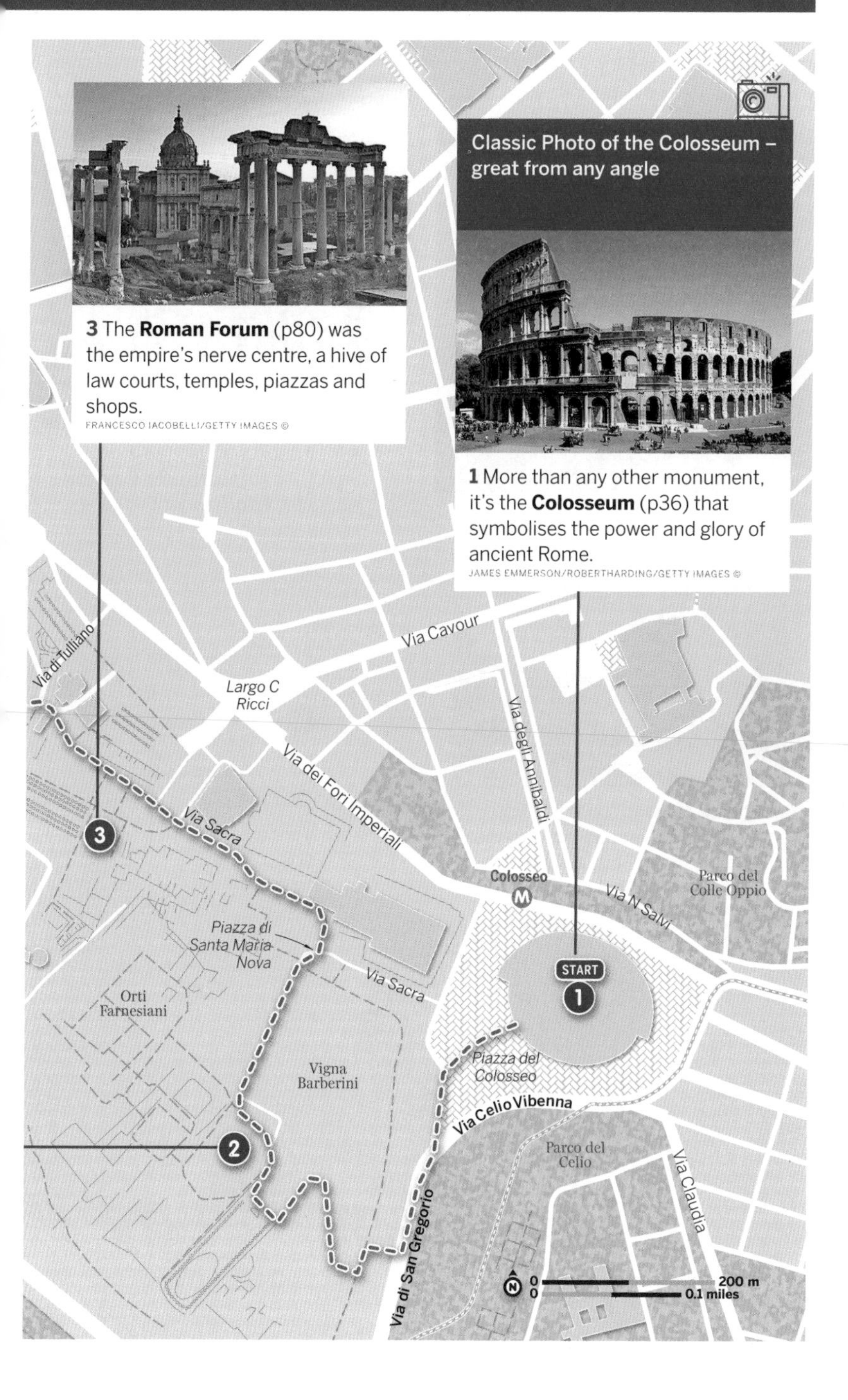
Classic Photo of the Colosseum – great from any angle
3 The **Roman Forum** (p80) was the empire's nerve centre, a hive of law courts, temples, piazzas and shops.
FRANCESCO IACOBELLI/GETTY IMAGES ©
1 More than any other monument, it's the **Colosseum** (p36) that symbolises the power and glory of ancient Rome.
JAMES EMMERSON/ROBERTHARDING/GETTY IMAGES ©
Via Cavour
Via di Tulliano
Largo C Ricci
Via degli Annibaldi
Via dei Fori Imperiali
Via Sacra
Colosseo
Via N Salvi
Parco del Colle Oppio
Piazza di Santa Maria Nova
START
Orti Farnesiani
Vigna Barberini
Piazza del Colosseo
Via Celio Vibenna
Parco del Celio
Via Claudia
Via di San Gregorio
200 m
0.1 miles

Centro Storico Piazzas

Rome's historic centre boasts some of the city's most celebrated piazzas and beautiful but lesser known squares. Together they encapsulate much of the city's beauty, history and drama.

Start: Via del Corso
Distance: 1.5km
Duration: 3.5 hours

Classic Photo of Piazza della Rotonda with the Pantheon in the background

4 It's a short walk along Via del Seminario to Piazza della Rotonda, where the **Pantheon** (p50) needs no introduction.

DAVID SOANES PHOTOGRAPHY/GETTY IMAGES ©

5 Piazza Navona (p64) is Rome's great showpiece square where you can compare the two giants of Roman baroque: Gian Lorenzo Bernini and Francesco Borromini.

BASIC ELEMENTS PHOTOGRAPHY/GETTY IMAGES ©

7 Just beyond the Campo, the more sober Piazza Farnese is overshadowed by the austere facade of the Renaissance **Palazzo Farnese** (p116).

1 Piazza Colonna (p111) is dominated by the 30m-high Colonna di Marco Aurelio and flanked by Palazzo Chigi, the official residence of the Italian PM.

FRED MATOS/GETTY IMAGES ©

2 Follow Via dei Bergamaschi to **Piazza di Pietra**, a refined space overlooked by the 2nd-century Tempio di Adriano.

3 Continue down Via de' Burro to Piazza di Sant'Ignazio, location of the **Chiesa di Sant'Ignazio di Loyola** (p111), with its celebrated *trompe l'œil* frescoes.

Take a Break

Caffè Sant'Eustachio (p170) is reckoned by many to serve the city's best coffee.

6 Over on the other side of Corso Vittorio Emanuele II, **Campo de' Fiori** (p67) hosts a noisy market and boisterous drinking scene.

CHRISTIAN MUELLER/SHUTTERSTOCK ©

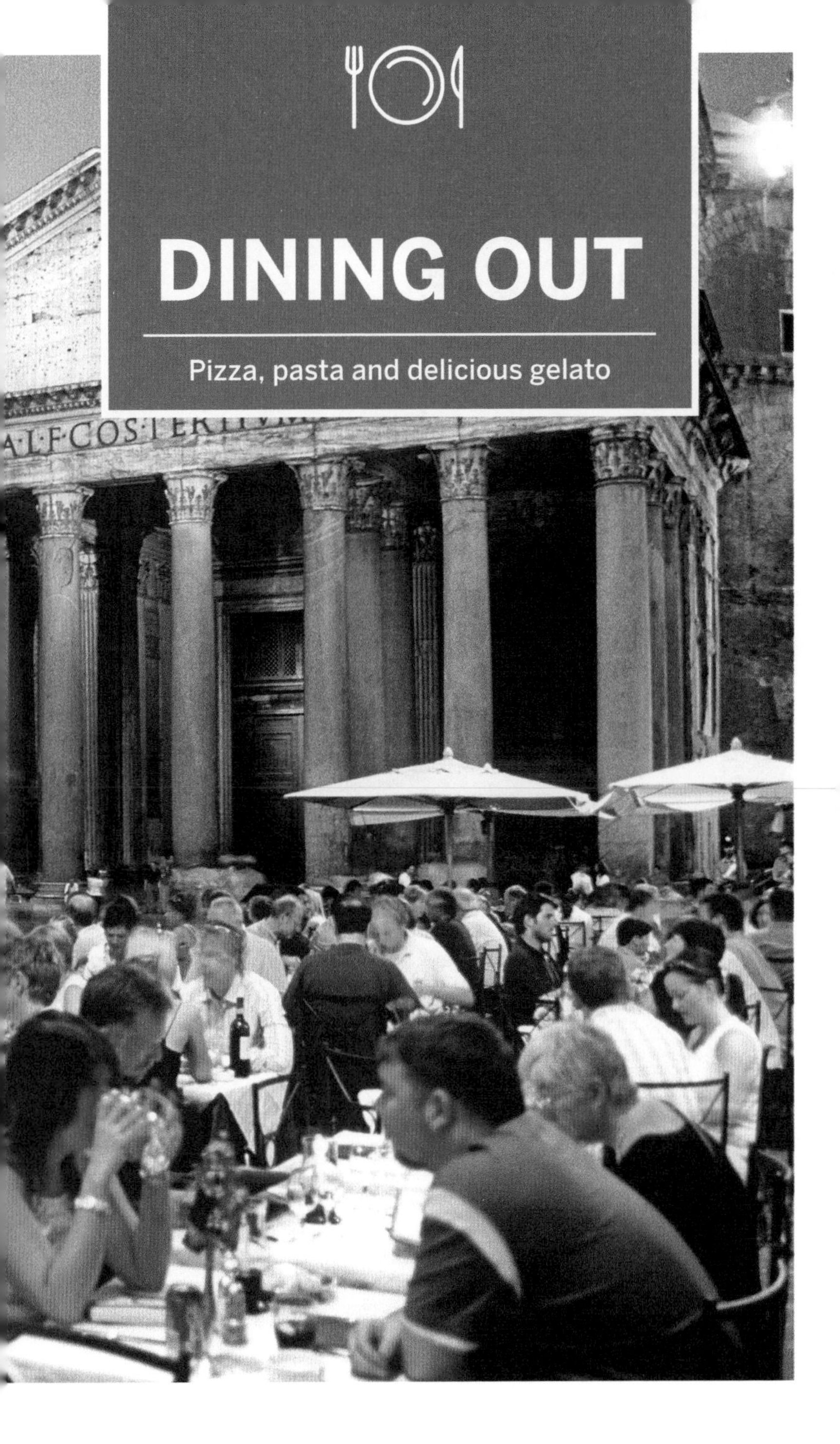
DINING OUT
Pizza, pasta and delicious gelato

Dining Out

Food is central to the Roman passion for life. Everyone has an opinion on it and the city teems with trattorias, pizzerias, fine-dining restaurants and gourmet gelaterie.

Over recent decades, Rome's dining scene has become increasingly sophisticated as cucina creativa *(creative cooking) has taken off and the city's gourmet restaurants have upped their game. But the bedrock of the Roman foodscape has always been, and still remains, the family-run trattorias that pepper the city's streets and piazzas. These simple eateries, often with rickety wooden tables and* nonna *(grandma) at the stove, have been feeding visitors for centuries and are still the best bet for classic, seasonal Roman fare.*

In This Section

Price Ranges

The following prices refer to a meal consisting of a *primo* (first course), *secondo* (second course) and *dolce* (dessert), plus a glass of wine.

€	less than €25
€€	€25–45
€€€	more than €45

Tipping

Although service is included, leave a tip: from 5% in a pizzeria to 10% in a restaurant.

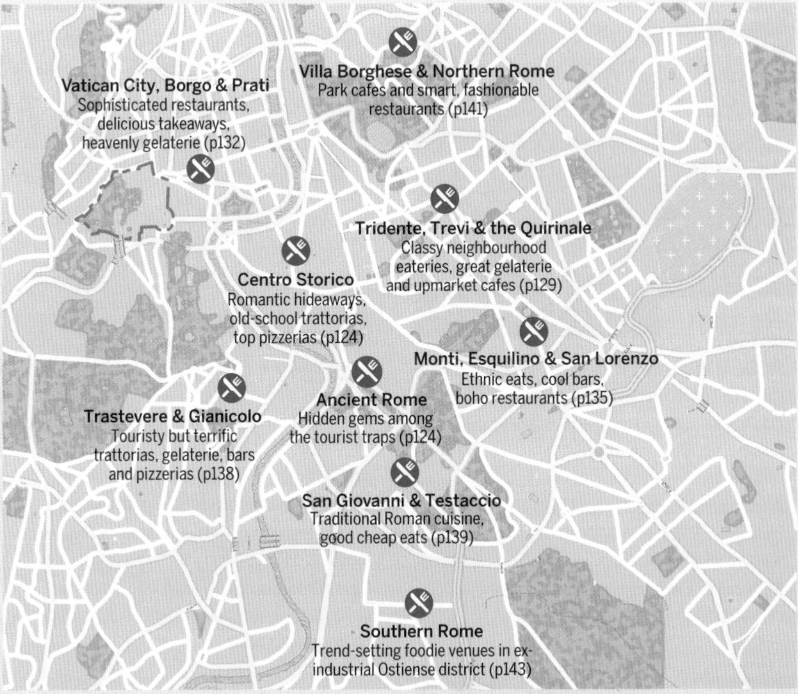

Vatican City, Borgo & Prati
Sophisticated restaurants, delicious takeaways, heavenly gelaterie (p132)

Villa Borghese & Northern Rome
Park cafes and smart, fashionable restaurants (p141)

Tridente, Trevi & the Quirinale
Classy neighbourhood eateries, great gelaterie and upmarket cafes (p129)

Centro Storico
Romantic hideaways, old-school trattorias, top pizzerias (p124)

Monti, Esquilino & San Lorenzo
Ethnic eats, cool bars, boho restaurants (p135)

Ancient Rome
Hidden gems among the tourist traps (p124)

Trastevere & Gianicolo
Touristy but terrific trattorias, gelaterie, bars and pizzerias (p138)

San Giovanni & Testaccio
Traditional Roman cuisine, good cheap eats (p139)

Southern Rome
Trend-setting foodie venues in ex-industrial Ostiense district (p143)

Useful Phrases

I'd like... vorrei (vo.*ray*...)

a table un tavolo (oon *ta*.vo.lo)

the menu il menù (eel me.*noo*)

two beers due birre (doo.e *bee*.re)

What would you recommend? Cosa mi consiglia? (*ko*.za mee kon.*see*.lya)

Can you bring me the bill, please? Mi porta il conto, per favore? (mee *por*.ta eel *kon*.to per fa.vo.re)

Classic Dishes

Bucatini all'amatriciana Thick spaghetti with tomato sauce, onions, pancetta, cheese and chilli.

Cacio e pepe Pasta mixed with pecorino romano cheese, black pepper and olive oil.

Carciofi alla giudia Deep-fried 'Jewish-style' artichokes.

Saltimbocca alla romana Veal cutlet with prosciutto (ham) and sage.

The Best...

Experience Rome's top restaurants and cafes

Roman

Flavio al Velavevodetto (p140) Classic *cucina romana,* served in huge portions.

Da Felice (p140) In the heartland of Roman cuisine, and sticking to a traditional weekly timetable.

Armando al Pantheon (p127) Family-run trattoria offering hearty Roman cuisine in the shadow of the Pantheon.

Ristorante L'Arcangelo (p133) A creative, contemporary take on Roman dishes.

Creative Cuisine

Metamorfosi (p142) Michelin-starred cuisine by wonder-chef Roy Carceres.

All'Oro (p132) Food as art at chef Riccardo Di Giacinto's glamorous Michelin-starred eatery.

Glass Hostaria (p139) Wonderful, innovative food in a contemporary setting in Trastevere.

Open Colonna (p137) Antonello Colonna's glass-roofed restaurant offers creative takes on Roman classics.

Pizzerias

Pizzeria Da Remo (p140) Spartan but stunning, the frenetic Roman pizzeria experience in Testaccio.

Pizza Ostiense (p143) New on the scene, but offering classic Roman neighbourhood pizza.

Pizzeria Ivo (p138) Always busy, fiercely traditional, loud-and-gruff Trastevere pizzeria.

For Ambiance

Aroma (p141) The Michelin-starred rooftop restaurant of the Palazzo Manfredi hotel has 'marry me' views over the Colosseum.

La Veranda (pictured above; p135) A location in Paolo Sorrentino's *The Great Beauty*; dine beneath 15th-century Pinturicchio frescoes.

Open Colonna (p137) Restaurant on a mezzanine under a soaring glass ceiling in the Palazzo degli Esposizioni.

Regional

Enoteca Regionale Palatium (p130) Wine bar showcasing the best of Lazio food and drink.

Colline Emiliane (p130) Fantastic roasted meats and hearty pastas.

Trattoria Monti (p135) Top-notch traditional cooking from the Marches, including heavenly fried things.

Terre e Domus (p124) All ingredients are sourced from the surrounding Lazio region.

Pastries

Pasticceria De Bellis (p125) Work-of-art cakes, pastries and *dolci* (sweets) at this chic *pasticceria* (pastry shop).

Andreotti (p143) Poem-worthy treats from buttery *crostate* (tarts) to the piles of golden *sfogliatelle romane* (ricotta-filled pastries).

Innocenti (p138) Classic old-school Trastevere bakery, with piled-high biscuits such as *brutti ma buoni* (ugly but good).

Eating with Locals

Pizzeria Da Remo (p140) The full neighbourhood Roman pizza experience, with lightning-fast waiters serving paper-thin pizzas.

Da Felice (p140) Traditional local cooking in Testaccio, the heartland of Roman cuisine.

Antico Forno Urbani (p126) Queue up with the locals for *pizza al taglio* (by the slice) in the Ghetto.

Pizza Ostiense (p143) Fabulous thin-crust pizza in Rome's ex-industrial, hip neighbourhood of Ostiense.

★ Lonely Planet's Top Choices

Metamorfosi (p142) Michelin-starred, chic yet informal.

Glass Hostaria (p139) Italian cuisine as a creative art in Trastevere.

Casa Coppelle (p126) Creative Italian- and French-inspired food in romantic surroundings.

Flavio al Velavevodetto (p140) For the real *cucina romana* (Roman kitchen).

L'Asino d'Oro (p135) Fantastic food, stunning value and Umbrian flavours.

Fatamorgana (p129) Incredible artisanal gelato, in Tridente, Vatican, Monti and Trastevere.

Ancient Rome

Caffè Capitolino Cafe €

(Map p252; Piazzale Caffarelli 4; 9am-7.30pm Tue-Sun; Piazza Venezia) The Capitoline Museums' charming terrace cafe is a good place to relax over a drink or light snack (*panini,* salads and pizza) and enjoy wonderful views across the city's rooftops. Although part of the museum complex, you don't need a ticket to come here as it's accessible via an independent entrance on Piazzale Caffarelli.

Terre e Domus Lazio Cuisine €€

(Map p252; 06 6994 0273; Via Foro Traiano 82-4; meals €30; 7.30am-12.30am Mon-Sat; Via dei Fori Imperiali) This modern white-and-glass restaurant is the best option in the touristy Forum area. Overlooking the Colonna di Traiano, it serves a menu of traditional staples, all made with ingredients sourced from the surrounding Lazio region, and a thoughtful selection of regional wines. Lunchtime can be busy but it quietens down in the evening.

Ristorante Roof Garden Circus Ristorante €€€

(06 678 78 16; www.fortysevenhotel.com; Hotel Forty Seven, Via Petroselli 47; meals €50; 12.30-3pm & 7.30-11.30pm; Via Petroselli) The rooftop of the Forty Seven hotel sets the romantic stage for chef Vito Grippa's menu of classic Roman dishes and contemporary Italian cuisine. With the Aventino hill rising in the background, you can tuck into stalwarts such as spaghetti *aglio e olio* (with garlic and olive oil) or push the boat out and opt for something richer like roast guinea fowl with black truffles.

Centro Storico

Forno Roscioli Pizza, Bakery €

(Map p250; Via dei Chiavari 34; pizza slices from €2, snacks from €1.50; 7am-7.30pm Mon-Sat; Via Arenula) This is one of Rome's top bakeries, much loved by lunching locals who crowd here for luscious sliced pizza, prize pastries and hunger-sating *supplì*. There's also a counter serving hot pastas and vegetable side dishes.

Spaghetti *aglio e olio*

Forno di Campo de' Fiori Pizza, Bakery €

(Map p250; Campo de' Fiori 22; pizza slices about €3; ⏲7.30am-2.30pm & 4.45-8pm Mon-Sat; 🚌Corso Vittorio Emanuele II) This buzzing bakery on Campo de' Fiori does a roaring trade in *panini* and delicious fresh-from-the-oven *pizza al taglio* (by the slice). Aficionados swear by the *pizza bianca* ('white' pizza with olive oil, rosemary and salt), but the *panini* and *pizza rossa* ('red' pizza, with olive oil, tomato and oregano) taste plenty good, too.

Supplizio Fast Food €

(Via dei Banchi Vecchi 143; supplì €3-5; ⏲noon-4pm Mon-Sat plus 5.30-10pm Mon-Thu, to 11pm Fri & Sat; 🚌Corso Vittorio Emanuele II) Rome's favourite snack, the *supplì* (a fried croquette filled with rice, tomato sauce and mozzarella), gets a gourmet makeover at this elegant new street-food joint. Sit back on the vintage leather sofa and dig into the classic article or throw the boat out and try something different, maybe a mildly spicy fish *supplì* stuffed with anchovies, tuna, parsley, and just a hint of orange.

Chiostro del Bramante Caffè Cafe €

(Map p250; www.chiostrodelbramante.it; Via Arco della Pace 5; meals €15-20; ⏲10am-8pm Mon-Fri, to 9pm Sat & Sun; 📶; 🚌Corso del Rinascimento) Many of Rome's galleries and museums have in-house cafes but few are as beautifully located as the Chiostro del Bramante Caffè on the 1st floor of Bramante's elegant Renaissance cloister. With outdoor tables overlooking the central courtyard and an all-day menu offering everything from cakes and coffee to baguettes, light lunches and aperitifs, it's a great spot for a break.

Gelateria del Teatro Gelateria €

(Map p250; Via dei Coronari 65; gelato from €2.50; ⏲11.30am-midnight; 🚌Corso del Rinascimento) All the ice cream served at this excellent gelateria is prepared on-site – look through the window and you'll see how. There are about 40 flavours to choose from, all made from thoughtfully sourced ingredients, including some excellent fruit combos and spicy chocolate.

Baguetteria del Fico Sandwiches €

(Map p250; Via della Fossa 12; panini €5-7; ⏲11am-2am; 🚌Corso del Risorgimento) A designer baguette bar ideal for a midday bite or late snack. Choose your bread, then select from the rich array of fillers – cured meats, cheeses, marinated vegetables, salads, homemade sauces. For liquid sustenance, there's a choice of bottled craft beers.

Pasticceria De Bellis Pasticceria €

(Map p250; Piazza del Paradiso 56-57; pastries €4; ⏲9am-8pm Tue-Sun; 🚌Corso Vittorio Emanuele II) The beautifully crafted cakes, pastries and *dolci* made at this chic pasticceria are miniature works of art. Curated in every detail, they look superb and taste magnificent. You'll find traditional offerings

Where to Eat

Take your pick according to your mood and your pocket, from the frenetic energy of a Roman pizzeria to the warm familiarity of a local trattoria, run for generations by the same family, or from a bar laden with sumptuous *aperitivi* to a restaurant where both presentation and the flavours are a work of art.

Romans rarely drink without eating, and you can eat well at many *enoteche*, wine bars that usually serve snacks (such as cheeses or cold meats, bruschette and crostini) and hot dishes. Some, such as Enoteca Regionale Palatium (p130) or Casa Bleve (p129), offer full-scale dining.

Traditionally, trattorias were family-run places that offered a basic, affordable local menu, while *osterie* usually specialised in one dish and *vino della casa* (house wine). There are still lots of these around. *Ristoranti* offer more choices and smarter service, and are more expensive.

Roman Cuisine

Like most Italian cuisines, Roman cooking was born of careful use of local ingredients – making use of the cheaper cuts of meat, like *guanciale* (pig's cheek), and greens that could be gathered wild from the fields.

There are a few classic dishes that are served by almost every trattoria and restaurant in Rome. These carb-laden comfort foods are seemingly simple, yet notoriously difficult to prepare well. Iconic Roman dishes include carbonara (pasta with lardons, egg and Parmesan), *alla gricia* (with pig's cheek and onions), *amatriciana* (invented when an enterprising chef from Amatrice added tomatoes to *alla gricia*) and *cacio e pepe* (with cheese and pepper).

The number of special-occasion, creative restaurants is ever rising in Rome, with a buzz around openings such as chef Riccardo Di Giacinto's All'Oro (p132). Roy Carceres continues to wow at Metamorfosi (p142), and Cristina Bowerman, director of Trastevere's Glass Hostaria (p139), is one of the few Italian female chefs to have received a Michelin star.

Another relatively new concept in Rome is all-day dining, with a few notable all-things-to-all-people restaurants including Porto Fluviale (p143) and the multistorey mall Eataly (p143), which has restaurants to suit almost every mood, from a hankering for *fritti* (fried things) to fine dining.

Guanciale

alongside unique creations such as the Assoluta, a decadent concoction combining several chocolate mousses.

Alfredo e Ada Trattoria €

(☎06 687 88 42; Via dei Banchi Nuovi 14; meals €25; ⊙noon-3pm & 7-10pm Tue-Sat; 🚌Corso Vittorio Emanuele II) For an authentic trattoria meal, search out this much-loved local eatery. It's distinctly no-frills with spindly, marble-topped tables and homey clutter, but there's a warm, friendly atmosphere and the traditional Roman food is filling and flavoursome.

Antico Forno Urbani Pizza, Bakery €

(Map p250; Piazza Costaguti 31; pizza slices from €1.50; ⊙7.40am-2.30pm & 5-8.45pm Mon-Fri, 9am-1.30pm Sat & Sun; 🚋Via Arenula) A popular kosher bakery, this Ghetto institution makes some of the best *pizza bianca* in town, as well as freshly baked bread, biscuits and focaccias. It gets very busy but once you catch a whiff of the yeasty odours wafting off the counter, there's nothing for it but to grab a ticket and wait your turn.

Casa Coppelle Ristorante €€

(Map p250; ☎06 6889 1707; www.casacoppelle.it; Piazza delle Coppelle 49; meals €45; ⊙12-3.30pm & 6.30-11.30pm; 🚌Corso del Rinascimento) Exposed brick walls, flowers and subdued lighting set the stage for creative Italian- and French-inspired food at this intimate, romantic restaurant. There's a full range of starters and pastas, but the real tour de force are the deliciously tender steaks and rich meat dishes. Service is attentive and the setting, on a small piazza near the Pantheon, memorable. Book ahead.

La Ciambella Italian €€

(Map p250; www.laciambellaroma.com; Via dell'Arco della Ciambella 20; fixed-price lunch menus €10-25, meals €30; ⊙7.30am-midnight; 🚌Largo di Torre Argentina) From breakfast pastries and lunchtime pastas to afternoon tea, Neapolitan pizzas and aperitif cocktails, this all-day eatery is a top find. Central but as yet undiscovered by the tourist

Pastries in Centro Storico (p124)

hordes, it's a spacious, light-filled spot set over the ruins of the Terme di Agrippa, visible through transparent floor panels. The mostly traditional food is spot on, and the atmosphere laid-back and friendly.

Armando al Pantheon Trattoria **€€**
(Map p250; 06 6880 3034; www.armandoalpantheon.it; Salita dei Crescenzi 31; meals €40; 12.30-3pm & 7-11pm Mon-Fri, 12.30-3pm Sat; Largo di Torre Argentina) An institution in these parts, Armando al Pantheon is a rare find – a genuine family-run trattoria in the touristy Pantheon area. It's been on the go for more than 50 years and has served its fair share of celebs, but it hasn't let fame go to its head and it remains one of the best bets for earthy Roman cuisine. Reservations essential.

Renato e Luisa Trattoria **€€**
(Map p250; 06 686 96 60; www.renatoeluisa.it; Via dei Barbieri 25; meals €45; 8pm-midnight Tue-Sun; Largo di Torre Argentina) Highly rated locally, this small backstreet trattoria is always packed. Chef Renato takes a creative approach to classic Roman cooking, resulting in dishes that are modern and seasonal yet undeniably local, such as his signature *cacio e pepe e fiori di zucca* (pasta with pecorino cheese, black pepper and courgette flowers). Bookings recommended.

Ditirambo Italian **€€**
(Map p250; 06 687 16 26; www.ristoranteditirambo.it; Piazza della Cancelleria 72; meals €40; 1-3pm & 7.20-10.30pm, closed Mon lunch; Corso Vittorio Emanuele II) Since opening in 1996, Ditirambo has won an army of fans with its informal trattoria vibe and seasonal, organic cuisine. Dishes cover many bases, ranging from old-school favourites to thoughtful vegetarian offerings and more exotic fare such as pasta with Sicilian prawns, basil and lime. Book ahead.

Grappolo D'Oro Italian **€€**
(Map p250; 06 689 70 80; www.hosteriagrappolodoro.it; Piazza della Cancelleria 80; tasting menu €28, meals €35-40; 12.45-3pm & 7-11.30pm, closed Wed lunch; Corso Vittorio Emanuele II) This informal eatery stands out among the many lacklustre options around Campo de' Fiori. The emphasis is on

"toppings loaded atop thin, crispy, light-as-air, slow-risen bread that verge on the divine"

traditional Roman cuisine, albeit with the occasional twist, so look out for artichoke starters, pastas littered with pecorino, pancetta and black pepper, and mains of no-nonsense braised and grilled meats.

Matricianella Trattoria **€€**

(☎06 683 21 00; www.matricianella.it; Via del Leone 2/4; meals €40; ⊙12.30-3pm & 7.30-11pm Mon-Sat; 🚌Via del Corso) With its gingham tablecloths, chintzy murals and fading prints, Matricianella is an archetypal trattoria, much loved for its traditional Roman cuisine. Its loyal clientele go crazy for ever-green crowd-pleasers like battered vegetables, artichoke *alla giudia* (fried, Jewish style), and *saltimbocca* (veal cutlet with ham and sage). Booking is essential.

Osteria dell'Ingegno Italian **€€**

(Map p250; ☎06 678 06 62; www.osteriaingegno.it; Piazza di Pietra 45; meals €40; ⊙noon-midnight; 🚌Via del Corso) A boho-chic restaurant–wine bar with a colourful art-filled interior and a prime location on a charming central piazza. The menu hits all the right notes with a selection of seasonal pastas, creative mains, salads and home-made desserts, while the 300-strong wine list boasts some interesting Italian labels. *Aperitivo* is served daily from 5pm to 8pm.

Salumeria Roscioli Italian **€€€**

(Map p250; ☎06 687 52 87; Via dei Giubbonari 21; meals €55; ⊙12.30-4pm & 7pm-midnight Mon-Sat; 🚊Via Arenula) The name Roscioli has long been a byword for foodie excellence, and this luxurious deli-restaurant is the place to experience it. Under a coffered ceiling, you'll find a display of mouth-watering Italian and foreign delicacies, while behind, in the small restaurant, diners sit down to sophisticated Italian food and some truly outstanding wines.

La Rosetta Seafood **€€€**

(Map p250; ☎06 686 10 02; www.larosetta.com; Via della Rosetta 8; lunch menu €65, meals €90-120; ⊙12.15-2.45pm & 7-10.45pm, closed 3 weeks

Aug; Corso del Rinascimento) Food fads might come and go but La Rosetta remains what it has long been, one of Rome's top fish restaurants. Hidden down a sidestreet near the Pantheon, it offers classic seafood dishes and a choice of raw delicacies alongside more elaborate modern creations. Bookings essential.

Casa Bleve Ristorante, Wine Bar **€€€**
(Map p250; 06 686 59 70; www.casableve.it; Via del Teatro Valle 48-49; meals €50-65; 12.30-3pm & 7.30-10.30pm Mon-Sat; Largo di Torre Argentina) Ideal for a special-occasion dinner, this palatial restaurant–wine bar dazzles with its column-lined dining hall and stained-glass roof. Its wine list, one of the best in town, accompanies a small but considered menu of hard-to-find cheeses, cold cuts, seasonal pastas and refined main courses.

Piperno Ristorante **€€€**
(Map p250; 06 6880 6629; www.ristorantepiperno.it; Via Monte de' Cenci 9; meals €50-55; 12.45-2.20pm & 7.45-10.20pm, closed Mon & Sun dinner; Via Arenula) This historic Ghetto restaurant, complete with its smart old-school look, is a top spot to get to grips with traditional Jewish-Roman cooking. Signature dishes include wonderful deep-fried *filetti di baccalà* (cod fillets) and *animelle di agnello con carciofi* (lamb sweetbreads with artichokes). To finish off, try the *tortino al cioccolato* (chocolate cake). Booking recommended.

Tridente, Trevi & the Quirinale

Fatamorgana Gelateria **€**
(Map p256; Via Laurina 10; noon-11pm; M Flaminio) The wonderful all-natural Fatamorgana, purveyors of arguably Rome's best artisanal ice cream, now has this handy central branch. Innovative and classic tastes of heaven abound, including flavours such as pear and caramel, all made from the finest seasonal ingredients.

Il Caruso Gelateria **€**
(Via Collina 15; noon-9pm; M Repubblica) Spot Il Caruso by the gelato-licking hordes outside. This best-kept-secret artisanal gelateria only does a few strictly seasonal flavours, but they're created to perfection. Try the incredibly creamy pistachio. It also offers two types of *panna:* the usual whipped cream or the verging-on-sublime *zabaglione* (egg and marsala custard) combined with whipped cream.

Dei Gracchi Gelateria **€**
(Map p256; Via di Ripetta 261; ice cream from €2; 11.30am-10pm, to midnight Jun-Sep; M Flaminio) A new outpost of the venerable Gelataria dei Gracchi, close to the Vatican, this serves up superb ice cream made from the best ingredients, with an excellent array of classic flavours. It's handily located just off Piazza del Popolo, so you can take your pick and then wander around the square as you revel in your excellent selection.

Gina Cafe **€**
(Map p255; 06 678 02 51; Via San Sebastianello 7a; snacks €8-16; 11am-8pm; M Spagna) Around the corner from the Spanish Steps, this is an ideal place to drop once you've shopped. Comfy white seats are strewn with powder-blue cushions, and it gets packed by a Prada-clad crowd, gossiping and flirting over sophisticated salads and perfect *panini*. You can also order a €40/60 regular/deluxe picnic-for-two to take up to Villa Borghese.

Canova Tadolini Cafe **€**
(Map p256; 06 3211 0702; Via del Babuino 150a/b; 9am-10.30pm Mon-Sat; M Spagna) In 1818 sculptor Canova signed a contract for this studio that agreed it would be forever preserved for sculpture. The place is still stuffed with statues and it's a unique experience to sit among the great maquettes and sup an upmarket tea or knock back some wine and snacks.

Caffè Greco Cafe **€**
(06 679 17 00; Via dei Condotti 86; 9am-9pm; M Spagna) Caffè Greco opened in 1760 and is still working the look: penguin

Pizza

Remarkably, pizza was only introduced to Rome post-WWII, by southern immigrants. It caught on. Every Roman's favourite casual (and cheap) meal is the gloriously simple pizza, with Rome's signature wafer-thin bases, covered in fresh, bubbling toppings, slapped down on tables by waiters on a mission. Pizzerias often only open in the evening, as their wood-fired ovens take a while to get going. Most Romans will precede their pizza with a starter of bruschetta or *fritti* (mixed fried foods, such as zucchini flowers, potato, olives etc) and wash it all down with beer. Some places in Rome serve pizza with a thicker, fluffier base, which is the Neapolitan style.

For a snack on the run, Rome's *pizza al taglio* (by the slice) places are hard to beat, with toppings loaded atop thin, crispy, light-as-air, slow-risen bread that verge on the divine. There's been an outbreak of gourmet pizza places in the last decade, with Gabriele Bonci's Pizzarium (p132) leading the way near the Vatican.

waiters, red flock and age-spotted gilt mirrors. Casanova, Goethe, Wagner, Keats, Byron, Shelley and Baudelaire were all once regulars. Now there are fewer artists and lovers and more shoppers and tourists. Prices reflect this, unless you do as the locals do and have a drink at the bar (*caffè* bar/seated €1.50/6).

Colline Emiliane Italian **€€**

(Map p255; 06 481 75 38; Via degli Avignonesi 22; meals €50; 12.45-2.45pm Tue-Sun & 7.30-10.45pm Tue-Sat, closed Aug; M Barberini) This welcoming, tucked-away restaurant just off Piazza Barberini flies the flag for Emilia-Romagna, the well-fed Italian province that has blessed the world with Parmesan, balsamic vinegar, bolognese sauce and Parma ham. This is a consistently excellent place to eat; there are delicious meats, homemade pasta and rich *ragù*. Try to save room for dessert too.

Al Gran Sasso Trattoria **€€**

(Map p256; 06 321 48 83; www.algransasso.com; Via di Ripetta 32; meals €35; 12.30-2.30pm & 7.30-11.30pm Sun-Fri; M Flaminio) A top lunchtime spot, this is a classic, dyed-in-the-wool trattoria specialising in old-school country cooking. It's a relaxed place with a welcoming vibe, garish murals on the walls (strangely often a good sign) and tasty, value-for-money food. The fried dishes are excellent, or try one of the daily specials, chalked up on the board outside.

Il Margutta RistorArte Vegetarian **€€**

(Map p256; 06 678 60 33; www.ilmargutta.it; Via Margutta 118; meals €40; 12.30-3pm & 7-11.30pm; ; M Spagna, Flaminio) Vegetarian restaurants in Rome are rarer than parking spaces, and this airy art gallery–restaurant is an unusually chic way to eat your greens. Dishes are excellent and most produce is organic, with offerings such as artichoke hearts with potato cubes and smoked provolone cheese. Best value is the weekday (€15 to €18) and weekend (€25) buffet brunch. There's a vegan menu, and live music, on weekends.

Enoteca Regionale Palatium Ristorante, Wine Bar **€€€**

(06 692 02 132; Via Frattina 94; meals €55; 11am-11pm Mon-Sat, closed Aug; Via del Corso) A rich showcase of regional bounty, run by the Lazio Regional Food Authority, this sleek wine bar serves excellent local specialities, such as *porchetta* (pork

PAOLO CORDELLI/GETTY IMAGES ©

★ The Culinary Calendar

According to the culinary calendar (initiated by the Catholic Church), fish is eaten on Friday and *baccalà* (salted cod) is often eaten with *ceci* (chickpeas), usually on Wednesday. Thursday is gnocchi day.

LEFT: PAOLO CORDELLI/GETTY IMAGES ©; RIGHT: MANUELVELASCO/GETTY IMAGES ©

Clockwise from top: Roman restaurant; Making pasta; Salami varieties roasted with herbs) or *gnocchi alla Romana con crema da zucca* (potato dumplings Roman-style with cream of pumpkin), as well as an impressive array of Lazio wines (try lesser-known drops such as Aleatico). *Aperitivo* is a good bet, too.

There's also a tantalising array of artisanal cheese and delicious salami and cold cuts.

Imàgo Italian €€€

(Map p255; 06 6993 4726; www.imagorestaurant.com; Piazza della Trinità dei Monti 6; tasting menus €120-140; 7-10.30pm; M Spagna;) Even in a city of great views, the panoramas from the Hassler Hotel's Michelin-starred romantic rooftop restaurant are special (request the corner table), extending over a sea of roofs to the great dome of St Peter's Basilica. Complementing the views are the bold, mod-Italian creations of culinary whizz, chef Francesco Apreda. Book ahead.

Babette Italian €€€

(Map p256; 06 321 15 59; Via Margutta 1; meals €45-55; 1-3pm Tue-Sun, 7-10.45pm daily, closed Jan & Aug; ; M Spagna, Flaminio) Babette is run by two sisters who used to produce a fashion magazine, which accounts for its effortlessly chic interior of exposed brick walls and vintage painted signs. You're in for a feast too, as the cooking is delicious, with a sophisticated, creative French twist (think *tortiglioni* with courgette and pistachio pesto). The *torta Babette* is the food of the gods, a light-as-air lemon cheesecake.

The weekend lunch buffet (adult/child €28/18) is a good deal, including water, bread, dessert and coffee.

All'Oro Italian €€€

(Map p256; Via del Vantaggio 14; tasting menu €98, meals €90; M Flaminio) A Michelin-starred fine-dining restaurant, All'Oro established itself under chef Riccardo Di Giacinto in the upmarket suburb of Parioli. It's now transferred to the contemporary art-styled First Luxury Art Hotel, with white surroundings and sophisticated dishes such as ravioli filled with mascarpone, duck ragout and red wine reduction and roasted suckling pig with potatoes and black truffle souce.

Vatican City, Borgo & Prati

Pizzarium Pizza €

(Via della Meloria 43; pizza slices from €3; 11am-10pm; M Cipro–Musei Vaticani) Pizzarium, or 'Bonci pizza rustica #pizzarium', as it has recently rebranded itself, serves some of Rome's best sliced pizza. Scissor-cut squares of meticulously crafted dough are topped with original combinations of seasonal ingredients and served on paper trays for immediate consumption. There's also a daily selection of freshly fried *supplì* (crunchy rice croquettes).

Gelarmony Gelateria €

(Via Marcantonio Colonna 34; gelato €1.50-3; 10am-late; M Lepanto) Sweet-tooths are spoiled for choice at this popular Sicilian gelateria. There's an ample selection of fruit and cream gelati but for a typically Sicilian flavour go for pistachio or cassata.

Fa-Bìo Sandwiches €

(06 6452 5810; www.fa-bio.com; Via Germanico 43; sandwiches €5; 10am-5.30pm Mon-Fri, to 4pm Sat) Sandwiches, salads and smoothies are all prepared with speed, skill and fresh organic ingredients at this tiny takeaway. Locals and in-the-know visitors come to grab a quick lunchtime bite, and if you can squeeze in the door you'd do well to follow suit.

Old Bridge Gelateria €

(www.gelateriaoldbridge.com; Viale dei Bastioni di Michelangelo 5; gelato €2-5; 9am-2am Mon-Sat, 2.30pm-2am Sun; Piazza del Risorgimento, Piazza del Risorgimento) Ideal for a pre- or post-Vatican pick-me-up, this tiny gelateria has been cheerfully dishing up huge portions of delicious gelato for over 20 years. Alongside all the traditional flavours, there are also yoghurts and refreshing sorbets.

Romeo Pizza, Ristorante **€€**

(☎06 3211 0120; www.romeo.roma.it; Via Silla 26a; pizza slices €2.50, meals €45; ⊙9am-midnight; MOttaviano–San Pietro) This chic, contemporary outfit is part bakery, part deli, part takeaway and part restaurant. For a quick bite, there's delicious sliced pizza or you can have a *panino* made up at the deli counter; for a full restaurant meal, the à la carte menu offers a mix of traditional Italian dishes and forward-looking international creations.

Il Sorpasso Italian **€€**

(www.sorpasso.info; Via Properzio 31-33; meals €20-35; ⊙7am-1am Mon-Fri, 9am-1am Sat; Piazza del Risorgimento) A bar-restaurant hybrid sporting a vintage cool look – vaulted stone ceilings, hanging hams, white bare-brick walls – Il Sorpasso is a hot ticket right now. Open throughout the day, it caters to a fashionable neighbourhood crowd, serving everything from pasta specials to aperitifs, *trappizini* (pyramids of stuffed pizza), and a full dinner menu.

Velavevodetto Ai Quiriti Lazio Cuisine **€€**

(☎06 3600 0009; www.ristorantevelavevodetto.it; Piazza dei Quiriti 5; meals €35; ⊙12.30-2.30pm & 7.30-11.30pm; MLepanto) This welcoming restaurant continues to win diners over with its unpretentious, earthy food and honest prices. The menu reads like a directory of Roman staples, and while it's all pretty good, standout choices include *fettuccine con asparagi, guanciale e pecorino* (pasta ribbons with asparagus, guanciale and pecorino cheese) and *polpette di bollito* (fried meat balls).

Del Frate Ristorante **€€**

(☎06 323 64 37; www.enotecadelfrate.it; Via degli Scipioni 122; meals €40; ⊙noon-3pm & 6-11.45pm Mon-Sat; MOttaviano–San Pietro) Locals love this upmarket wine bar with its simple wooden tables and high-ceilinged brick-arched rooms. There's a formidable wine and cheese list with everything from Sicilian ricotta to Piedmontese Gorgonzola, and a small but refined menu of tartars, salads, fresh pastas and main courses.

Ristorante L'Arcangelo Ristorante **€€€**

(☎06 321 09 92; www.larcangelo.com; Via Belli 59-61; tasting menus lunch/dinner €25/55, meals €60; ⊙12.30-2.30pm Mon-Fri, 8-11pm Mon-Sat; Piazza Cavour) Styled as an informal bistro with wood-panelling, leather banquettes

Fast Food

Fast food is a long-standing Roman tradition, with plenty of street-food favourites.

A *tavola calda* (hot table) offers cheap, pre-prepared pasta, meat and vegetable dishes, while a *rosticceria* sells mainly cooked meats. Neither make for a romantic meal, but they're often very tasty.

Another favourite on the run are *arancini*, fried risotto balls that have fillings such as mozzarella and ham. These originate from Sicily, but are much loved in Rome too, where they're known as *supplì*.

Fast food is the latest Roman tradition to be reinvented, with a newfangled offering of gourmet snacks that riff on family favourites. These days you'll find hip new places serving *supplì* or *fritti* (fried things) with a twist. And these are no victory of style over substance – the new guard takes their gastronomy just as seriously as the old.

Arancini

QANAT SOCIETÀ COOPERATIVA/GETTY IMAGES ©

Rome on a Plate

Forget extravagant toppings, but *panna* (cream) is fine.

Aficionados go for gelato that's *artigianale* (artisanal) or *produzione propria* (owner produced).

Fruit flavours must be seasonal and a natural colour.

Some purists claim cones detract from the flavour.

Gelato Romano

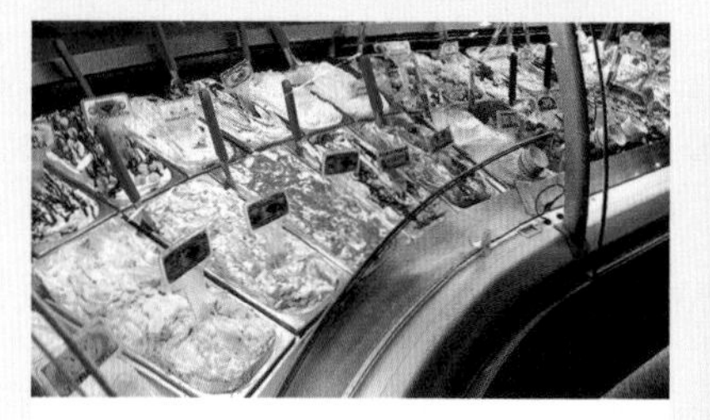

Get Your Ice Cream!

Gelato is one of Rome's great social unifiers. Everyone, from elegant matrons to politicians and toddlers, adores the stuff and the city's gelaterie cater to locals as much as tourists. The best are often small with a limited choice of freshly made seasonal flavours. Once you've got yours, hit the streets and stroll as you slurp, trying not to drip the ice cream all over yourself.

Gelato display
XUAN CHE/GETTY IMAGES ©

★ Top Five Gelaterie

Fatamorgana (p129) Rome's finest artisanal flavours, now in multiple central locations.

Gelateria del Teatro (p125) Around 40 choices of delicious ice cream, all made on-site.

Il Caruso (p129) A small but perfect selection of creamy flavours.

Gelarmony (p132) A Sicilian gelataria with many great tastes, including typically Sicilian pistachio or cassata.

Dei Gracchi (p129) A taste of heaven in several locations across Rome.

and casual table settings, L'Arcangelo enjoys a stellar local reputation. The highlight for many are the classic Roman staples such as carbonara and *amatriciana*, but there's also a limited selection of more innovative modern dishes. The wine list is a further plus, boasting some interesting Italian labels.

Enoteca La Torre Ristorante **€€€**
(06 4566 8304; www.enotecalatorreroma.com; Villa Laetitia, Lungotevere delle Armi 22; fixed-price lunch menu €55, meals €110; 12.30-2.30pm Tue-Sat, 7.30-10pm Mon-Sat; Lungotevere delle Armi) The art-nouveau Villa Laetitia provides the romantic setting for this refined Michelin-starred restaurant. A relative newcomer to the capital's fine-dining scene, chef Danilo Ciavattino has quickly established himself with his original culinary style and love of authentic country flavours.

Settembrini Ristorante **€€€**
(06 323 26 17; www.viasettembrini.it; Via Settembrini 25; menus lunch €28-38, dinner €48-65; 12.30-3pm Mon-Fri, 8-11pm Mon-Sat; Piazza Giuseppe Mazzini) All labels, suits and media gossip, this fashionable restaurant is part of the ever-growing Settembrini empire. Next door is a stylish all-day cafe, while over the way, Libri & Cucina is a laid-back bookshop eatery, and L'Officina an upscale food store. At the casually chic main restaurant expect contemporary Italian cuisine and quality wine to match.

La Veranda Ristorante **€€€**
(06 687 29 73; www.laveranda.net; Borgo Santo Spirito 73; meals €60-70, brunch €18-27; 12.30-3pm & 7.30-11pm Tue-Sun; Piazza Pia) A location in Paolo Sorrentino's Oscar-winning film *The Great Beauty*, this fine-dining restaurant sets a memorable stage for quality Italian cuisine. Inside, you can dine under 15th-century Pinturicchio frescoes, while in the warmer months, you can opt for a shady table in the garden. To enjoy the atmosphere for a snip of the regular price, stop by for Sunday brunch.

Monti, Esquilino & San Lorenzo

Panella l'Arte del Pane Bakery, Cafe **€**
(06 487 24 35; Via Merulana 54; snacks about €3.50; 8am-11pm Mon-Thu, to midnight Fri & Sat, 8.30am-4pm Sun; M Vittorio Emanuele) With a magnificent array of *pizza al taglio*, *arancini*, focaccia, fried croquettes and pastries, this smart bakery-cum-cafe is good any time of the day. The outside tables are ideal for a leisurely breakfast or chilled evening drink, or you can perch on a high stool and lunch on something from the sumptuous counter display.

Roscioli Pizza, Bakery **€**
(Via Buonarroti 48; pizza slices €3.50; 7am-8pm Mon-Sat; M Vittorio Emanuele) Off-the-track branch of this splendid deli-bakery-pizzeria, with delish *pizza al taglio*, pasta dishes and other goodies that make it ideal for a swift lunch or picnic stock-up. It's on a road leading off Piazza Vittorio Emanuele II.

L'Asino d'Oro Italian **€€**
(Map p255; 06 4891 3832; Via del Boschetto 73; meals €45; 12.30-2.30pm Sat, 7.30-11pm Tue-Sat; M Cavour) This fabulous restaurant was transplanted from Orvieto and its Umbrian origins resonate in Lucio Sforza's delicious, exceptional cooking. It's unfussy yet innovative, with dishes featuring lots of flavourful contrasts, such as lamb meatballs with pear and blue cheese. Save room for the amazing desserts. For such excellent food, this intimate, informal yet classy place is one of Rome's best deals. Hours are changeable so call ahead.

Trattoria Monti Ristorante **€€**
(06 446 65 73; Via di San Vito 13a; meals €40-45; 1-3pm Tue-Sun, 8-11pm Tue-Sat, closed Aug; M Vittorio Emanuele) The Camerucci family runs this elegant brick-arched place, proffering top-notch traditional cooking from the Marches region. There are wonderful *fritti* (fried things), delicate pastas and ingredients such as *pecorino*

di fossa (sheep's cheese aged in caves), goose, swordfish and truffles. Try the egg-yolk *tortelli* pasta. Desserts are delectable, including apple pie with *zabaglione*. Word has spread, so book ahead.

Temakinho Sushi **€€**

(Map p252; www.temakinho.com; Via dei Serpenti 16; meals €40; 12.30-3.30pm & 7pm-midnight; Cavour) In a city where food is still mostly resolutely (though deliciously) Italian, this Brazilian-Japanese hybrid serves up sushi and ceviche, and makes for a refreshing, sensational change. As well as delicious, strong *caipirinhas,* which combine Brazilian *cachaça,* sugar, lime and fresh fruit, there are '*sakehinhas*' made with sake. It's very popular; book ahead.

Tram Tram Osteria **€€**

(06 49 04 16; www.tramtram.it; Via dei Reti 44; meals around €45; 12.30-3.30pm & 7.30-11.30pm Tue-Sun; Via Tiburtina) This trendy yet old-style lace-curtained trattoria takes its name from the trams that rattle past outside. It's a family-run concern whose menu is an unusual mix of Roman and Pugliese (southern Italian) dishes, featuring taste sensations such as *tiella riso, patata e cozze* (baked rice dish with rice, potatoes and mussels). Book ahead.

Da Valentino Trattoria **€€**

(Map p255; 06 488 06 43; Via del Boschetto; meals €25-30; 1-2.45pm & 7.30-11.30pm Mon-Sat; Cavour) The vintage 1930s sign outside says 'Birra Peroni', and inside the lovely old-fashioned feel indicates that not much has changed here for years, with black-and-white photographs on the walls, white tablecloths and tiled floors. Come here when you're in the mood for grilled *scamorza* (a type of Italian cheese, similar to mozzarella), as this is the main focus of the menu, with myriad variations: served with tomato and rocket, tomato and Gorgonzola, cheese and artichokes, grilled meats, hamburgers and so on.

Primo Italian **€€**

(06 701 38 27; www.primoalpigneto.it; Via del Pigneto 46; meals around €45; 7.30pm-2am Tue-Sat, lunch Sun; ; Via Prenestina) Flagship of the Pigneto scene, Primo is still buzzing after several years, with outdoor

ANGELO CAVALLI/ROBERT HARDING/GETTY IMAGES ©

tables and an industrial brasserie-style interior. Service is slow, though.

Doozo Japanese **€€**

(Map p255; ☎06 481 56 55; Via Palermo 51; set lunch €16-26, set dinner €15-28; ⏰12.30-3pm & 7.30-11pm Tue-Sat, 7.30-10.30pm Sun; 🚌Via Nazionale) Doozo (meaning 'welcome') is a spacious, Zen restaurant-bookshop and gallery that offers tofu, sushi, *soba* (buckwheat noodle) soup and other Japanese delicacies, plus beer and green tea in wonderfully serene surroundings. It's a little oasis, particularly the shady courtyard garden.

Open Colonna Italian **€€€**

(Map p255; ☎06 4782 2641; www.antonellocolonna.it; Via Milano 9a; meals €20-80; ⏰12.30-3.30pm Tue-Sun, 8-11.30pm Tue-Sat; ❄; 🚌Via Nazionale) Spectacularly set at the back of Palazzo delle Esposizioni, superchef Antonello Colonna's superb restaurant is tucked onto a mezzanine floor under an extraordinary glass roof. The cuisine is new Roman: innovative takes on traditional dishes, cooked with wit and flair. The best thing? There's a more basic but still delectable fixed two-course lunch for €16, and Saturday and Sunday brunch is €30, served in the dramatic, glass-ceilinged hall, with a terrace for sunny days.

Agata e Romeo Italian **€€€**

(☎06 446 61 15; Via Carlo Alberto 45; meals €120; ⏰12.30-2.30pm & 7.30-11.30pm Mon-Fri; Ⓜ Vittorio Emanuele) This elegant, restrained place was one of Rome's gastronomic pioneers and still holds its own as one of the city's most gourmet takes on Roman cuisine. Chef Agata Parisella prepares the menus and runs the kitchen, offering creative uses of Roman traditions; husband Romeo curates the wine cellar; and daughter Maria Antonietta chooses the cheeses. Bookings essential.

Pastificio San Lorenzo Italian **€€€**

(☎06 9727 3519; Via Tiburtina 135; meals €50; ⏰from 7pm; 🚌Via Tiburtina, 🚋Via dei Reti) The biggest buzz in San Lorenzo is to be found at this brasserie-style restaurant, housed in a corner of the former pasta factory that is now Rome's contemporary art hub as it's also home to a collective of artists' studios. The place is packed, the

GARY YEOWELL/GETTY IMAGES ©

★ Top Five for Street Food

Pizzarium (p132)

Supplizio (p125)

Forno Roscioli (p124)

Forno di Campo de' Fiori (p125)

Trapizzino (p140)

From left: Restaurants on Piazza della Rotonda; Al fresco dining in a Roman laneway; Food stall, Piazza di Spagna

RACHEL LEWIS/GETTY IMAGES ©

vibe is 'this is where it's at', and the food....is fine – nothing to shout about, but perfectly scrumptious old favourites with pappadelle and *ragù*, served up in a stylish fashion with equivalent prices.

Trastevere & Gianicolo

Fatamorgana – Trastevere Gelateria €

(Via Roma Libera 11, Piazza San Cosimato; cones/tubs from €2; noon-midnight summer, to 10.30pm winter; Viale di Trastevere, Viale di Trastevere) One of several Fatamorgana outlets across Rome, this is one of the finest among the city's gourmet gelatarie. Quality natural ingredients are used to produce creative flavour combos such as pineapple and ginger or pear and Gorgonzola. Gluten-free.

Innocenti Bakery €

(06 580 39 26; Via delle Luce 21; 8am-8pm Mon-Sat, 9.30am-2pm Sun; Viale di Trastevere, Viale di Trastevere) It's at reassuring spots like this that you can feel that the world never changes, in some corners at least. Here you can buy light-as-air *crostata*, and stock up on biscuits such as *brutti ma buoni* (ugly but good).

Panetteria Romana Pizza €

(Via della Lungaretta 28-31; pizza slices from around €2.50; 8am-3.30pm & 4.30-7.30pm Mon-Fri, 9am-3pm Sat) Run by the Conti family since the 19th century, this venerable bakery is a handy address for pizza by the slice, cakes and biscuits in Trastevere, especially as you can grab a quick lunch here, eating in at the high tables and stools.

Pianostrada Laboratorio di Cucina Italian €

(Vicolo del Cedro; meals €25; 1-4pm & 7.30-11.30pm Tue-Sun; Piazza Trilussa) A diminutive, tucked-away place, this all-female-run foodie stop has been attracting attention with its delicious meals such as parmigiana with aubergine and pumpkin, meatballs, burgers, pasta with swordfish and wild fennel, and gourmet sandwiches. It's all exquisitely made and conceived, so squeeze in along the bar or take one of the tiny tables with barstools.

Da Corrado Roman €

(Via della Pelliccia 39; meals €25; 12.30-2.30pm & 7-11pm Mon-Sat; Viale di Trastevere, Viale di Trastevere) Don't expect refined service or a fancy interior. This is a proper Roman old-school trattoria, with no outdoor seating, but an unfussy, rough-and-ready atmosphere. It's packed with locals, feasting on hearty Roman soul food, such as *amatriciana*.

Pizzeria Ivo Pizza €

(06 581 70 82; Via di San Francesco a Ripa 158; pizzas from €7; 7pm-midnight Wed-Mon; Viale di Trastevere, Viale di Trastevere) One of Trastevere's most famous pizzerias, Ivo's has been slinging pizzas for some 40 years, and still the hungry come. With the TV on in the corner and the tables full (a few outside on the cobbled street), Ivo is a noisy and vibrant place, and the waiters fit the gruff-and-fast stereotype.

Da Olindo Trattoria €

(06 581 88 35; Vicolo della Scala 8; meals €25; 7.30-11pm Mon-Sat; Viale di Trastevere) This is your classic family affair; the menu is short, cuisine robust, portions huge, and the atmosphere lively. Expect *baccalà con patate* on Fridays and gnocchi on Thursdays, but other dishes – such as *coniglio all cacciatore* (rabbit, hunter-style) or *polpette al sugo* (meatballs in sauce) – whichever day you like.

La Gensola Sicilian €€

(06 581 63 12; Piazza della Gensola 15; meals €45; 12.30-3pm & 7.30-11.30pm, closed Sun mid-Jun–mid-Sep; Viale di Trastevere, Viale di Trastevere) This tranquil, classy yet unpretentious trattoria thrills foodies with delicious food that has a Sicilian slant and emphasis on seafood, including an excellent tuna tartare, linguine with fresh anchovies and divine *zuccherini* (tiny fish) with fresh mint.

Le Mani in Pasta Ristorante **€€**

(☎06 581 60 17; Via dei Genovesi 37; meals €35; ⏰12.30-3pm & 7.30-11pm Tue-Sun; 🚌Viale di Trastevere, 🚋Viale di Trastevere) Popular and lively, this rustic, snug place has arched ceilings and an open kitchen that serves up delicious fresh pasta dishes such as *fettucine con ricotta e pancetta*. The grilled meats are great, too.

Glass Hostaria Italian **€€€**

(☎06 5833 5903; Vicolo del Cinque 58; meals €90; ⏰7.30-11.30pm Tue-Sun; 🚌Piazza Trilussa) Trastevere's foremost foodie address, the Glass is a modernist-styled, sophisticated setting decorated in warm wood and contemporary gold, with fabulous cooking to match. Chef Cristina Bowerman creates inventive, delicate dishes that combine with fresh ingredients and traditional elements to delight and surprise the palate. There are tasting menus at €75, €80 and €100.

Paris Ristorante **€€€**

(☎06 581 53 78; www.ristoranteparis.it; Piazza San Calisto 7a; meals €45-55; ⏰7.30-11pm Mon, 12.30-3pm & 7.30-11pm Tue-Sun; 🚌Viale di Trastevere, 🚋Viale di Trastevere) An old-school restaurant set in a 17th-century building with tables on a small piazza, Paris – named for its founder, not the French capital – is the best place outside the Ghetto to sample Roman-Jewish cuisine. Signature dishes include *gran fritto vegetale con baccalà* (deep-fried vegetables with salt cod) and *carciofi alla giudia* (fried artichoke).

San Giovanni & Testaccio

Cafè Cafè Bistro **€**

(☎06 700 87 43; www.cafecafebistrot.it; Via dei Santissimi Quattro Coronati 44; meals €15-20; ⏰9.30am-11pm; 🚌Via di San Giovanni in Laterano) Cosy, relaxed and welcoming, this cafe-bistro is a far cry from the usual impersonal eateries in the Colosseum area. With its rustic wooden tables, butternut walls and wine bottles, it's a charming spot to recharge your batteries over tea and

Roman-Jewish Cuisine

If you thought fried fish was exclusive to British takeaways, think again. Deep-frying is a staple of *cucina ebraico-romanesca* (Roman-Jewish cuisine), a form of cooking that developed between the 16th and 19th centuries when Rome's Jews were confined to the city's Ghetto. To add flavour to their limited ingredients they began to fry everything from mozzarella to *baccalà* (salted cod). To delicious effect. Particularly addictive are locally grown artichokes, which are flattened out to form a kind of flower shape and then deep-fried and salted.

For the best place to try these *carciofi alla guidia* (crisp fried artichokes) and other Roman-Jewish staples, head to Via del Portico d'Ottavia, the main strip through the Jewish Ghetto. Lined with trattorias and restaurants, it's a lively hangout, especially on hot summer nights when diners crowd the many sidewalk tables. For a taste of typical Ghetto cooking, try **Nonna Betta** (Map p250; ☎06 6880 6263; www.nonnabetta.it; Via del Portico d'Ottavia 16; meals €30-35; ⏰noon-4pm & 6-11pm, closed Fri dinner & Sat lunch; 🚋Via Arenula), a small tunnel of a trattoria serving traditional kosher food. Further down the road, the unmarked **Cremeria Romana** (Map p250; Via del Portico d'Ottavia 1b; gelato €2-5; ⏰8am-11pm Sun-Thu, to 4pm Fri, 6pm-midnight Sat; 🚋Via Arenula) has a small selection of tasty kosher gelato.

Carciofi alla guidia

Offal Specialities

The hallmark of an authentic Roman menu is the presence of offal. The Roman love of nose-to-tail eating arose in Testaccio, a traditionally working-class district clustered around the city's former slaughterhouse. In the past, butchers who worked there would often be paid in cheap cuts of meat as well as money. As a result, local cooks came up with recipes to make the best possible use of these unfashionable scraps. The Roman staple *coda alla vaccinara*, which translates as 'oxtail cooked butcher's style', is cooked for hours to create a rich sauce with tender slivers of meat. Another famous dish that's not for the faint hearted is pasta with *pajata*, made with the entrails of young veal calves, considered a delicacy since they contain the mother's congealed milk.

Other offal delicacies to look out for include *trippa* (tripe), *coratella* (heart, lung and liver), *animelle* (sweetbreads), *testarella* (head), *lingua* (tongue) and *zampe* (trotters).

homemade cake, a light lunch or laid-back dinner. There's also brunch on Sundays.

Pizzeria Da Remo Pizza €

(☎06 574 62 70; Piazza Santa Maria Liberatrice 44; pizzas from €5.50; 7pm-1am Mon-Sat; Via Marmorata) For an authentic Roman experience, join the noisy crowds at this, one of the city's best-known and most popular pizzerias. It's a spartan-looking place, but the thin-crust Roman pizzas are the business, and there's a cheerful, boisterous vibe. Expect to queue after 8.30pm.

Li Rioni Pizza €

(☎06 7045 0605; Via dei Santissimi Quattro Coronati 24; meals €15-20; 7pm-midnight Thu-Tue, closed Aug; Via di San Giovanni in Laterano) Locals swear by Li Rioni, arriving for the second sitting around 9pm after the tourists have left. A classic neighbourhood pizzeria, it buzzes most nights as diners squeeze into the kitschy interior – set up as a Roman street scene – and tuck into wood-fired thin-crust pizzas and crispy fried starters.

Trapizzino Fast Food €

(www.trapizzino.it; Via Branca 88; trapizzini from €3.50; noon-1am Tue-Sun; Via Marmorata) This pocket-size joint is the birthplace of the *trapizzino*, a kind of hybrid sandwich made by stuffing a small cone of doughy bread with fillers like *polpette al sugo* (meatballs in tomato sauce) or *pollo alla cacciatore* (stewed chicken). They're messy to eat but quite delicious.

Flavio al Velavevodetto Trattoria €€

(☎06 574 41 94; www.ristorantevelavevodetto.it; Via di Monte Testaccio 97-99; meals €30-35; 12.30-3pm & 7.45-11pm; Via Galvani) Housed in a rustic Pompeian-red villa, this welcoming eatery specialises in earthy, no-nonsense *cucina romana* (Roman cuisine). Expect *antipasti* of cheeses and cured meats, huge helpings of homemade pastas, and uncomplicated meat dishes.

Da Felice Lazio Cuisine €€

(☎06 574 68 00; www.feliceatestaccio.it; Via Mastro Giorgio 29; meals €35-40; 12.30-3pm & 7.30-10.30pm; Via Marmorata) Foodies swear by this historic stalwart, famous for its unwavering dedication to local culinary traditions. In contrast to the light-touch modern decor, the menu is pure old-school with a classic weekly timetable: *pasta e fagioli* (pasta and beans) on Tuesdays, *bollito di manzo* (boiled beef) on Thursdays, seafood on Fridays. Reservations essential.

Trattoria da Bucatino Trattoria €€

(☎06 574 68 86; www.bucatino.com; Via Luca della Robbia 84; meals €30-35; noon-3pm & 7pm-midnight Tue-Sun; Via Marmorata) This genuine neighbourhood trattoria is hugely popular. It's far from refined with its dated decor and brusque service, but the typical Roman food is excellent – try its trademark *bucatini all'amatriciana* – and helpings are generous.

Il Bocconcino Trattoria €€
(☎06 7707 9175; www.ilbocconcino.com; Via Ostilia 23; meals €35; ⌚12.30-3.30pm & 7.30-11.30pm Thu-Tue, closed Aug; 🚌Via Labicana) One of the better options in the touristy pocket near the Colosseum, this laid-back trattoria stands out for its authentic regional cooking and use of locally sourced seasonal ingredients. Daily specials are chalked up on blackboards or there's a regular menu of classic Roman pastas, meaty mains and imaginative desserts.

Aroma Ristorante €€€
(☎06 9761 5109; www.aromarestaurant.it; Via Labicana 125; tasting menu €130; ⌚12.30-3pm & 7.30-11.30pm; 🚌Via Labicana) One for a special-occasion dinner, the rooftop restaurant of the Palazzo Manfredi hotel offers 'marry-me' views of the Colosseum and Michelin-starred food that rises to the occasion. Overseeing the kitchen is chef Giuseppe Di Iorio, whose brand of luxurious, forward-thinking Mediterranean cuisine has won widespread applause from critics and diners alike.

Checchino dal 1887 Lazio Cuisine €€€
(☎06 574 63 18; www.checchino-dal-1887.com; Via di Monte Testaccio 30; tasting menus €40-65; ⌚12.30-3pm & 8pm-midnight Tue-Sat; 🚌Via Galvani) A pig's whisker from the city's former slaughterhouse, Checchino is one of the grander restaurants specialising in the *quinto quarto* (fifth quarter – or insides of the animal). Signature dishes include *coda all vaccinara* (oxtail stew) and *rigatoni alla pajata* (pasta tubes with a sauce of tomato and veal intestines).

Villa Borghese & Northern Rome

Neve di Latte Gelateria €
(Via Poletti 6; gelato €2.50-5; ⌚noon-10pm Sun-Thu, to 11pm Fri & Sat) Near the MAXXI gallery, this out-of-the-way gelateria is reckoned one of Rome's best. There are few exotic flavours, rather the onus is on the classics, all prepared with high-quality organic ingredients. The pistachio, made with nuts from

Trippa alla Roma

CULTURA RM EXCLUSIVE/PHILIP LEE HARVEY/GETTY IMAGES ©

Cheese display in a delicatessen's window

the Sicilian town of Bronte, is outstanding, as is the creme caramel.

Bar Pompi Pastries €

(Via Cassia 8; tiramisu €3.50; 7am-midnight Wed-Mon, 4pm-midnight Tue; Ponte Milvio) This renowned *pasticceria* serves Rome's most celebrated tiramisu. Alongside the classic coffee, liqueur and cocoa combination, there are several other versions including strawberry, pistachio, and banana and chocolate.

Caffè delle Arti Cafe, Ristorante €€

(Map p256; 06 3265 1236; www.caffedelleartiroma.com; Via Gramsci 73; meals €45; 12.30-3.30pm daily & 7.30-11pm Tue-Sun; Piazza Thorvaldsen) The cafe-restaurant of the Galleria Nazionale d'Arte Moderna (p58) sits in neoclassical splendour in a tranquil corner of Villa Borghese. An elegant venue, it's at its best on sultry summer evenings when you can sit on the terrace and revel in the romantic atmosphere over coffee, cocktails or an al fresco dinner of classic Italian cuisine.

Metamorfosi Ristorante €€€

(06 807 68 39; www.metamorfosiroma.it; Via Giovanni Antonelli 30-32; tasting menus €80-110, lunch menu €45; 12.30-2.30pm & 8-10.30pm, closed Sat lunch & Sun; Via Giovanni Antonelli) Since opening in 2011, chef Roy Carceres' Michelin-starred restaurant has established itself as one of Rome's top dining tickets, offering innovative, contemporary cuisine, impeccable service, and a chic but informal setting. Various tasting menus are available, including a three-course lunch option.

Molto Ristorante €€€

(06 808 29 00; www.moltoitaliano.it; Viale Parioli 122; meals €45-50; 1-3pm & 8-11pm; Viale Parioli) Fashionable and quietly glamorous, Molto is a Parioli favourite. The discreet entrance gives onto an elegant, modern interior and open-air terrace, while the menu offers everything from simple, homemade pastas to more decadent truffle-flavoured dishes and succulent roast meats.

Southern Rome

Pizza Ostiense Pizza €
(Via Ostiense 56; 6.30pm-1am; M Pyramide) Run by folk formerly of the much-lauded classic Roman pizzeria Remo, in Testaccio, Pizza Ostiense offers similarly paper-thin, crispy bases and delicious fresh toppings and scrumptious *fritti* (fried things) in unfussy surroundings, with a friendly vibe.

Doppiozeroo Italian €
(06 5730 1961; www.doppiozeroo.com; Via Ostiense 68; 7am-2am Mon-Sat; M Piramide) This easy-going bar was once a bakery, hence the name ('double zero' is a type of flour). But today the sleek, modern interior attracts hungry, trendy Romans like bees to honey, especially for the cheap lunches (*primo/secondo* €4.50/6.50) and famously lavish, dinnertastic *aperitivo* between 6pm and 9pm.

Andreotti Pastries €
(06 575 07 73; Via Ostiense 54; 7.30am-9.30pm; Via Ostiense) Film director and Ostiense local Ferzan Ozpetek is such a fan of the pastries here he's known to cast them in his films. They're all stars, from the buttery *crostate* (tarts) to the piles of golden *sfogliatelle romane* (ricotta-filled pastries). You can also eat a cheap lunch or dinner here from the tasty *tavola calda*, with pasta dishes ringing in at €5.

Eataly Italian €
(06 9027 9201; www.eataly.net/it_en; Air Terminal Ostiense, Piazzale XII Ottobre 1492; shop 10am-midnight, restaurants noon-11.30pm; M Piramide) Eataly is an enormous, mall-like complex, a glittering, gleaming, somewhat confusing department store, entirely devoted to Italian food. As well as foodstuffs from all over the country, books and cookery implements, the store is also home to 19 cafes and restaurants, including excellent pizzas, pasta dishes, ice cream and more.

Porto Fluviale Trattoria, Pizza €€
(06 574 31 99; www.portofluviale.com; Via del Porto Fluviale 22; meals €35; 10.30am-2am; M Piramide) A hip, buzzing restaurant in the industrial-chic vein, Porto Fluviale is a great space and a good place to go with families: it's lively, spacious and good value, offering pasta, pizza and *cicchetti* (tapas-style appetisers, eg artichoke and ham bruschetta) that are served until late.

Trattoria Priscilla Trattoria €€
(06 513 63 79; Via Appia Antica 68; meals €30; 1-3pm daily, 8-11pm Mon-Sat; Via Appia Antica) Set in a 16th-century former stable, this intimate family-run trattoria has been feeding hungry travellers along the Appian Way for more than a hundred years, serving up traditional *cucina Romana*, so think carbonara, *amatriciana* and *cacio e pepe*. The tiramisu wins plaudits.

Qui Non se More Mai Italian €€
(06 780 3922; Via Appia Antica 198; meals around €40; 12.30-3pm & 6.30-11.30pm Tue-Sat; Via Appia Antica) This small, charismatically rustic restaurant has an open fire for grilling, plus a small terrace for when the weather's good. The menu offers Roman classics such as pasta *amatriciana*, carbonara, *gricia*, *cacio e pepe*, and so on. Just the thing to set you up for the road ahead.

LEONARDO PATRIZI/GETTY IMAGES ©

TREASURE HUNT

Begin your shopping adventure

Treasure Hunt

Rome boasts the usual cast of flagship chain stores and glitzy designer outlets, but what makes shopping here so special is its legion of small, independent shops: historic, family owned-delis, picture-framers, dusty furniture workshops, small-label fashion boutiques and artists' studios. Adding to the fun are the much-frequented neighbourhood markets selling everything from secondhand jeans to bumper produce from local farms.

Italy's reputation for quality is deserved and Rome is a top place to shop for designer clothes, shoes and leather goods. Foodie treats are another obvious choice and you'll find no end of heavenly delis, bakeries, pasticcerie (pastry shops) and chocolate shops. Homewares are another Italian speciality, and many shops focus on covetable stainless-steel kitchenware and sleek interior design.

In This Section

Useful Phrases

I'd like to buy... Vorrei comprare... (vo.*ray* kom.*pra*.re)

I'm just looking. Sto solo guardando. (sto *so*.lo gwar.*dan*.do)

Can I look at it? Posso dare un'occhiata? (po.so *da*.re oo.no.*kya*.ta)

How much is this? Quanto costa questo? (*kwan*.to *kos*.ta *kwe*.sto)

Vatican City, Borgo & Prati
Fill up on foodie treasures, accessories and vintage clothes (p158)

Villa Borghese & Northern Rome
Explore an antique market and a historic wine shop (p163)

Tridente, Trevi & the Quirinale
From high-fashion designer boutiques to flagship chain stores (p154)

Centro Storico
Boutiques, one-off designers, antiques, vintage threads, jewellery and swoon-worthy delis (p150)

Monti, Esquilino & San Lorenzo
Centre for independent fashion, homeware and vintage boutiques (p159)

Trastevere & Gianicolo
Gifts and one-off shops in one of Rome's prettiest neighbourhoods (p161)

San Giovanni & Testaccio
Browse a colourful food market and glorious delis (p163)

Opening Hours

Most city-centre shops 9am to 7.30pm (or 10am to 8pm) Monday to Saturday; some close Monday morning

Smaller shops 9am to 1pm and 3.30pm to 7.30pm (or 4pm to 8pm) Monday to Saturday

Sales

To grab a bargain, time your visit to coincide with the *saldi* (sales). Winter sales run from the first Saturday in January to mid-February, and summer sales from the first Saturday in July to early September.

The Best...

Experience Rome's best shopping

Handicrafts

Bottega di Marmoraro (p154) Commission a marble inscription to remind you of Rome.

Le Artigiane (p151) Maintaining Italy's artisanal traditions with a collection of handmade clothes, costume jewellery, ceramics, design objects and lamps.

Officina della Carta (p162) Beautiful hand-decorated notebooks, paper and cards.

Ibiz – Artigianato in Cuoio (p150) Wallets, bags and sandals in a variety of soft leathers.

Bookshops

Feltrinelli International (p161) An excellent range of the latest releases in English, Spanish, French, German and Portuguese.

Almost Corner Bookshop (pictured above; p162) A crammed haven full of rip-roaring reads.

Open Door Bookshop (p162) Many happy moments browsing secondhand books in English, Italian, French and Spanish.

Libreria l'Argonauta (p163) Travel bookshop great for sparking dreams of your next trip.

Clothing

Luna & L'Altra (pictured above; p150) Fashion-heaven, with clothes by Comme des Garçons, Issey Miyake and Yohji Yamamoto.

Tina Sondergaard (p159) This Monti boutique adjusts retro-inspired dresses to fit perfectly.

daDADA 52 (p152) Cocktail dresses and summer frocks to make you stand out from the crowd (in a good way).

Gifts

Vertecchi Art (p155) Classy stationers, selling different hues of paper, notebooks and gifts appropriate to the season.

Arion Esposizioni (p161) Art, architecture and children's books, plus design-conscious presents.

Fabriano (p155) Leather-bound diaries, funky notebooks and products embossed with street maps of Rome.

A S Roma Store (p152) Trastevere treasure-trove perfume store, with hundreds of choices from niche labels.

Ai Monasteri (p154) Exquisite herbal unguents made by monks, plus wines, liqueurs and biscuits.

Foodstuffs

Volpetti (p163) Bulging with delicious delicacies, and notably helpful staff.

Eataly (p143) Mall-scale food shop, filled with products from all over Italy, plus books, cooking utensils and more.

Salumeria Roscioli (p151) The name is a byword for foodie excellence, with mouth-watering Italian and foreign delicacies.

Pio La Torre (p152) Unpretentious, and every cent you spend helps in the fight against the mafia.

Homewares

Spot (p159) A careful selection of beautiful midcentury furnishings.

Mercato Monti Urban Market (p159) Vintage homeware finds cram this weekend market.

C.U.C.I.N.A (p157) Gastronomic gadgets to enhance your culinary life.

Shoes

Borini (p151) An unfussy shop filled to the brim with the latest women's footwear fashions.

Danielle (p155) A fast-changing collection of whatever is in right now, in a rainbow palette of colours, at affordable prices.

Barrilà Boutique (p155) With hundreds of different women's styles, it's a good chance to find the perfect shoe.

Giacomo Santini (p161) Pick up an exquisite, Fausto-designed bargain at this outlet shop.

★ Lonely Planet's Top Choices

Confetteria Moriondo & Gariglio (p150) A magical chocolate shop.

Vertecchi Art (p155) Art emporium with beautiful paper and notebooks.

Bottega di Marmoraro (p154) Have the motto of your choice carved into a marble slab at this delightful shop.

Pelletteria Nives (p155) Leather artisans make bags, wallets and more to your specifications.

Ancient Rome

Mercato di Circo Massimo — Food & Drink

(Map p252; www.mercatocircomassimo.it; Via di San Teodoro 74; ⏲9am-6pm Sat, to 4pm Sun, closed Sun Jul & all Aug; 🚌Via dei Cerchi) Rome's best and most popular farmers market is a colourful showcase for seasonal, zero-kilometre produce. As well as fresh fruit and veg, you can stock up on *pecorino Romano* cheese, milky mozzarella (known locally as *fior di latte*), olive oils, preserves, and *casareccio* bread from the nearby town of Genzano.

Centro Storico

Confetteria Moriondo & Gariglio — Food

(Map p250; Via del Piè di Marmo 21-22; ⏲9am-7.30pm Mon-Sat; 🚌Via del Corso) Roman poet Trilussa was so smitten with this historic chocolate shop – established by the Torinese confectioners to the royal house of Savoy – that he dedicated several sonnets to it. And we agree, it's a gem. Many of the bonbons and handmade chocolates laid out in ceremonial splendour in the glass cabinets are still prepared according to original 19th-century recipes.

Ibiz – Artigianato in Cuoio — Accessories

(Map p250; Via dei Chiavari 39; ⏲9.30am-7.30pm Mon-Sat; 🚌Corso Vittorio Emanuele II) In their diminutive workshop, Elisa Nepi and her father craft exquisite, well-priced leather goods, including wallets, bags, belts and sandals, in simple but classy designs and myriad colours. You can pick up a belt for about €35, while for a bag you should bank on at least €110.

Luna & L'Altra — Fashion

(Map p250; Piazza Pasquino 76; ⏲10am-2pm Tue-Sat, 3.30-7.30pm Mon-Sat; 🚌Corso Vittorio Emanuele II) An address for fashionistas with their fingers on the pulse, this is one of a number of independent boutiques on and around Via del Governo Vecchio. In its austere, gallery-like interior, clothes by Comme

Shoppers on Via del Corso, Centro Storico

LONELY PLANET/GETTY IMAGES ©

des Garçons, Issey Miyake, Yohji Yamamoto and others are exhibited in reverential style.

Le Artigiane Clothing, Handicrafts
(Map p250; www.leartigiane.it; Via di Torre Argentina 72; 10am-7.30pm; Largo di Torre Argentina) A space for local artisans to showcase their wares, this eclectic shop is the result of an ongoing project to sustain and promote Italy's artisanal traditions. It's a browser's dream, with an eclectic range of handmade clothes, costume jewellery, ceramics, design objects and lamps.

Salumeria Roscioli Food
The rich scents of fine Italian produce, cured meats and cheeses intermingle in this top-class deli.

SBU Fashion
(Map p250; www.sbu.it; Via di San Pantaleo 68-69; 10am-7.30pm Mon-Sat; Corso Vittorio Emanuele II) The flagship store of hip fashion label SBU, aka Strategic Business Unit, occupies a 19th-century workshop near Piazza Navona, complete with cast-iron columns and wooden racks. Pride of place goes to the jeans, superbly cut from top-end Japanese denim, but you can also pick up shirts, jackets, hats, sweaters and T-shirts.

Borini Shoes
(Map p250; Via dei Pettinari 86-87; 9am-1pm Tue-Sat, 3.30-7.30pm Mon-Sat; Via Arenula) Don't be fooled by the discount, workaday look – those in the know head to this seemingly down-at-heel shop for the latest footwear fashions. Women's styles, ranging from ballet flats to heeled boots, are displayed in the functional glass cabinets, alongside a small selection of men's leather shoes.

Arsenale Fashion
(Map p250; www.patriziapieroni.it; Via del Pellegrino 172; 10am-7.30pm Tue-Sat, 3.30-7.30pm Mon; Corso Vittorio Emanuele II) Arsenale, the atelier of Roman designer Patrizia Pieroni, is a watchword for original, high-end women's fashion. The virgin white interior creates a clean, contemporary showcase for beautifully cut clothes ranging from winter coats in warm, earthy tones to wispy, free-flowing summer dresses.

High Fashion

Big-name designer boutiques gleam in the grid of streets between Piazza di Spagna and Via del Corso. The great Italian and international names are represented here, as well as many lesser-known designers, selling clothes, shoes, accessories and dreams. The immaculately clad spine is Via dei Condotti, but there's also lots of high fashion in Via Borgognona, Via Frattina, Via della Vite and Via del Babuino.

Downsizing a euro or two, Via Nazionale, Via del Corso, Via dei Giubbonari and Via Cola di Rienzo are good for midrange clothing stores, with some enticing small boutiques set amid the chains.

Rachele Children's clothing
(Map p250; www.racheleartchildrenswear.it; Vicolo del Bollo 6; 10.30am-2pm & 3.30-7.30pm Tue-Sat; Corso Vittorio Emanuele II) Mums looking to update their kids' (under 12s) wardrobe would do well to look up Rachele in her delightful shop just off Via del Pellegrino. With everything from hats and mitts to romper suits and jackets, all brightly coloured and all handmade, this sort of shop is a dying breed. Most items are around the €40 to €50 mark.

Tempi Moderni Jewellery, Clothing
(Map p250; Via del Governo Vecchio 108; 9am-1.30pm & 3-7.30pm Mon-Sat; Corso Vittorio Emanuele II) Klimt prints sit side by side with pop-art paintings and cartoon ties at this kooky curiosity shop on Via del Governo Vecchio. It's packed with vintage costume jewellery, Bakelite pieces from the '20s and '30s, art nouveau and art deco trinkets, 19th-century resin brooches and pieces by couturiers such as Chanel, Dior and Balenciaga.

Help Fight the Mafia

To look at it there's nothing special about **Pio La Torre** (Map p250; www.liberaterra.it; Via dei Prefetti 23; ⏰10.30am-7.30pm Tue-Sat, 10.30am-2.30pm Sun, 3.30-7.30pm Mon; 🚌Via del Corso), a small, unpretentious food store near Piazza del Parlamento. But shop here and you're making a small but concrete contribution to the fight against the mafia. All the gastronomic goodies on sale, including organic olive oils, pastas, flours, honeys and wine, have been produced on land confiscated from organised crime outfits in Calabria and Sicily. The shop is one of several across the country set up by Libera Terra, a grassroots movement of agricultural cooperatives working on terrain that was once owned by the mob.

I Colori di Dentro — Arts

(www.mgluffarelli.com; Via dei Banchi Vecchi 29; ⏰11am-7pm Mon-Sat; 🚌Corso Vittorio Emanuele II) Take home some Mediterranean sunshine. Artist Maria Grazia Luffarelli's paintings are a riotous celebration of Italian colours, with sunny yellow landscapes, blooming flowers, Roman cityscapes and comfortable-looking cats. You can buy original watercolours or prints, as well as postcards, T-shirts, notebooks and calendars.

A S Roma Store — Sports

(Map p250; Piazza Colonna 360; ⏰10am-7.30pm Mon-Sat, 10.30am-7pm Sun; 🚌Via del Corso) An official club store of A S Roma, one of Rome's two top-flight football teams. There's an extensive array of Roma-branded kit, including replica shirts, caps, T-shirts, scarves, hoodies, keyrings and a whole lot more. You can also buy match tickets here.

daDADA 52 — Fashion

(Map p250; www.dadada.eu; Via dei Giubbonari 52; ⏰noon-8pm Mon, 11am-2pm & 3-8pm Tue-Sat, 11.30am-7.30pm Sun; 🚋Via Arenula) Girls with an eye for what works should make a beeline for this small boutique. Here you'll find a selection of eye-catching cocktail dresses that can be dressed up or down, print summer frocks, eclectic coats and colourful hats. There's a second **branch** (Map p256; ☎06 6813 9162; www.dadada.eu; Via del Corso 500; Ⓜ Flaminio, Spagna) at Via del Corso 500.

Officina Profumo Farmaceutica di Santa Maria Novella — Beauty

(Map p250; www.smnovella.it; Corso del Rinascimento 47; ⏰10am-7.30pm Mon-Sat; 🚌Corso del Rinascimento) This, the Roman branch of one of Italy's oldest pharmacies, stocks natural perfumes and cosmetics as well as herbal infusions, teas and potpourri, all shelved in wooden, glass-fronted cabinets under a Murano chandelier. The original pharmacy was founded in Florence in 1612 by the Dominican monks of Santa Maria Novella, and many of its cosmetics are based on 17th-century herbal recipes.

Nardecchia — Arts

(Map p250; Piazza Navona 25; ⏰10am-1pm Tue-Sat, 4.30-7.30pm Mon-Sat; 🚌Corso del Rinascimento) Famed for its antique prints, this historic Piazza Navona shop sells everything from 18th-century etchings by Giovanni Battista Piranesi to more affordable 19th-century panoramas. Bank on at least €150 for a small framed print.

Casali — Arts

(Via dei Coronari 115; ⏰10am-1pm Mon-Sat plus 3.30-7.30pm Sat; 🚌Corso del Rinascimento) On lovely Via dei Coronari, Casali deals in original and reproduction etchings and old prints, many delicately hand-coloured. The shop is small but the choice is not, ranging from 16th-century botanical manuscripts to postcard prints of Rome.

Le Tele di Carlotta — Handicrafts

(Map p250; Via dei Coronari 228; ⏰10.30am-1pm & 3.30-7pm Mon-Fri; 🚌Corso del Rinascimento) Search out this tiny sewing box of a shop for hand-embroidered napkins, cushion covers, bags and antique jewellery. If you're stopping long enough in Rome,

★ Artisans

Rome's shopping scene has a surprising number of artists and artisans who create their goods on the spot in hidden workshops. There are several places in Tridente where you can get a custom made bag, wallet or belt.

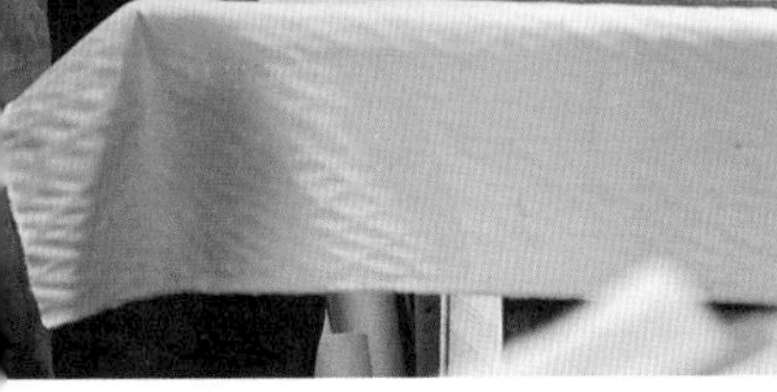

Clockwise from top: A S Roma Store; Market, Trastevere (p161); Frame crafter you can have pieces embroidered to your specifications.

Aldo Fefè Handicrafts
(Map p250; Via della Stelletta 20b; 8am-7.30pm Mon-Sat; Corso del Rinascimento) This authentic artisanal workshop produces beautifully hand-painted paper as well as leather-bound notebooks (€32), picture frames and photo albums (from €15). You can also buy Florentine wrapping paper and calligraphic pens.

Bartolucci Toys
(Map p250; www.bartolucci.com; Via dei Pastini 98; 10am-10pm; Via del Corso) It's difficult to resist going into this magical toy shop where everything is carved out of wood. It's guarded by a cycling Pinocchio and a full-sized motorbike, and within are all manner of ticking clocks, rocking horses, planes and more Pinocchios than you'll have ever seen in your life.

Alberta Gloves Accessories
(Map p250; Corso Vittorio Emanuele II 18; 10am-6.30pm; Largo di Torre Argentina) From elbow-length silk gloves to tan-coloured driving mitts, this tiny family-run shop sells a wide range of handmade gloves for every conceivable occasion. Scarves and woolly hats too.

Ai Monasteri Beauty
(Map p250; www.aimonasteri.it; Corso del Rinascimento 72; 10.30am-7.30pm, closed Thu afternoon & Sun; Corso del Rinascimento) With balms for the body and food for the soul, this monastic apothecary stocks a range of herbal essences, lotions and cosmetics, all made by monks from across Italy, as well as wines, liqueurs and biscuits. There are even elixirs promising love, happiness, and eternal youth.

De Sanctis Ceramics
(Map p250; www.desanctis1890.com; Piazza di Pietra 24; 10am-1.30pm & 3-7.30pm Mon-Sat, closed Tue morning; Via del Corso) De Sanctis – in business since 1890 – is full of impressive Sicilian and Tuscan ceramics, with sunbursts of colour decorating crockery, kitchenware and objets d'art. If your purchases are too heavy to carry, it ships worldwide.

Leone Limentani Kitchenware
(Map p250; www.limentani.com; Via del Portico d'Ottavia 47; 9am-1pm & 3.30-7pm Mon-Fri, 10am-8pm Sat; Via Arenula) Family-run for seven generations, this well-stocked basement store has a huge, rambling choice of kitchenware and tableware, expensive porcelain and knick-knacks, crockery, cutlery and crystal, many by top brands and all at bargain prices.

Loco Shoes
(Map p250; Via dei Baullari 22; 10.30am-7.30pm Tue-Sat, 3.30-7.30pm Mon; Corso Vittorio Emanuele II) Sneaker fetishists should hotfoot it to Loco for the very latest in big-statement trainers. It's a small shop, but full of attitude, with a jazzy collection of original sneakers (for boys and girls), boots and pumps by international and Italian designers. It also sells bags and costume jewellery.

Tartarughe Fashion
(Map p250; www.letartarughe.eu; Via del Piè di Marmo 17; 10am-7.30pm Tue-Sat, noon-7.30pm Mon; Via del Corso) Fashionable, versatile and elegant, Susanna Liso's seasonal designs adorn this relaxed, white-walled boutique. Her clothes, which include understated woollen coats, strikingly cut jackets, sweaters, and trousers, provide a vibrant modern update on classic styles. You'll also find a fine line in novel accessories.

Tridente, Trevi & the Quirinale

Bottega del Marmoraro Arts
(Map p256; Via Margutta 53b; 8am-7.30pm Mon-Sat; M Flaminio) A particularly charismatic hole-in-the-wall shop lined with marble carvings, where you can get marble tablets engraved with any inscription you like (€15). Peer inside at lunchtime and you might see the cheerfully quizzical *mar-*

moraro, Enrico Fiorentini, cooking pasta for his lunch next to the open log fire.

Danielle Shoes
(☎06 679 24 67; Via Frattina 85a; ⊙10.30am-7.30pm; Ⓜ Spagna) If you're female and in need of an Italian shoe fix, this is an essential stop on your itinerary. It sells both classic and fashionable styles – foxy heels, boots and ballet pumps – at extremely reasonable prices. Shoes are soft leather and come in myriad colours.

Pelletteria Nives Accessories
(☎333 3370831; Via delle Carrozze 16; ⊙9am-1pm & 4-8pm Mon-Sat; Ⓜ Spagna) Take the rickety lift to this workshop, choose from the softest leathers, and you will shortly be the proud owner of a handmade, designer-style bag, wallet, belt or briefcase. Bags cost €200 to €350 and take around a week to make.

Vertecchi Art Arts
(Via della Croce 70; ⊙3.30-7.30pm Mon, 10am-7.30pm Tue-Sat; Ⓜ Spagna) Ideal for last-minute gift buying, this large paperware and art shop has beautiful printed paper, cards and envelopes that will inspire you to bring back the art of letter writing, plus an amazing choice of notebooks, art stuff and trinkets.

Fabriano Arts, Crafts
(Map p256; ☎06 3260 0361; www.fabrianoboutique.com; Via del Babuino 173; ⊙10am-8pm; Ⓜ Flaminio, Spagna) Fabriano makes stationery sexy, with deeply desirable leather-bound diaries, funky notebooks and products embossed with street maps of Rome. It's perfect for picking up a gift, with other items including beautifully made leather key rings (€10) and quirky paper jewellery by local designers.

Mercato delle Stampe Market
(Largo della Fontanella di Borghese; ⊙7am-1pm Mon-Sat; 🚌 Piazza Augusto Imperatore) The Mercato delle Stampe (Print Market) is well worth a look if you're a fan of vintage books and old prints. Squirrel through the permanent stalls and among the tired posters and dusty back editions, and you might turn up some interesting music scores, architectural engravings or chromolithographs of Rome.

Delis & Markets

Rome's well-stocked delis and fresh-produce markets are a fabulous feature of the city's shopping-scape. Most neighbourhoods have a few local delis and their own daily food market, which typically operates from around 7am to 1.30pm Monday to Saturday. The most high profile of these is on **Campo de' Fiori** (Map p250; ⊙6am-2pm Mon-Sat; 🚌 Corso Vittorio Emanuele II), but you'll also find plenty to peruse at **Piazza San Cosimato** (⊙7am-2pm Mon-Sat; 🚌 Viale di Trastevere, 🚋 Viale di Trastevere) in Trastevere and the Nuovo Mercato di Testaccio (p163). There are also some excellent farmers markets, such as the Mercato di Circo Massimo (p150), which provides a colourful weekend showcase for seasonal, zero-kilometre produce.

Rome's top delis include the sumptuously stocked Salumeria Roscioli (p151) and historic Volpetti (p163). For the full gamut of foodie purchases, Eataly (p143) is a mall-sized state-of-the-art food emporium with produce from all over Italy and several restaurants.

Nuovo Mercato di Testaccio (p163)

Barrilà Boutique Shoes
(Map p256; Via del Babuino 34; ⊙10am-8pm; Ⓜ Flaminio, Spagna) For classic, handmade Italian women's shoes that won't crack the credit card, head to Barrilà. This boutique

5 Must-Buy Mementos

BARRY WINIKER/GETTY IMAGES ©

Antique Print

Photos are fine but for a timeless depiction of Rome consider a historic etching or print from the Piazza Navona specialist Nardecchia (p152) or nearby Casali (p152).

COMPOST_PHOTO/GETTY IMAGES ©

Leather Bag

Nothing screams Roman style as much as a silky soft leather bag. There are several shops where you can get one made, including Pelletteria Nives (p155).

Rosary Beads

Rosary beads and other assorted religious paraphernalia, ranging from papal key rings to original hand-painted icons, can be found in the Vatican and Borgo area.

LONELY PLANET/GETTY IMAGES ©

Cheese & Chocolate

Buy *pecorino Romano* cheese at Volpetti (p163) and chocolate at Confetteria Moriondo & Gariglio (p150).

Espresso Maker

A design classic and easy-to-carry icon, a caffettiera will remind you of Rome every time you make a coffee. Try C.U.C.I.N.A.

stocks myriad styles in soft leather. From the window they all look a bit traditional, but you're bound to find something you'll like in the jam-packed interior.

L'Olfattorio Beauty

(Map p256; 06 361 23 25; Via di Ripetta 34; 10.30am-7.30pm Mon-Sat, 11am-7pm Sun; M Flaminio) This is like an *enoteca* (wine bar), but with perfume instead of drinks: scents are concocted by names such as Artisan Parfumeur, Diptyque, Les Parfums de Rosine and Coudray. The assistants will guide you through different combinations of scents to work out your ideal fragrance. Exclusive perfumes are available to buy. Smellings are free but you should book ahead.

Lucia Odescalchi Jewellery

(Map p255; 06 6992 5506; Palazzo Odescalchi, Piazza dei Santissimi Apostoli 81; 9.30am-2pm Mon-Fri; M Spagna) If you're looking for a unique piece of statement jewellery that will make an outfit, this is the place to head. Housed in the evocative archives of the family *palazzo* (mansion), the avant-garde pieces often have an almost-medieval beauty, and run from incredible polished steel and chain mail to pieces created out of pearls and fossils. Beautiful. Prices start at around €140.

C.U.C.I.N.A. Homewares

(06 679 12 75; Via Mario de' Fiori 65; 3.30-7.30pm Mon, 10am-7.30pm Tue-Fri, 10.30am-7.30pm Sat; M Spagna) If you need a foodie gadget, C.U.C.I.N.A. is the place. Make your own *cucina* (kitchen) look the part with the designerware from this famous shop, with myriad devices you'll decide you simply must have, from jelly moulds to garlic presses.

Furla Accessories

(Map p255; 06 6920 0363; Piazza di Spagna 22; 10am-8pm Mon-Sat, 10.30am-8pm Sun; M Spagna) Simple, good-quality bags in soft leather and a brilliant array of colours is why the handbagging hordes keep flocking to Furla, where all sorts of accessories, from sunglasses to shoes, are made. There are many other branches dotted across Rome.

Sermoneta Accessories

(Map p255; 06 679 19 60; www.sermonetagloves.com; Piazza di Spagna 61; 9.30am-8pm Mon-Sat, 10am-7pm Sun; M Spagna) Buying leather gloves in Rome is a rite of passage for some, and its most famous glove-seller is the place to do it. Choose from a kaleidoscopic range of quality leather and suede gloves lined with silk and cashmere. An expert assistant will size up your hand in a glance. Just don't expect them to crack a smile.

Fendi Clothing

(06 69 66 61; Largo Goldoni 420; 10am-7.30pm Mon-Sat, 11am-2pm & 3-7pm Sun; M Spagna) A temple to subtly blinging accessories, this multistorey art deco building is the Fendi mother ship: this is the global headquarters, as the brand was born in Rome. Fendi is particularly famous for its products made of leather and (more controversially) fur.

Focacci Food

(06 679 12 28; Via della Croce 43; 8am-8pm; M Spagna) One of several smashing delis along this pretty street, this is the place to buy cheese, cold cuts, smoked fish, caviar, pasta, olive oil and wine.

Bulgari Jewellery

(06 679 38 76; Via dei Condotti 10; 10am-5pm Tue-Sat, 11am-7pm Sun & Mon; M Spagna) If you have to ask the price, you can't afford it. Sumptuous window displays mean you can admire the world's finest jewellery without spending a *centesimo*.

Fausto Santini Shoes

(06 678 41 14; Via Frattina 120; 11am-7.30pm Mon, 10am-7.30pm Tue-Sat, 11am-2pm & 3-7pm Sun; M Spagna) Rome's best-known shoe designer, Fausto Santini is famous for his beguilingly simple, architectural shoe designs, with beautiful boots and shoes made from butter-soft leather. Colours are beautiful, the quality impeccable. Seek out

Ecclesiastical Threads

Even if you're not in the market for a bishop's mitre or a ceremonial cassock, Rome's religious outfitters harbour some good deals. South of the Pantheon, a string of ecclesiastical shops has clerics from all over the world trying out elaborate capes and classic dog collars. That might not be you, but if you're after a sober V-neck sweater, an icon or a pair of glorious cardinal's socks in poppy red or deep purple, head to Via dei Cestari.

the end-of-line discount shop (p161) if this looks out of your price range.

Tod's Shoes

(☎06 6821 0066; Via della Fontanella di Borghese 56; 🚌Via del Corso) Tod's trademark is its rubber-studded loafers (the idea was to reduce those pesky driving scuffs), perfect weekend footwear for kicking back at your country estate.

Galleria Alberto Sordi Shopping Centre

(Map p250; Piazza Colonna; ⏰10am-10pm; 🚌Via del Corso) This elegant stained-glass arcade appeared in Alberto Sordi's 1973 classic film, *Polvere di stelle,* and has since been renamed for Rome's favourite actor, who died in 2003. It's a serene place to browse stores such as Zara and Feltrinelli, and there's an airy cafe ideal for a quick coffee break.

Underground Market

(Map p255; ☎06 3600 5345; Via Francesco Crispi 96, Ludovisi underground car park; ⏰3-8pm Sat & 10.30am-7.30pm Sun, 2nd weekend of the month Sep-Jun; Ⓜ Barberini) Monthly market held underground in a car park near Villa Borghese. There are more than 150 stalls selling everything from antiques and collectables to clothes and toys.

Vatican City, Borgo & Prati

Enoteca Costantini Wine

(www.pierocostantini.it; Piazza Cavour 16; ⏰9am-1pm Tue-Sat, 4.30-8pm Mon-Sat; 🚌Piazza Cavour) If you're after a hard-to-find grappa or something special for your wine collection, this historic *enoteca* is the place to try. Opened in 1972, Piero Costantini's superbly stocked shop is a point of reference for aficionados across town with its 800-sq-metre basement cellar and a colossal collection of Italian and world wines and more than 1000 spirits.

Antica Manufattura Cappelli Accessories

(☎06 3972 5679; www.antica-cappelleria.it; Via degli Scipioni 46; ⏰9am-7pm Mon-Fri; Ⓜ Ottaviano–San Pietro) A throwback to a more elegant age, the atelier-boutique of milliner Patrizia Fabri offers a wide range of beautifully crafted hats. Choose from the off-the-peg line of straw Panamas, vintage cloches, felt berets and tweed deerstalkers, or have one made to measure. Prices range from about €70 to €300 and ordered hats can be delivered within the day.

Rechicle Vintage

(Piazza dell'Unità 21; ⏰11am-1.30pm Tue-Sat, 2.30-7.30pm Mon-Sat; 🚌Via Cola di Rienzo) Search out this discreet boutique behind the covered market on Piazza dell'Unità for secondhand styles and vintage fashions. Designer labels are in obvious evidence among the racks of women's clothes, shoes, bags and accessories displayed alongside the occasional vintage piece.

Castroni Food & Drink

(www.castronicoladirienzo.com; Via Cola di Rienzo 196; ⏰7.45am-8pm Mon-Sat, 9.30am-8pm Sun; 🚌Via Cola di Rienzo) This is a real Aladdin's cave of gourmet treats. Towering, ceiling-high shelves groan under the weight of Italian wines and foodie specialities, classic foreign delicacies, and all manner of sweets and chocolates. Adding to the atmosphere are the coffee odours that waft up from the in-store bar.

Piazza dell' Unità Market
(Piazza del Risorgimento) Near the Vatican, perfect for stocking up for a picnic.

Monti, Esquilino & San Lorenzo

Mercato Monti Urban Market Market
(Map p252; www.mercatomonti.com; Via Leonina 46; 10am-8pm Sat & Sun; MCavour) Vintage clothes, accessories, one-off pieces by local designers, this market in the hip hood of Monti is well worth a rummage.

Tina Sondergaard Clothing
(Map p255; 334 3850799; Via del Boschetto 1d; 3-7.30pm Mon, 10.30am-7.30pm Tue-Sat, closed Aug; MCavour) Sublimely cut and whimsically retro-esque, these handmade threads are a hit with female fashion cognoscenti, including Italian rock star Carmen Consoli and the city's theatre and TV crowd. You can have adjustments made (included in the price), and dresses cost around €140.

Spot Homewares
(Map p255; 338 9275739; Via del Boschetto; 10.30am-7.30pm Mon-Sat; MCavour) This small shop has an impeccable collection of midcentury furnishings, plus glassware designed by the owners and papier mâché vases designed by their friends. It's frequented by the likes of Paolo Sorrentino (who directed *La Grande Belleza*).

La Bottega del Cioccolato Food
(Map p252; 06 482 14 73; Via Leonina 82; 9.30am-7.30pm Oct-Aug; MCavour) Run by the younger generation of Moriondo & Gariglio, this is a magical world of scarlet walls and old-fashioned glass cabinets set into black wood, with irresistible smells wafting in from the kitchen and rows of lovingly homemade chocolates on display.

Fabio Piccioni Jewellery
(Map p252; 06 474 16 97; Via del Boschetto 148; 10.30am-1pm Tue-Sat, 2-8pm Mon-Sat;

> *"stalls selling everything from rare books...to Peruvian shawls and MP3 players"*

Porta Portese Market (p161)

★ Top Five For Foodie Treats

Salumeria Roscioli (p151)

Confetteria Moriondo & Gariglio (p150)

Volpetti (p163)

Eataly (p143)

Podere Vecciano (see below)

From left: Chocolates, Confetteria Moriondo & Gariglio (p150); Vinegar tasting, Volpetti (p163)

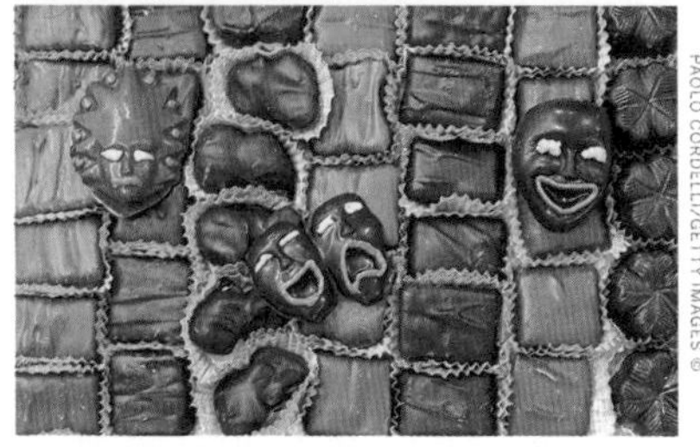

PAOLO CORDELLI/GETTY IMAGES ©

M Cavour) A sparkling Aladdin's cave of decadent, one-of-a-kind costume jewellery; artisan Fabio Piccioni recycles old trinkets to create remarkable art deco–inspired jewellery.

101 Clothing
(Via Urbana; 10am-1.30pm & 2-8pm; M Cavour) The collection at this individual boutique might include gossamer-light jumpers, broad-brimmed hats, chain-mail earrings and silk dresses: it's always worth a look to discover a special something.

Podere Vecciano Food
(Map p252; 06 4891 3812; Via dei Serpenti 33; 10am-8pm; M Cavour) Selling produce from its Tuscan farm, this shop is a great place to pick up presents, such as different varieties of pesto, honey and marmalade, selected wines, olive-oil-based cosmetics and beautiful olive wood chopping boards. There's even an olive tree growing in the middle of the shop.

Creje Clothing
(Map p255; 06 4890 5227; Via del Boschetto 5A; 10am-2.30pm & 3-8pm; M Cavour) This eclectic, inexpensive Monti boutique sells a mix of clothing sourced from exotic places, including Indian dresses, plus dramatic silver costume jewellery and soft leather bags.

Abito Clothing
(Map p255; 06 488 10 17; http://abito61.blogspot.co.uk; Via Panisperna 61; 10.30am-8pm Mon-Sat, noon-8pm Sun; M Cavour) Wilma Silvestre designs elegant clothes with a difference. Choose from the draped, chic, laid-back styles on the rack, and you can have one made up just for you in a day or just a few hours – customise the fabric and the colour. There's usually one guest designer's clothes also being sold at the shop.

La Grande Officina Jewellery
(06 445 03 48; http://lagrandeofficinagioielli.blogspot.co.uk; Via dei Sabelli 165B; 11am-7.30pm Tue-Fri, 11am-2pm Sat, 1-7.30pm Mon; Via Tiburtina) Under dusty workshop lamps, husband-and-wife team Giancarlo Genco and Daniela Ronchetti turn everything from old clock parts and Japanese fans into beautiful work-of-art jewellery. Head here for something truly unique.

Arion Esposizioni Books
(Map p255; 06 4891 3361; Via Milano 15-17; 10am-8pm Sun-Thu, to 10.30pm Fri & Sat; Via Nazionale) In cool, gleaming white rooms designed by Firouz Galdo, Arion Esposizioni – the bookshop attached to Palazzo delle Esposizioni – is just made for browsing. There are books on art, architecture and photography, DVDs, CDs, vinyl, children's books and gifts for the design lover in your life.

Giacomo Santini Shoes
(Map p255; 06 488 09 34; Via Cavour 106; 3.30-7.30pm Mon, 10am-1pm & 3.30-7.30pm Tue-Sat; Cavour) Close to the Basilica di Santa Maria Maggiore, this Fausto Santini outlet store is named after the accessory designer's father, Giacomo. It sells end-of-line and discounted Fausto Santini boots, shoes and bags, and has bargain signature architectural designs in butter-soft leather at a fraction of the retail price. Sizes are limited, however.

Feltrinelli International Books
(Map p255; 06 482 78 78; Via VE Orlando 84; 9am-8pm Mon-Sat, 10.30am-1.30pm & 4-8pm Sun; Repubblica) The international branch of Italy's ubiquitous bookseller has a splendid collection of books in English, Spanish, French, German and Portuguese. You'll find everything from recent best sellers to dictionaries, travel guides, DVDs and an excellent assortment of maps.

Nuovo Mercato Esquilino Market
(Via Lamarmora; 5am-3pm Mon-Sat; Vittorio Emanuele) Cheap, colourful food market, and the best place to find exotic herbs and spices.

Trastevere & Gianicolo

Porta Portese Market Market
(Piazza Porta Portese; 6am-2pm Sun; Viale di Trastevere, Viale di Trastevere) To see another side of Rome, head to this mammoth flea market. With thousands of stalls selling everything from rare books and fell-off-a-lorry bikes to Peruvian shawls and MP3 players, it's crazily busy and a lot of fun. Keep your valuables safe and wear your haggling hat.

Boutiques & Vintage Clothes

The bohemian enclave of Monti is a hotbed of hip shopping action. Via del Boschetto, Via Urbana and Via dei Serpenti are lined with independent-label clothing boutiques and small-scale artisanal jewellery makers. The area also boasts a number of vintage clothes shops, as well as a weekend vintage market, Mercato Monti Urban Market (p159).

Other good areas for cutting-edge designer boutiques and vintage clothes include Via del Governo Vecchio, a delightful cobbled street that runs from a small square just off Piazza Navona towards the river, Via del Pellegrino, and around Campo de' Fiori.

Vintage shop on Via del Governo Vecchio

Roma-Store Beauty

(☎06 581 87 89; Via della Lungaretta 63; ⏰10am-8pm; 🚌Viale di Trastevere, 🚋Viale di Trastevere) An enchanting perfume shop crammed full of deliciously enticing bottles of scent, including lots of small, lesser-known brands that will have perfume lovers practically fainting with joy.

Antica Caciara Trasteverina Food & Drink

(Via San Francesco a Ripa 140; ⏰7am-2pm & 4-8pm Mon-Sat; 🚌Viale di Trastevere, 🚋Viale di Trastevere) The fresh ricotta is a prized possession at this century-old deli, and usually snapped up by lunch. If you're too late, take solace in the famous *pecorino Romano* or the *burrata pugliese* (a creamy cheese from the Puglia region), or simply lust after the fragrant hams, bread, Sicilian anchovies and local wines.

Officina della Carta Gifts

(☎06 589 55 57; Via Benedetta 26b; ⏰10.30am-7.30pm Mon-Sat; 🚌Piazza Trilussa) A perfect present pit stop, this tiny workshop produces attractive hand-painted paper-bound boxes, photo albums, recipe books, notepads, photo frames and diaries.

Almost Corner Bookshop Books

(☎06 583 69 42; Via del Moro 45; ⏰10am-7.30pm Mon-Thu, 10am-8pm Fri & Sat, 11am-8pm Sun; 🚌Piazza Trilussa) This is how a bookshop should look: a crammed haven full of rip-roaring reads, with every millimetre of wall space containing English-language fiction and nonfiction (including children's) and travel guides.

Open Door Bookshop Books

(Via della Lungaretta 23; ⏰10am-8pm Mon-Sat; 🚌Viale di Trastevere, 🚋Viale di Trastevere) A lovely crammed secondhand bookshop, this is a great place to browse and happen on a classic, with novels and nonfiction in English, Italian, French and Spanish.

La Cravatta su Misura Accessories

(☎06 890 69 41; Via di Santa Cecilia 12; ⏰10am-7pm Mon-Sat; 🚌Viale di Trastevere, 🚋Viale di Trastevere) With ties draped over the wooden furniture, this inviting shop resembles the study of an absent-minded professor. But don't be fooled: these guys know their ties. Only the finest Italian silks and English wools are used in neckwear made to customers' specifications. At a push, a tie can be ready in a few hours.

Scala Quattorodici Clothing Clothing

(Villa della Scala 13-14; ⏰10am-1.30pm & 4-8pm Tue-Sat, 4-8pm Mon; 🚌Piazza Trilussa) Make yourself over à la Audrey Hepburn with these classically tailored clothes in beautiful fabrics – either made to measure or off the peg. Pricey (a frock will set you back €600 plus) but oh so worth it.

San Giovanni to Testaccio

Volpetti Food & Drink
(www.volpetti.com; Via Marmorata 47; 8am-2pm & 5-8.15pm Mon-Sat; Via Marmorata) This superstocked deli, considered by many the best in town, is a treasure trove of gourmet delicacies. Helpful staff will guide you through the extensive selection of smelly cheeses, homemade pastas, olive oils, vinegars, cured meats, vegie pies, wines and grappas. It also serves excellent sliced pizza.

Nuovo Mercato di Testaccio Market
(entrances Via Galvani, Via Beniamino Franklin; 6am-3pm Mon-Sat; Via Marmorata) Even if you don't need to buy anything, a trip to Testaccio's daily food market is fun. Occupying a modern, purpose-built site, it hums with activity as locals go about their daily shopping, picking, prodding and sniffing the brightly coloured produce and cheerfully shouting at all and sundry.

Calzature Boccanera Shoes
(Via Luca della Robbia 36; 9.30am-1.30pm Tue-Sat & 3.30-7.30pm Mon-Sat; Via Marmorata) From just-off-the-runway heels to classic driving shoes, high-end trainers and timeless lace-ups, this historic shoe shop stocks a wide range of footwear by top international brands, as well as bags, belts and leather accessories.

Soul Food Music
(www.haterecords.com; Via di San Giovanni in Laterano 192; 10.30am-1.30pm & 3.30-8pm Tue-Sat; Via di San Giovanni in Laterano) Run by Hate Records, Soul Food is a laid-back record store with an eclectic collection of vinyl that runs the musical gamut, from '60s garage and rockabilly to punk, indie, new wave, folk, funk and soul. You'll also find retro-design T-shirts, fanzines and other groupie clobber.

Via Sannio Market
(9am-1.30pm Mon-Sat; M San Giovanni) This clothes market in the shadow of the Aurelian Walls is awash with wardrobe staples. It has a good assortment of new and vintage clothes, bargain-price shoes, jeans and leather jackets.

Villa Borghese & Northern Rome

Libreria l'Argonauta Books
(www.librerialargonauta.com; Via Reggio Emilia 89; 10am-8pm Mon-Fri, 10am-1pm & 4-8pm Sat; Via Nizza) Near the Museo d'Arte Contemporanea di Roma (MACRO) museum, this travel bookshop is a lovely place to browse. With its serene atmosphere and shelves of travel literature, guides, maps and photo tomes, it can easily spark daydreams of far-off places.

Bulzoni Wine
(www.enotecabulzoni.it; Viale Parioli 36; 8.30am-2pm & 4.30-8.30pm; Viale Parioli) This historic wine shop has been supplying Parioli with wine since 1929. It has a formidable collection of Italian regional wines, as well as European and New World labels, and a carefully curated selection of champagnes, liqueurs, olive oils and gourmet delicacies.

Bagheera Fashion
(www.bagheeraboutique.com; Piazza Euclide 30; 9.30am-1pm Tue-Sat, 3.30-7.30pm Mon-Sat; Piazza Euclide) This modish boutique has long been a local go-to for the latest fashions. Alongside sandals and vampish high heels you'll find dresses by Dries Van Noten and a selection of bags and accessories by big-name international designers.

Anticaglie a Ponte Milvio Market
(Via Capoprati; 9am-8pm 1st & 2nd Sun of month, closed Aug; Ponte Milvio) The 2nd-century-BC Ponte Milvio forms the backdrop to this monthly antique market. On the first and second Sunday of every month up to 200 stalls spring up on the riverbank laden with antiques, objets d'art, vintage clothes, period furniture and all manner of collectable clobber.

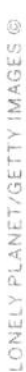

BAR OPEN

Afternoon beers, evening wines and midnight cocktails

Bar Open

Often the best way to enjoy nightlife in Rome is to wander from restaurant to bar, getting happily lost down picturesque cobbled streets. There's simply no city with better backdrops for a drink: you can savour a Campari overlooking the Roman Forum or sample artisanal beer while watching the light bounce off baroque fountains.

Night-owl Romans tend to eat late, then drink at bars before heading off to a club at around 1am. Like most cities, Rome is a collection of districts, each with its own character, which is often completely different after dark. The centro storico *(historic centre) and Trastevere pull in a mix of locals and tourists as night falls, while Ostiense and Testaccio are the grittier clubbing districts.*

In This Section

Opening Hours

Cafes 7.30am to 8pm

Traditional bars 7.30am to 1am or 2am

Bars, pubs & enoteche (wine bars) lunchtime or 6pm to 2am

Nightclubs 10pm to 4am

Villa Borghese & Northern Rome
From a cool aperitif bar to hip alternative venues (p181)

Vatican City, Borgo & Prati
Low-key scene with a sprinkling of quiet wine bars (p173)

Tridente, Trevi & the Quirinale
Historic cafes and swanky, good-looking cocktail bars (p172)

Centro Storico
Bars and a few clubs, a mix of touristy and sophisticated (p170)

Monti, Esquilino & San Lorenzo
Boho bars in Monti and Pigneto, underground clubs in San Lorenzo (p173)

Ancient Rome
A couple of popular retreats near the Roman ruins (p170)

Trastevere & Gianicolo
Buzzing area riddled with bars, pubs and cafes (p177)

San Giovanni & Testaccio
Nightowls swarm to Testaccio's strip of poptastic clubs (p180)

Southern Rome
Serious clubbing territory with cool venues in Ostiense's ex-industrial venues (p183)

Costs

Expect to pay somewhere around the following for a drink:

Espresso €0.80

Cappuccino €1.20

Glass of wine from €3

Beer €5

Tipping

Tipping in bars is not necessary, although many people leave small change, perhaps a €0.20 coin, if standing at the bar.

Websites

Roma 2 Night (http://2night.it)

Zero (http://roma.zero.eu)

The Best...

Experience Rome's finest drinking establishments

Enoteche

Il Tiaso (p174) With a hip, living-room vibe, plentiful wines and live music.

Fafiuché (p174) A charming warm-orange space with wine and artisanal beers.

La Barrique (p174) Inviting Monti address, with great wines and accompanying meals.

Il Goccetto (p171) An old-school *vino e olio* (wine and oil) shop that makes for a great neighbourhood wine bar.

Aperitivo

La Meschita (p180) Delicious nibbles in this tiny *enoteca* (wine bar) adjoining La Ferrara restaurant.

Momart (p182) Students and local professionals love its expansive array of pizza and other snacks.

Freni e Frizioni (p179) Perenially cool bar with lavish nightly buffet of snacks.

Alternative Venues

Lanificio 159 (p182) Cool underground venue that hosts live gigs and club nights.

Big Bang (p180) Reggae, dancehall, dub and techno in a graffitti-sprayed former slaughterhouse.

Big Star (p179) Backstreet Trastevere bar, with regular DJs and a laid-back crowd.

Yeah! Pigneto (p176) Cool bar hosting live gigs and DJs in Rome's most boho district.

For a Lazy Drink

Ombre Rosse (p179) Lovely, relaxed Trastevere bar, with outside seating.

Stravinskij Bar (p172) Hotel de Russie's elegant bar, with its courtyard garden backed by Borghese gardens.

Yeah! Pigneto (p176) Boho bar that has plenty of places to sit and chat and DJs and regular live gigs.

Gay

Coming Out (pictured above; p181) A friendly gay bar near the Colosseum, open all day, with gigs, drag shows and karaoke later on.

L'Alibi (p181) Kitsch shows and house, techno and dance pumping up a mixed gay and straight crowd.

My Bar (p181) A mixed crowd by day, and gay-er by night, in the shadow of the Colosseum.

See & Be Seen

Etablì (p171) Chic bar near Campo de'Fiori, filled with vintage French furniture and laid-back cool.

Salotto 42 (p172) A sitting-room-style bar, offering cocktails facing the ancient Roman Stock Exchange.

Co.So (p174) Opened by the Hotel de Russie's former mixologist, this is Pigneto's hippest haunt.

Rec 23 (p180) With NY style and locally inspired cocktails, this is the place to be seen in Testaccio.

★ Lonely Planet's Top Choices

Ai Tre Scalini (p173) Buzzing *enoteca* that feels as convivial as a pub.

Ma Che Siete Venuti a Fà (p179) Tiny pub that's the heart of Rome's artisanal-beer explosion.

Co.So (p174) A real buzz in this Pigneto hotspot, serving up out-there cocktails on bubble-wrap coasters.

Sciascia Caffè (p173) Classy joint serving the unparalleled *caffè eccellente,* a velvety smooth espresso in a chocolate-lined cup.

Barnum Cafe (p170) Cool vintage armchairs to sink into by day and dressed-up cocktails by night.

Ancient Rome

0,75 Bar

(Map p252; www.075roma.com; Via dei Cerchi 65; ⌚11am-2am; 📶; 🚌Via dei Cerchi) This welcoming bar on the Circo Massimo is good for a lingering drink, an *aperitivo* (6.30pm onwards) or a light meal (mains €6 to €13.50, salads €5.50 to €7.50). It's a friendly place with a laid-back vibe, an attractive exposed-brick look and cool tunes.

Cavour 313 Wine Bar

(Map p252; ☎06 678 54 96; www.cavour313.it; Via Cavour 313; ⌚12.30-2.45pm & 7.30pm-12.30am, closed Sun summer; Ⓜ Cavour) Close to the Forum, wood-panelled Cavour 313 attracts everyone from tourists to actors and politicians. It serves a daily food menu and a selection of salads, cold cuts and cheeses (€8 to €12), but the headline act is the wine. And with more than 1200 labels to choose from you're sure to find something to tickle your palate.

"it's well worth trying a few local drops while you're in Rome"

Centro Storico

Barnum Cafe Cafe

(Map p250; www.barnumcafe.com; Via del Pellegrino 87; ⌚9am-10pm Mon, 8.30am-2am Tue-Sat; 📶; 🚌Corso Vittorio Emanuele II) A relaxed, friendly spot to check your email over a freshly squeezed orange juice or spend a pleasant hour reading a newspaper on one of the tatty old armchairs in the white bare-brick interior. Come evenings and the scene is cocktails, smooth tunes and coolly dressed-down locals.

Caffè Sant'Eustachio Cafe

(Map p250; www.santeustachioilcaffe.it; Piazza Sant'Eustachio 82; ⌚8.30am-1am Sun-Thu, to 1.30am Fri, to 2am Sat; 🚌Corso del Rinascimento) This small, unassuming cafe, generally three deep at the bar, is reckoned by many to serve the best coffee in town. Created by beating the first drops of espresso and several teaspoons of sugar into a frothy

paste, then adding the rest of the coffee, it's superbly smooth and guaranteed to put some zing into your sightseeing.

La Casa del Caffè Tazza d'Oro Cafe

(Map p250; www.tazzadorocoffeeshop.com; Via degli Orfani 84-86; 7am-8pm Mon-Sat, 10.30am-7.30pm Sun; Via del Corso) A busy, stand-up cafe with burnished 1940s fittings, this is one of Rome's best coffee houses. Its espresso hits the mark nicely and there's a range of delicious coffee concoctions, including a cooling *granita di caffè,* a crushed-ice coffee drink served with whipped cream. There's also a small shop and, outside, a coffee vending machine for those out-of-hours caffeine emergencies.

Open Baladin Bar

(Map p250; www.openbaladinroma.it; Via degli Specchi 6; noon-2am; ; Via Arenula) A hip, shabby-chic lounge bar near Campo de' Fiori, Open Baladin is a leading light in Rome's craft-beer scene, with more than 40 beers on tap and up to 100 bottled brews, many from Italian artisanal microbreweries. There's also a decent food menu with *panini,* gourmet burgers and daily specials.

Etablì Bar

(Map p250; 06 9761 6694; www.etabli.it; Vicolo delle Vacche 9a; 11am-2am, closed Mon in winter, Sun in summer; ; Corso del Rinascimento) Housed in a lofty 16th-century *palazzo,* Etablì is a rustic-chic lounge-bar-restaurant where you can drop by for a morning coffee, have a light lunch or chat over an *aperitivo*. It's laid-back and good-looking, with original French-inspired country decor – think leather armchairs, rough wooden tables and a crackling fireplace. It also serves weekend brunch and full restaurant dinners (€45), and hosts the occasional jam session.

Circus Bar

(Map p250; www.circusroma.it; Via della Vetrina 15; 10.30am-2am; ; Corso del Rinascimento) A great little bar, tucked around the corner from Piazza Navona. It's a relaxed place popular with out-of-town students who come here to catch up on the news – wi-fi is free and there are international newspapers to read – and hang out over a drink. The atmosphere hots up in the evening when cocktails and shots take over from tea and cappuccino.

Bars & Pubs

Bars range from regular Italian cafe-bars that have seemingly changed little over the centuries, to chic, contemporarily styled places serving esoteric cocktails – such as Co.So (p174) and Salotto 42 (p172) – and laid-back, perennially popular haunts – such as Freni e Frizioni (p179) – that have a longevity rarely seen in other cities. Pubs are also popular, with several long-running Irish-style pubs filled with chattering Romans, and more pub-like bars opening on the back of the artisanal beer trend.

No.Au Bar

(Map p250; Piazza Montevecchio 16; 6pm-1am Tue-Thu, noon-1am Fri-Sun; Corso del Rinascimento) Opening onto a charming *centro storico* piazza, No.Au – pronounced know how – is a cool bistro-bar set-up. Like many fashionable bars, it's big on beer and offers a knowledgeable list of artisanal craft brews, as well as local wines and a small but select food menu.

Il Goccetto Wine Bar

(Via dei Banchi Vecchi 14; 11.30am-2pm Tue & Sat, 6.30pm-midnight Mon-Sat, closed Aug; Corso Vittorio Emanuele II) This old-school *vino e olio* (wine and oil) shop has everything you could want in a neighbourhood wine bar: a colourful cast of regulars, a cosy, bottle-lined interior, a selection of cheeses and cold cuts, and a serious, 800-strong wine list.

Clubbing & Centri Sociali

Rome has a range of clubs, with DJs spinning everything from lounge and jazz to dancehall and hip-hop. The scene is centred on Testaccio (mainstream clubs) and Ostiense (industrial, warehouse vive for serious clubbers), although you'll also find popular places in Trastevere and the *centro storico* (historics centre).

Clubs tend to get busy after midnight, or even after 2am. Admission is often free, but drinks are expensive. Cocktails can cost from €10 to €20, but you can drink much more cheaply in the studenty clubs of San Lorenzo, Pigneto and the *centri sociali* (social centres).

Rome's *centri sociali* were originally hubs of left-wing activism but many have now resurrected themselves as arts centres hosting live music and contemporary arts events. They offer Rome's most unusual, cheap and alternative nightlife options. Top spots include Brancaleone (p182) and Esc Atelier (p177).

You'll need to dress the part for the big clubs, some of which have a seemingly whimsical door policy, and men, whether single or in groups, will often find themselves turned away.

Jerry Thomas Project Cocktail Bar

(06 9684 5937; www.thejerrythomasproject.it; Vicolo Cellini 30; 10pm-4am; Corso Vittorio Emanuele II) A self-styled speakeasy with a 1920s look and a password to get in – check the website and call to book – this hidden bar is setting the standards for the cocktail trend currently sweeping Rome. Its hipster mixologists know their stuff and the retro decor lends the place a real Prohibition-era feel.

L'Angolo Divino Wine Bar

(Map p250; www.angolodivino.it; Via dei Balestrari 12; 10.30am-3pm Tue-Sat, 5pm-1.30am daily; Corso Vittorio Emanuele II) A hop and a skip from Campo de' Fiori, this is a warm, woody wine bar. It's an oasis of genteel calm, with a carefully selected wine list, mostly Italian but a few French and New World labels, and a small daily menu of hot and cold dishes such as creamy Andria *burrata* (cheese made from mozzarella and cream) with sun-dried tomatoes.

Salotto 42 Bar

(Map p250; www.salotto42.it; Piazza di Pietra 42; 10.30am-2am Tue-Sun; Via del Corso) On a picturesque piazza, facing the columns of the Temple of Hadrian, this is a glamorous lounge bar, complete with subdued lighting, vintage 1950s armchairs, Murano lamps and a collection of heavyweight design books. Come for the daily lunch buffet or to hang out with the 'see-and-be-seen' crowd over an evening cocktail.

Tridente, Trevi & the Quirinale

La Scena Bar

(Map p256; Via della Penna 22; 7am-1am; Flaminio) Part of the art deco Hotel Locarno, this bar has a faded Agatha Christie–era feel, and a greenery-shaded outdoor terrace bedecked in wrought-iron furniture. Cocktails cost €13 to €15, or you can partake of afternoon tea from 3pm to 6pm and *aperitivo* from 7pm to 10pm.

Stravinskij Bar – Hotel de Russie Bar

(Map p256; 06 328 88 70; Via del Babuino 9; 9am-1am; Flaminio) Can't afford to stay at the celeb-magnet Hotel de Russie? Then splash out on a drink at its swish bar. There are sofas inside, but best is a drink in the sunny courtyard, with sunshaded tables overlooked by terraced gardens. Impossibly romantic in the best dolce vita style, it's perfect for a cocktail (from €20) or beer (€13) and some posh snacks.

Canova Bar

(Map p256; 06 361 22 31; Piazza del Popolo 16; 8am-midnight; Flaminio) While left-wing authors Italo Calvino and Alberto Moravia used to drink at Rosati, over the square, their right-wing counterparts came here. Today tourists are the main clientele, and the views are as good as ever

Micca Club Club

(Map p255; ☎393 3236244; www.miccaclub.com; Via degli Avignonesi; MBarberini) No longer in its brick-arched cellar in southern Rome, but now close to Piazza Barberini, Micca Club now has a less arresting interior but still retains its vintage, quirky vibe. This is Rome's burlesque club, where you can sip cocktails while watching shimmying acts upping the kitsch factor. Reserving a table by phone is advised.

Vatican City, Borgo & Prati

Sciascia Caffè Cafe

(Via Fabio Massimo 80/A; 7.30am-6.30pm Mon-Sat; MOttaviano-San Pietro) The timeless elegance of this polished cafe is perfectly suited to the exquisite coffee it makes. There are various options but nothing can beat the *caffè eccellente,* a velvety smooth espresso served in a delicate cup that has been lined with melted chocolate. The result is nothing short of magnificent.

Makasar Wine Bar, Teahouse

(www.makasar.it; Via Plauto 33; noon-midnight Tue-Thu, to 2am Fri & Sat, 5.30-11.30pm Sun; Piazza del Risorgimento) Recharge your batteries with a quiet drink at this oasis of bookish tranquillity. Pick your tipple from the nine-page tea menu or opt for an Italian wine and sit back in the casually stylish, softly lit interior. For something to eat, there's a small menu of salads, bruschette, baguettes and healthy hot dishes.

Passaguai Wine Bar

(☎06 8745 1358; www.passaguai.it; Via Leto 1; 10am-2am Mon-Fri, 6pm-2am Sat & Sun; ; Piazza del Risorgimento) A cosy basement bar with tables in a vaulted interior and on a quiet side street, Passaguai feels pleasingly off the radar. It's a great spot for a beer or glass of wine – there's an excellent choice of both – accompanied by cheese and cold cuts, or even a full meal from the limited menu. Free wi-fi.

Grattachecca

It's summertime, the living is easy, and Romans like nothing better in the sultry evening heat than to amble down to the river and partake of some *grattachecca* (crushed ice covered in fruit and syrup). It's the ideal way to cool down and there are kiosks along the riverbank satisfying this very Roman need. Try Sora Mirella Caffè (grattachecce €3-6; 11am-3am May-Sep), next to Ponte Cestio.

Mint *grattachecca*

Monti, Esquilino & San Lorenzo

Ai Tre Scalini Wine Bar

(Map p252; Via Panisperna 251; 12.30pm-1am; MCavour) The 'Three Steps' is always packed, with crowds spilling out into the street. Apart from a tasty choice of wines, it sells the damned fine Menabrea beer, brewed in northern Italy. You can also tuck into a heart-warming array of cheeses, salami and dishes such as *polpette al sugo* (meatballs with sauce; €7.50).

Il Tiaso Bar

(☎06 4547 4625; www.iltiaso.com; Via Perugia 20; ; Circonvallazione Casilina) Think living room with zebra-print chairs, walls of indie art, Lou Reed biographies shelved between wine bottles, and 30-something owner Gabriele playing his latest New York Dolls album to neobeatnik chicks, corduroy professors and the odd neighbourhood dog. Well-priced wine, an intimate chilled vibe and regular live music.

Co.So Bar

(Via Braccio da Montone 80; cocktails €10; 7pm-3am Mon-Sat; Via Prenestina) The chicest bar in the Pigneto district, this tiny place, opened by Massimo D'Addezio, former master mixologist at Hotel de Russie, is buzzing and is hipster to the hilt, with its Carbonara Sour cocktail (with vodka infused with pork fat), bubble-wrap coasters, and popcorn and M&M bar snacks.

Birra Piu Bar

(06 7061 3106; Via del Pigneto 105; beer €5; 5pm-midnight Mon-Thu, 5pm-2am Fri & Sat, 7pm-midnight Sun; Circonvallazione Casilina) A small, relaxed bar, with a laid-back crowd draped over blonde-wood bar stools and tables. To a soundtrack of the Doors, Blur and so on, you can drink a wide variety of craft beers, with names such as 'Total Insanity'.

Fafiuché Wine Bar

(Map p252; 06 699 09 68; www.fafiuche.it; Via della Madonna dei Monti 28; 5.30pm-1am Mon-Sat; M Cavour) Fafiuché means 'light-hearted fun' in the Piedmontese dialect, and this place lives up to its name. The narrow, bottle-lined warm-orange space exudes charm: come here to enjoy wine and artisanal beers, eat delicious dishes originating from Puglia to Piedmont or buy delectable foodstuffs. *Aperitivo* is from 6.30pm to 9pm.

La Barrique Wine Bar

(Map p255; Via del Boschetto 41b; 12.30-3.30pm & 5.30pm-1am Mon-Sat; M Cavour) This appealing *enoteca*, with wooden furniture and whitewashed walls, is a classy yet informal place to hang out and sample excellent French, Italian and German wines; a choice of perfectly cooked, delicious main courses provide a great accompaniment, or you can stick to artisanal cheeses and cold cuts.

★ Enoteche

An *enoteca* was originally where the old boys from the neighbourhood would gather to drink rough local wine poured straight from the barrel. Times have changed: nowadays they tend to be sophisticated but still atmospheric places, offering Italian and international vintages, delicious cheeses and cold cuts.

Clockwise from top left: Via della Scala, Trastevere (p179); Trastevere cafe; Frascati Superiore wine (p177); Irish-style pub

Wine on display

Al Vino al Vino Wine Bar

(Map p252; Via dei Serpenti 19; 6pm-1am, shop open all day; MCavour) A rustic *enoteca* that's a favourite with the locals, mixing ceramic tabletops and contemporary paintings, this is an attractive spot to linger over a fine collection of wines, particularly *passiti* (sweet wines). The other speciality is *distillati* – grappa, whisky and so on – and there are snacks to help it all go down, including some Sicilian delicacies.

Bohemien Bar

(Map p252; Via degli Zingari 36; 6pm-2am Wed-Sun; MCavour) This little bar lives up to its name; it feels like something you might stumble on in Left Bank Paris. It's small, with mismatched chairs and tables and an eclectic crowd drinking wine by the glass, and tea and coffee.

Yeah! Pigneto Bar

(06 6480 1456; www.yeahpigneto.com; Via Giovanni de Agostini 41; small beer €3, aperitivo €7; 7pm-2am Mon-Fri, 8pm-2am Sat & Sun) We say si! to Yeah! Pigneto. A relaxed boho-feeling bar with DJs playing jazz and the walls covered in collages and classic album covers, this is a good place for lingering over not-too-expensive beer. Regular live gigs.

Locanda Atlantide Club

(06 4470 4540; www.locandatlantide.it; Via dei Lucani 22b; free or €3-5; 9pm-2am Oct-Jun; MVia Tiburtina, Scalo San Lorenzo) Come, tickle Rome's grungy underbelly. Descend through a door in a graffiti-covered wall into this cavernous basement dive, packed to the rafters with studenty, alternative crowds and featuring everything from prog-folk to DJ-spun electro music. It's good to know that punk is not dead.

Gente di San Lorenzo Bar

(06 445 44 25; Via degli Aurunci 42; 7am-2am; Via dei Reti) On the corner of San Lorenzo's Piazza dell'Immacolata, which gets thronged with students on balmy nights, this is a relaxed place for a drink and snack or meal. The interior is airy, with warm wooden floors and brick arches, and there are some outdoor tables as well as regular DJs and occasional live music.

Esc Atelier Club

(www.escatelier.net; Via dei Volsci 159; varies; Via Tiburtina, Via dei Reti) This left-wing alternative arts centre hosts live gigs and club nights: expect electronica DJ sets featuring live sax, discussions, exhibitions, political events and more. Admission and drinks are cheap.

Ice Club Bar

(Map p252; www.iceclubroma.it; Via della Madonna dei Monti 18; 6pm-2am; M Colosseo) Novelty value is what the Ice Club is all about. Pay €15 (you get a free vodka cocktail served in a glass made of ice), put on a thermal cloak and mittens, and enter the bar, in which everything is made of ice (temperature: −5°C). Most people won't chill here for long – the record is held by a Russian (four hours).

Vicious Club Club

(06 7061 4349; www.viciousclub.com; Via Achille Grandi 3a; admission varies; 10pm-4.30am Tue & Thu-Sat, to 4am Sun; Roma Laziali) Vicious is a gay-friendly club that welcomes all to dance and chatter to a soundtrack of electro, no wave, deep techno, glam indie, and deep house. It's small enough to feel intimate; try Alchemy every Saturday.

Vini e Olii Bar

(Via del Pigneto 18; Circonvallazione Casilina) Forget the other bars that line Pigneto's main pedestrianised drag, with their scattered outside tables and styled interiors. This is where the locals head, turning their noses up at newer interlopers. This traditional 'wine and oil' shop sells cheap beer and wine (bottles from €7.50), and you can snack on platefuls of antipasti and *porchetta* (pork roasted in herbs). It's outside seating only.

Bar Zest at the Radisson Blu Es Bar

(Via Filippo Turati 171; 9am-1am; Via Cavour) In need of a cocktail in the Termini district? Pop up to the 7th-floor bar at the slinkily designed Radisson Blu Es. Waiters are cute, chairs are by Jasper Morrison, views are through plate glass and there's a sexy rooftop pool to look at.

Regional Lazio Wines

Wines produced in the surrounding region of Lazio may not be household names yet, but it's well worth trying a few local drops while you're in Rome. Although whites dominate Lazio's production – 95% of the region's Denominazione di origine controllata (DOC; the second of Italy's four quality classifications) wines are white – there are a few notable reds as well. To sample Lazio wines, Palatium (p130) and Terre e Domus (p124) are good places to go.

Most of the house white in Rome will be from the Castelli Romani area to the southeast of Rome, centred on Frascati and Marino. New production techniques have led to a lighter, drier wine that is beginning to be taken seriously. Frascati Superiore is now an excellent tipple, Castel de Paolis' Vigna Adriana wins plaudits, while the emphatically named Est! Est!! Est!!!, produced by the renowned wine house Falesco, based in Montefiascone on the volcanic banks of Lago Bolsena, is increasingly drinkable.

Falesco also produces the excellent Montiano, blended from Merlot grapes. Colacicchi's Torre Ercolana from Anagni is another opulent red, which blends local Cesanese di Affile with cabernet sauvignon and merlot. Velvety, complex and fruity, this is a world-class wine.

Trastevere & Gianicolo

Ma Che Siete Venuti a Fà Pub

(www.football-pub.com; Via Benedetta 25; 11am-2am; Piazza Trilussa) Named after a football chant, which translates politely as 'What did you come here for?', this pint-sized Trastevere pub is a beer-buff's paradise, packing in at least 13 international

Rome in a Glass

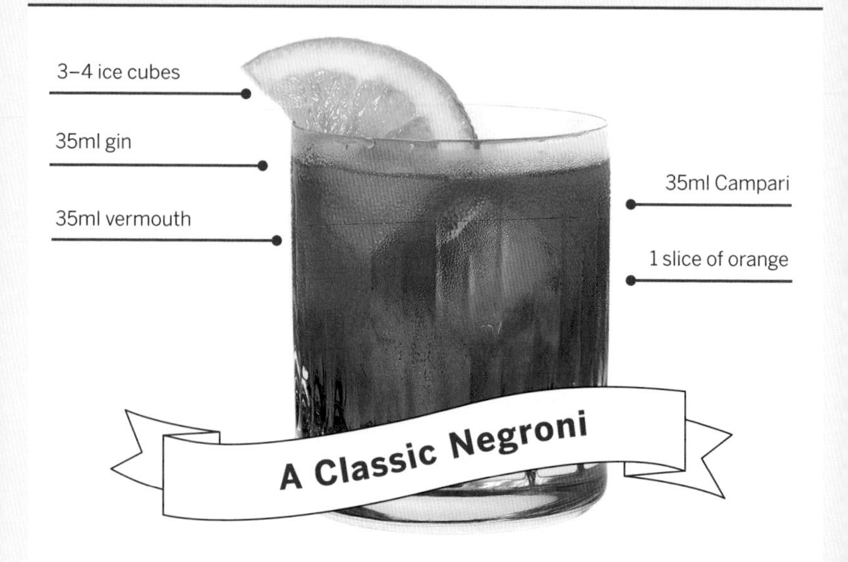

Make a Negroni

- Tip the ice cubes into a short whisky glass and pour on the gin, vermouth and Campari.
- Stir quickly with a cocktail spoon before topping with the orange slice. *Salute!*

Story Behind the Cocktail

Equal parts gin, vermouth and Campari, the ruby-red negroni is one of Rome's great cocktails. According to drinking lore, it was created in Florence in 1919, after Count Camillo Negroni asked a bartender to fortify his Americano by adding gin in place of soda water. Modern variations include the fizzier, lighter *negroni sbagliato,* made with *prosecco* (a type of sparkling wine) instead of gin.

★ Top 3 Bars for Negronis

Barnum Cafe (p170)

Stravinskij Bar – Hotel De Russie (p172)

Jerry Thomas Project (p172)

Friends at a bar
LEONARDO PATRIZI/GETTY IMAGES ©

craft beers on tap and even more by the bottle.

Bar San Calisto Cafe

(06 589 56 78; Piazza San Calisto 3-5; 6am-1.45am Mon-Sat; Viale di Trastevere, Viale di Trastevere) Those in the know head to the down-at-heel 'Sanca' for its basic, stuck-in-time atmosphere and cheap prices (beer €1.50). It attracts everyone from intellectuals to keeping-it-real Romans, alcoholics and American students. It's famous for its chocolate – hot with cream in winter, ice cream in summer. Try the *sambuca con la mosca* ('with flies' – raw coffee beans). Expect occasional late-night jam sessions.

Da Biagio Wine Bar

(www.dabiagio.it; Via della Scala 64; 10am-1.30pm & 5pm-midnight; Piazza Sonnino) With the sign 'Vini & Olio' scrawled above the door, this is a hole-in-the-wall Trastevere institution, lined by bottles of grappa and wine-for-sale, but also offering wine and spirits by the glass, shots and beer on tap. The owner is a funny guy, and has been serving up tipples since 1972. In the evening, drinkers spill out on the cobbled Trastevere street.

Freni e Frizioni Bar

(Map p250; 06 4549 7499; www.freniefrizioni.com; Via del Politeama 4-6; 6.30pm-2am; Piazza Trilussa) This perennially cool Trastevere bar is housed in a former mechanic's workshop – hence its name (Brakes and Clutches). It draws a young *spritz*-loving crowd that swells onto the small piazza outside to sip well-priced cocktails (from €7) and to snack on the daily *aperitivo* (€6 to €10, 7pm to 10pm).

Ombre Rosse Bar

(06 588 41 55; Piazza Sant'Egidio 12; 8am-2am Mon-Sat, 11am-2am Sun; Piazza Trilussa) A seminal Trastevere hang-out; grab a table on the terrace and watch the world go by amid a clientele ranging from elderly Italian wide boys to wide-eyed tourists. Tunes are slinky and there's live music (jazz, blues, world) on Thursday evenings from September to April.

Cocktails & Digestives

Cocktail bars are the current buzz in Rome, some featuring special local creations such as the Carbonara Sour at Co.So (p174), which has vodka infused with pork fat in a homage to the classic Roman pasta sauce.

Popular aperitifs are based on bitter alcoholic liqueurs, such Campari soda or Aperol *spritz*, which mixes Aperol with *prosecco*. Crodino is a herbal, medicinal-tasting nonalcoholic aperitif. Italians love to finish off a meal with a digestif. The best of these aren't shop bought, so if it's '*fatta in casa*' (homemade), give it a try.

Crodino

Big Star Bar

(Via Goffredo Mameli 25; 6pm-2am; Viale de Trastevere, Viale de Trastevere) Off the beaten Trastevere track, this is a cool back-street bar set away from the main action, with an alternative feel and the drink prices scrawled up on a blackboard. It's a small yet airy interior, where you can drink a wide range of beers and cocktails while listening to the hipster DJs, with a laid-back, appealing vibe.

Bar Stuzzichini Bar

(Piazzale Giuseppe Garibaldi; 7.30am-1am or 2am; Passeggiata del Gianicolo) This little kiosk nestles on the top of Gianicolo, and serves up coffees and drinks, including cocktails. There are a few tables to perch at and the views are unmatchable. On New Year's Eve it opens all night.

Coffee

Drinking an espresso standing at a bar shoulder to shoulder with the locals is a quintessential Roman experience. But to enjoy it you'll need to know what to order.

For an espresso (a shot of strong black coffee), ask for *un caffè;* if you want it with a drop of hot or cold milk, order *un caffè macchiato* ('stained' coffee). Long black coffee (as in a watered-down version) is known as *caffè lungo* (an espresso with more water) or *caffè all'american* (a filter coffee).

Then, of course, there's the cappuccino (served warm rather than hot), a staple of many Roman breakfasts. Italians drink cappuccino only during the morning and never after meals.

In summer, *cappuccino freddo* (iced coffee with milk, usually already sugared), *caffè freddo* (iced espresso) and *granita di caffè* (frozen coffee, usually with cream) top the charts.

A *caffè latte* is a milkier version of the cappuccino with less froth; a *latte macchiato* is even milkier (warmed milk 'stained' with a spot of coffee).

La Mescita Wine Bar

(Map p250; ☎06 5833 3920; Piazza Trilussa 41; ⏰5pm-midnight Sun-Thu, to 1am Fri & Sat; 🚌Piazza Trilussa) This tiny bar inside the entrance to upmarket restaurant Enoteca Ferrara serves delectable *aperitivo* and has a wide range of wines by the glass, from €7. Fancy an intimate tête-à-tête, with fine wines and yummy snacks? This is your place.

Il Baretto Bar

(☎06 5836 5422; Via Garibaldi 27; ⏰6am-2am Mon-Sat, 5pm-2am Sun; 🚌Piazza Sonnino, 🚋Piazza Sonnino) Venture a little way up the Gianicolo, up a steep flight of steps from Trastevere. Go on, it's worth it: you'll discover this well-kept-secret cocktail bar. The basslines are meaty, the bar staff hip, and the interior mixes vintage with pop art.

Bar le Cinque Bar

(Vicolo del Cinque 5; ⏰6.30am-2am Mon-Sat; 🚌Piazza Sonnino) There's no sign outside, and it looks like a rundown ordinary bar, but this is a long-standing Trastevere favourite and always has a small crowd clustered around outside; they're here for the pivotal location, easygoing vibe and cheap drinks.

San Giovanni & Testaccio

Rec 23 Bar

(☎06 8746 2147; www.rec23.com; Piazza dell'Emporio 2; ⏰6.30pm-2am daily & 12.30-3.30pm Sat & Sun; 🚌Via Marmorata) All plate glass and exposed brick, this popular, NY-inspired venue caters to all moods, serving aperitifs, restaurant meals and a weekend brunch. Arrive thirsty to take on the Testaccio Mule, one of a long list of cocktails, or get to grips with the selection of Scottish whiskies and Latin American rums. It also hosts regular live gigs.

Big Bang Club

(www.bigbangroma.org; Via di Monte Testaccio 22; ⏰10pm-4.30am Fri & Sat; 🚌Via Galvani) For one of the capital's best reggae parties, head to the Bababoomtime Friday-night session at Big Bang. The club, housed in Rome's graffiti-sprayed former slaughterhouse, draws a casual, music-loving crowd who know their reggae, dancehall, dub and techno.

L'Oasi della Birra Bar

(☎06 574 61 22; Piazza Testaccio 41; ⏰4.30pm-2am; 🚌Via Marmorata) In a local bottle shop, this cramped cellar bar is exactly what it says it is – an Oasis of Beer. With up to 500 labels, from Teutonic heavyweights to boutique brews, as well as wines, cheeses, cold cuts and the like, it's ideally set up for elbow-to-elbow quaffing.

Coming Out Bar
(www.comingout.it; Via di San Giovanni in Laterano 8; 7am-2am; Via Labicana) On warm evenings, with lively crowds on the street and the Colosseum as a backdrop, there are few finer places to sip than this friendly, gay bar. It's open all day but at its best in the evening when the the atmosphere hots up and the gigs, drag shows and karaoke nights get under way.

Il Pentagrappolo Wine Bar
(Via Celimontana 21b; noon-3pm & 6pm-1am Tue-Fri, 6pm-1am Sat & Sun; MColosseo) This vaulted, softly lit bar is the perfect antidote to sightseeing overload. Join the mellow crowd for an evening of wine and jazz courtesy of the frequent live gigs. There's also lunch and a daily *aperitif*.

L'Alibi Club
(Via di Monte Testaccio 44; 11.30pm-5am Thu-Sun; Via Galvani) A historic gay club, L'Alibi does high camp with style, putting on kitsch shows and playing house, techno and dance to a mixed gay and straight crowd. It's spread over three floors and if the sweaty atmosphere on the dance floors gets too much, head up to the spacious summer terrace. Saturday's Tommy Night is the hot date right now.

My Bar Bar
(Via di San Giovanni in Laterano 12; 9am-2am; Via Labicana) On the nearest thing that Rome has to a gay street, this cafe near the Colosseum serves mainly tourists during the day and a gay clientele in the evenings. DJs and live music fuel the fun.

Villa Borghese & Northern Rome

Momart Bar
(www.momartcafe.it; Viale XXI Aprile 19; noon-2am Mon-Fri, 6pm-2am Sat & Sun; Via XXI Aprile) A spacious modern bar in the university district near Via Nomentana, Momart serves one of Rome's most popular

> *drinking an espresso standing at a bar is a quintessential Roman experience*

Barista preparing coffee

KATHRIN ZIEGLER/GETTY IMAGES ©

PAOLO CORDELLI/GETTY IMAGES ©

aperitifs. A mixed crowd of students and local professionals flocks here to fill up on the pizza-led buffet and kick back over cocktails on the pavement terrace.

Brancaleone Club
(www.brancaleone.eu; Via Levanna 11; hours vary, typically 10.30pm-late; Via Nomentana) From its anti-establishment roots as a *centro sociale* (social centre), Brancaleone has grown to become one of Rome's top clubs, drawing blockbuster DJs and a young clubbing crowd. Rap, hip-hop, drum and bass, and electronica feature heavily, and there's a regular calendar of events and one-off evenings. The club is in the outlying Montesacro district.

Lanificio 159 Club
(www.lanificio159.com; Via Pietralata 159a; hours vary, typically 11pm-4am; Via Val Brembana) Occupying an ex-wool factory in Rome's northeastern suburbs, this cool underground venue hosts live gigs and hot clubbing action, led by top Roman crews and international DJs. The club is part of a larger complex which stages more reserved events such as Sunday markets, exhibitions and aperitifs.

Chioschetto di Ponte Milvio Bar
(Ponte Milvio 44; 6pm-2am summer only; Ponte Milvio) A local landmark, this green kiosk next to the Ponte Milvio bridge is perennially popular with the young crowd from Rome's wealthy northern suburbs. It might look like a shack – it is a shack – but the mojitos are the business and it does an excellent thirst-quenching *grattachecca* (shaved ice flavoured with fruit syrup).

Piper Club Club
(www.piperclub.it; Via Tagliamento 9; 11pm-5am Fri & Sat; Viale Regina Margherita) To Rome what Studio 54 was to New York, Piper has been a nightlife fixture for 50 years, and it just keeps on going. Fridays it hosts themed parties, everything from Latin nights to '90s house celebrations, while on Saturdays resident DJs drive the rhythms.

Southern Rome

Porto Fluviale Bar

(☎06 574 31 99; www.portofluviale.com; Via del Porto Fluviale 22; ⌚10.30am-2am; Ⓜ Piramide) A large bar in a converted factory, this has an ex-industrial look – dark-green walls and a brickwork floor – and is a relaxing, appealing place for morning coffee, *aperitivo* or an evening drink to a soundtrack of plinky jazz. In line with Rome's current love of artisanal brews, it serves its own Porto Fluviale craft beer (medium €5.50).

Goa Club

(☎06 574 82 77; www.goaclub.com; Via Libetta 13; ⌚11.30pm-4.30am Thu-Sat; Ⓜ Garbatella) Goa is Rome's serious superclub, with international names, ethnic styling, a fashion-forward crowd, podium dancers and heavies on the door.

Neo Club Club

(Via degli Argonauti 18; ⌚11pm-4am Fri & Sat; Ⓜ Garbatella) This small, dark two-level club has an underground feel and is one of the funkiest choices in the zone, featuring a dancetastic mishmash of breakbeat, techno and old-skool house.

La Saponeria Club

(☎06 574 69 99; Via degli Argonauti 20; ⌚11pm-4.30am Tue-Sun Oct-May; Ⓜ Garbatella) Formerly a soap factory, nowadays La Saponeria is a cool space that's all exposed brick and white walls and brain-twisting light shows. It lathers up the punters with guest DJs spinning everything from nu-house to nu-funk, minimal techno, dance, hip-hop and 1950s retro.

Craft Beer

In recent years beer drinking has really taken off in Italy, and especially in Rome, with specialised bars and restaurants offering microbrewed beers. Local favourites include Birradamare in Fiumicino, Porto Fluviale in Ostiense, and Birra Del Borgo in Rieti (on the border between Lazio and Abruzzo), which opened local beer haunts **Bir & Fud** (Via Benedetta 23; meals €25; ⌚7.30pm-midnight, to 2am Fri & Sat; ❄; 🚌Piazza Trilussa) and Open Baladin (p171). Local beers reflect the seasonality that's so important in Rome – for example, look for winter beers made from chestnuts.

Other important addresses on the artisanal beer trail include Porto Fluviale, Ma Che Siete Venuti a Fà (p179) and Birra Piu (p174).

Rashomon Club

(www.rashomonclub.com; Via degli Argonauti 16; ⌚11pm-4am Fri & Sat Oct-May; Ⓜ Garbatella) Rashomon is sweaty, not posey, and where to head when you want to dance your ass off. Shake it to a music-lovers' feast of the sound of the underground, especially house, techno and electronica.

SHOWTIME

Opera, jazz, theatre and more

Showtime

Watching the world go by in Rome is often entertainment enough, but there's plenty more to having a good time in Rome than people-watching. The city's music scene is a hive of activity with gigs and concerts of all musical genres drawing knowledgeable and enthusiastic audiences. Theatres put on everything from Shakespearian drama to avant-garde dance, while cinemas screen art-house flicks and arts festivals turn the city into a stage, particularly in summer when al fresco performances play out against backdrops of spectacular Roman ruins.

So whether you're an opera buff or a rapper, a cinephile or a theatregoer, you're sure to find something to suit your taste.

In This Section

Tickets

Tickets for concerts, live music and theatrical performances are widely available across the city. Hotels can often reserve tickets for guests, or you can contact the venue or organisation directly. Otherwise you can try the following:

Hellò Ticket (www.helloticket.it)

Orbis (Map p255; Piazza dell'Esquilino 37)

Opera performers in Giuseppe Verdi's *Nabucco*, Terme di Caracalla (p98)

The Best...

Classical Venues

Auditorium Parco della Musica (p192) Top international musicians and multiple concert halls.

Teatro dell'Opera di Roma (p189) Red-velvet and gilt interior for Rome's opera and dance companies.

Terme di Caracalla (p98) Wonderful outdoor setting for Rome's opera and ballet companies.

Auditorium Conciliazione (p189) Classical and contemporary concerts, cabarets, dance, theatre, film and exhibitions.

Teatro Olimpico (p192) Home to the Accademia Filarmonica Romana.

Jazz

Alexanderplatz (p189) Rome's premier jazz club stages international and local musicians.

Charity Café (p190) An intimate space, hosting regular live gigs.

Big Mama (p190) An atmospheric Trastevere venue for jazz, blues, funk, soul and R & B.

Gregory's (p188) Popular with local musicians, a smooth venue close to the Spanish Steps.

Fonclea (p189) Pub venue regularly hosting live jazz, moving riverside in the summer.

☆ Centro Storico

Teatro Argentina Theatre

(Map p250; 06 684 00 03 11; www.teatrodiroma.net; Largo di Torre Argentina 52; tickets €16-29; Largo di Torre Argentina) Founded in 1732, Rome's top theatre is one of the two official homes of the Teatro di Roma (the other is the Teatro India in the southern suburbs). Rossini's *Barber of Seville* premiered here in 1816; today it stages a wide-ranging program of drama (mostly in Italian), high-profile dance performances and classical-music concerts.

Isola del Cinema Cinema

(www.isoladelcinema.com) Independent films in the romantic outdoor setting of the Isola Tiberina in July and August. This runs in conjunction with the riverside Lungo il Tevere festival.

Teatro dell'Orologio Theatre

(06 687 55 50; www.teatroorologio.com; Via dei Filippini 17a; Corso Vittorio Emanuele II) A well-known experimental theatre, the three-stage Orologio offers a varied program of contemporary and classic works, including occasional performances in English.

☆ Tridente, Trevi & the Quirinale

Gregory's Live Music

(Map p255; 06 679 63 86; www.gregorysjazz.com; Via Gregoriana 54d; 7pm-2am Tue-Sun Sep-Jun; M Barberini, Spagna) If Gregory's were a tone of voice, it'd be husky: unwind in the downstairs bar, then unwind some more on squashy sofas upstairs to some slinky live jazz and swing, with quality local performers, who also like to hang out here.

Teatro Quirino Theatre

(Map p255; 06 679 45 85; www.teatroquirino.it; Via delle Vergini 7; Via del Tritone) Within splashing distance of the Trevi Fountain, this grand 19th-century theatre produces the odd new work and a stream of well-known classics – expect to see works (in Italian) by Arthur Miller, Tennessee

Teatro Argentina

Williams, Shakespeare, Seneca and Luigi Pirandello.

Teatro Sistina Theatre
(Map p255; 06 420 07 11; www.ilsistina.com; Via Sistina 129; M Barberini) Big-budget theatre spectaculars, musicals, concerts and comic star turns are the staples of the Sistina's ever-conservative, ever-popular repertoire.

☆ Vatican City, Borgo & Prati

Alexanderplatz Jazz
(06 3972 1867; www.alexanderplatzjazzclub.com; Via Ostia 9; 8.30pm-2am, concerts 9.45pm; M Ottaviano-San Pietro) Small, intimate and underground, Rome's most celebrated jazz club draws top Italian and international performers and a respectful cosmopolitan crowd. Book a table for the best stage views or if you want to dine to the tunes. Check the website for upcoming gigs.

Auditorium Conciliazione Live Performance
(06 3281 0333; www.auditoriumconciliazione.it; Via della Conciliazione 4; Piazza Pia) On the main approach road to St Peter's Basilica, this large auditorium plays host to a wide range of events – classical and contemporary concerts, cabarets, dance spectacles, theatre productions, film screenings and exhibitions.

Fonclea Live Music
(06 689 63 02; www.fonclea.it; Via Crescenzio 82a; 7pm-2am Sep-May; Piazza del Risorgimento) Fonclea is a great little pub venue, serving up nightly gigs by bands playing everything from jazz and soul to funk, rock and Latin (concerts start at around 9.30pm). Get in the mood with a drink during happy hour (7pm to 8.30pm daily). From June to August, the pub ups sticks and moves to a site by the Tiber.

Jazz, Rock & Rap

Jazz, introduced by US troops during WWII, grew in popularity during the postwar period and took off in the 1960s. Since then, it has gone from strength to strength and the city now boasts some fabulous jazz clubs, including Alexanderplatz, Big Mama (p190) and the Casa del Jazz (p192).

Major music concerts are staged at the Auditorium Parco della Musica as well as arenas and stadiums, such as the Stadio Olimpico (p198) and the racetrack on Via Appia Nuova, the Ippodromo La Capannelle.

The *centri sociali*, alternative arts centres set up in venues across Rome, are also good places to catch a gig, especially Brancaleone (p182) in northern Rome, with music policies encompassing hip-hop, electro, dubstep, reggae and dancehall.

Alexanderplatz

☆ Monti, Esquilino & San Lorenzo

Teatro dell'Opera di Roma Opera
(Map p255; 06 481 70 03; www.operaroma.it; Piazza Beniamino Gigli; ballet €12-80, opera €17-150; 9am-5pm Tue-Sat, to 1.30pm Sun; M Repubblica) Rome's premier opera house boasts a plush and gilt interior, a Fascist 1920s exterior and an impressive history: it premiered Puccini's *Tosca,* and Maria Callas once sang here. Opera and ballet

Opera & Classical Music

Rome's abundance of beautiful settings makes it a superb place to catch a concert. Many international stars play at the Auditorium Parco della Musica (p192), a state-of-the-art, Renzo Piano–designed complex. However, there are often creative uses of other spaces, and in recent years there have been major gigs on the ancient racetrack **Circo Massimo** (Circus Maximus; Map p252; Via del Circo Massimo; MCirco Massimo).

Music in Rome is not just about the Auditorium Parco della Musica (p192). There are concerts by the Accademia Filarmonica Romana at the Teatro Olimpico (p192); the Auditorium Conciliazione (p189), Rome's premier classical-music venue before the newer Auditorium was opened; and the Istituzione Universitaria dei Concerti, holds concerts in the Aula Magna of La Sapienza University.

Classical-music performances – often free – are regularly held in churches, especially around Easter, Christmas and the New Year.

Rome's opera house, the Teatro dell'Opera di Roma (p189), is a magnificent, grandiose venue, but productions can be a bit hit-and-miss. It's also home to Rome's official Corps de Ballet and has a ballet season running in tandem with its opera performances. Both ballet and opera move outdoors for the summer season to the ancient Terme di Caracalla (p98), an even more spectacular setting.

performances are staged between September and June.

Blackmarket Live Music
(Map p255; www.black-market.it; Via Panisperna 101; 5.30pm-2am; MCavour) A bit outside the main Monti hub, this charming, living-room-style bar filled with eclectic vintage furniture is a small but rambling place, great for sitting back on mismatched armchairs for a leisurely, convivial drink. It hosts regular acoustic indie and folk gigs, which feel a bit like having a band in your living room.

Charity Café Live Music
(Map p255; 328 8452915; www.charitycafe.it; Via Panisperna 68; 7pm-2am; MCavour) Think narrow space, spindly tables, dim lighting and a laid-back vibe: this is a place to snuggle down and listen to some slinky live jazz. Civilised, relaxed, untouristy and very Monti. Gigs usually take place from 10pm, with live music and *aperitivo* on Sundays. There's open mic from 7pm on Monday and Tuesday.

Istituzione Universitaria dei Concerti Live Music
(IUC; 06 361 00 51; www.concertiiuc.it; Piazzale Aldo Moro 5; MCastro Pretorio) The IUC organises a season of concerts in the Aula Magna of La Sapienza University, including many visiting international artists and orchestras. Performances cover a wide range of musical genres, including baroque, classical, contemporary and jazz.

Teatro Ambra Jovinelli Theatre
(06 8308 2884; www.ambrajovinelli.org; Via G Pepe 43-47; MVittorio Emanuele) A home from home for many famous Italian comics, the Ambra Jovinelli is a historic venue for alternative comedians and satirists. Between government-bashing, the theatre hosts productions of classics, musicals, opera, new works and the odd concert.

☆ Trastevere & Gianicolo

Big Mama Blues
(06 581 25 51; www.bigmama.it; Vicolo di San Francesco a Ripa 18; 9pm-1.30am, shows 10.30pm, closed Jun-Sep; Viale di Trastevere, Viale di Trastevere) Head to this cramped Trastevere basement for a mellow night of Eternal City blues. A long-standing venue,

Teatro dell'Opera di Roma (p189)

it also stages jazz, funk, soul and R & B, as well as popular Italian cover bands.

Lettere Caffè Gallery Live Music
(06 9727 0991; www.letterecaffe.org; Vicolo di San Francesco a Ripa 100/101; 7pm-2am, closed mid-Aug–mid-Sep; Viale di Trastevere, Viale di Trastevere) Like books? Poetry? Blues and jazz? Then you'll love this place – a clutter of bar stools and books, where there are regular live gigs, poetry slams, comedy and gay nights, plus DJ sets playing indie and new wave.

Alcazar Cinema Cinema
(06 588 00 99; Via Merry del Val 14; Viale di Trastevere, Viale di Trastevere) This old-style cinema with plush red seats occasionally shows films in their original language with Italian subtitles.

Nuovo Sacher Cinema
(06 581 81 16; www.sacherfilm.eu; Largo Ascianghi 1; Viale di Trastevere, Viale di Trastevere) Owned by cult Roman film director Nanni Moretti, this small, red-velvet-seated cinema is the place to catch the latest European art-house offering, with regular screenings of films in their original language.

Teatro Vascello Theatre
(06 588 10 21; www.teatrovascello.it; Via Giacinto Carini 72, Monteverde; Via Giacinto Carini) Left-field in vibe and location, this independent, fringe theatre stages interesting, cutting-edge new work, including avant-garde dance, multimedia events and works by emerging playwrights.

☆ San Giovanni & Testaccio

ConteStaccio Live Music
(www.contestaccio.com; Via di Monte Testaccio 65b; 7pm-4am Tue-Sun; Via Galvani) With an under-the-stars terrace and cool, arched interior, ConteStaccio is one of the top venues on the Testaccio clubbing strip. It's something of a multipurpose outfit, with a cocktail bar, a pizzeria and a restaurant but, is best known for its daily concerts. Gigs by emerging groups set the tone, spanning indie, rock, acoustic, funk and electronic.

☆ Villa Borghese & Northern Rome

Auditorium Parco della Musica Concert Venue

(☎06 8024 1281; www.auditorium.com; Viale Pietro de Coubertin 30; 🚋Viale Tiziano) The hub of Rome's thriving cultural scene, the Renzo Piano-designed Auditorium is the capital's premier concert venue and one of Europe's most popular arts centres. Its three concert halls offer superb acoustics, and, together with a 3000-seat open-air arena, stage everything from classical-music concerts to jazz gigs, public lectures, and film screenings.

The Auditorium is also home to Rome's world-class Orchestra dell' Accademia Nazionale di Santa Cecilia (www.santacecilia.it).

Teatro Olimpico Theatre

(☎06 326 59 91; www.teatroolimpico.it; Piazza Gentile da Fabriano 17; 🚌Piazza Mancini, 🚋Piazza Mancini) The Teatro Olimpico is home to the Accademia Filarmonica Romana (www.filarmonicaromana.org), a classical-music organisation whose past members have included Rossini, Donizetti and Verdi. The theatre offers a varied program of classical and chamber music, opera, ballet, one-man shows and comedies.

Silvano Toti Globe Theatre Theatre

(Map p256; ☎06 06 08; www.globetheatreroma.com; Largo Aqua Felix, Villa Borghese; tickets €10-23; 🚌Piazzale Brasile) Like London's Globe Theatre but with better weather, Villa Borghese's open-air Elizabethan theatre serves up Shakespeare (performances mostly in Italian) from July through to September.

☆ Southern Rome

La Casa del Jazz Jazz

(☎06 70 47 31; www.casajazz.it; Viale di Porta Ardeatina 55; admission varies; ⊙gigs around 8-9pm; Ⓜ Piramide) In the middle of a 2500-sq-metre park in the southern suburbs, the Casa del Jazz is housed in a three-storey 1920s villa that once belonged

Auditorium Parco della Musica

to a Mafia boss. When he was caught, the Comune di Roma (Rome Council) converted it into a jazz-fuelled complex, with a 150-seat auditorium, rehearsal rooms, a cafe and a restaurant. Some events are free.

Caffè Letterario Live Music
(☎06 5730 2842, 340 3067460; www.caffeletterarioroma.it; Via Ostiense 83, 95; ⌚10am-2am Tue-Fri, 4pm-2am Sat & Sun; 🚌Via Ostiense) Caffè Letterario is an intellectual hang-out housed in the funky converted, post-industrial space of a former garage. It combines designer looks, a bookshop, a gallery, performance space and a lounge bar. There are regular gigs from 10pm to midnight, ranging from soul and jazz to Indian dance.

XS Live Live Music
(☎06 5730 5102; www.xsliveroma.com; Via Libetta 13; ⌚11.30pm-4am Thu-Sun Sep-May; MGarbatella) A rocking live-music and club venue, hosting regular gigs. Big names playing here in recent times range from Peter Doherty to Jefferson Starship, and club nights range from Cool Britanni–themed trips to '80s odysseys.

Teatro India Theatre
(☎06 8400 0311; www.teatrodiroma.net; Lungotevere dei Papareschi; tickets €10-30; 🚌Via Enrico Fermi) Inaugurated in 1999 in the postindustrial landscape of Rome's southern suburbs, the India is the younger sister of Teatro Argentina. It's a stark modern space in a converted industrial building, a fitting setting for its cutting-edge program, with a calendar of international and Italian works.

Theatre

Rome has a thriving local theatre scene, with theatres including both traditional places and smaller experimental venues. Performances are usually in Italian.

Particularly wonderful are the summer festivals that make use of Rome's archaeological scenery. Performances take place in settings such as Ostia Antica's Roman theatre and the Teatro di Marcello. In summer the **Miracle Players** (☎06 7039 3427; www.miracleplayers.org) perform classic English drama or historical comedy in English next to the Roman Forum and other open-air locations. Performances are usually free.

Teatro di Marcello
KEN WELSH/GETTY IMAGES ©

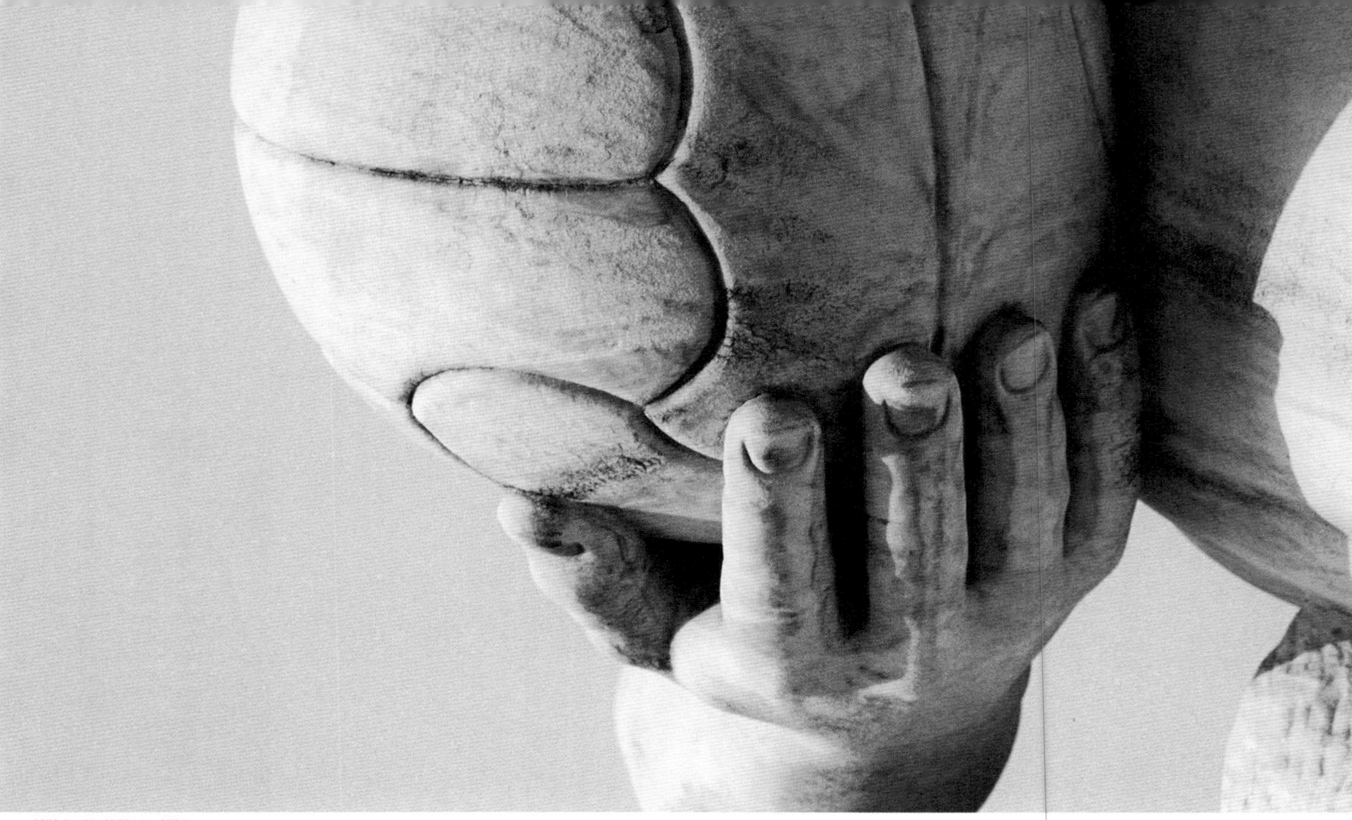

ACTIVE ROME

From football to cooking courses

Active Rome

The Romans have long been passionate about sport. Ever since crowds flocked to the Colosseum to support their favourite gladiators and to the Circo Massimo to cheer on chariot riders, the locals have enjoyed a good performance. Rome's modern-day Colosseum is the Stadio Olimpico, where footballing rivalries are played out in front of thousands of fans and major-league rock stars strut their stuff.

If you prefer your pursuits more hands-on, there is a whole range of courses you can take, ranging from cooking in historic palazzi *(mansions) to wine tasting and language learning. Tours are also a popular activity and a good way of covering a lot of ground in a short time.*

In This Section

Sports Seasons

Football is the big sport in Rome with the season running from September to May. Rugby can be seen on weekends in February and March, while tennis fans can enjoy world-class games at the Italian Open Tennis Tournament in May.

Foro Italico (p198)

The Best...

Sports Venues

Foro Italico (p198) Magnificent Fascist-era sports complex.

Stadio Olimpico (p198) Rome's 70,000-seat football stadium, part of the fascist-era Foro Italico.

Piazza di Siena (p58) Lovely race-course in the heart of the Villa Borghese park.

Courses

Roman Kitchen (p200) Learn about the intricacies of Italian cuisine with a hands-on cooking course.

Vino Roma (p200) Sniff, sip and spit your way through Italian vintages on a wine course.

Art Studio Café (p201) Put your creativity to the test, learning the ancient art of mosaic-making.

Torre di Babele Centro di Lingua e Cultura Italiana (p201) Learn to speak the language in like-minded company.

Spectator Sports

Stadio Olimpico Stadium

(☎06 3685 7520; Viale dei Gladiatori 2, Foro Italico) A trip to Rome's impressive Stadio Olimpico offers an unforgettable insight into Rome's sporting heart. Throughout the football season (September to May) there's a game on most Sundays. Tickets cost from around €16, depending on the match, and can be bought at Lottomatica (lottery centres), the stadium, ticket agencies, www.listicket.it, or one of the many Roma or Lazio stores around the city. Note that ticket purchase regulations are far stricter than they used to be. Tickets have to bear the holder's name and passport or ID number, and you must present a photo ID at the turnstiles when entering the stadium.

Foro Italico Stadium

(☎800 622662; www.foroitalicoticketing.it; Viale del Foro Italico) The Fascist-era Foro Italico was built between 1928 and 1938 and originally named the Foro Mussolini. Foro Italico hosts the Italian Open, one of the most important events on the European tennis circuit, attracting the world's top players to the clay courts. Tickets can usually be bought at the Foro Italico each day of the tournament, except for the final days, which are sold out weeks in advance.

Stadio Flaminio Stadium

(☎06 3685 7309; www.federugby.it; Viale Maresciallo Pilsudski) Large concerts take place at Stadio Flaminio, a sports stadium.

Tours

Taking a tour is a good way to see a lot in a short time or investigating a sight in depth. There are several outfits running hop-on hop-off bus tours, typically costing about €20 per person. Both the Colosseum and Vatican Museums offer official guided tours, but for a more personalised service you'll be better off with a private guide.

A Friend in Rome Tour

(☎340 501 92 01; www.afriendinrome.it) Silvia Prosperi organises private tailor-made tours (on foot, by bike or scooter) to suit

Cyclists on Piazza della Chiesa Nuova, Centro Storico

your interests. She covers the Vatican and main historic centre as well as areas outside the capital. Rates are €50 per hour, with a minimum of three hours for most tours. She can also arrange kid-friendly tours, cooking classes, vintage-car tours and more.

Roman Guy Tour
(http://theromanguy.com) A professional set-up that organises a wide range of group and private tours. Packages, led by English-speaking experts, include early-bird visits to the Vatican Museums (US$84), foodie tours of Trastevere and the Jewish Ghetto (US$84), and a bar-hop through the historic centre's cocktail bars.

Dark Rome Tour
(☎06 8336 0561; www.darkrome.com) Runs a range of themed tours, costing from €25 to €150, including skip-the-line visits to the Colosseum and Vatican Museums, and semiprivate visits to the Sistine Chapel. Other popular choices include a Crypts and Catacombs tour, which takes in Rome's buried treasures, and a day trip to Pompeii.

Through Eternity Cultural Association Walking Tour
(☎06 700 93 36; www.througheternity.com) A reliable operator offering private and group tours led by English-speaking experts. Popular packages include a twilight tour of Rome's piazzas and fountains (€39, 2½ hours), a skip-the-line visit to the Vatican Museums by day/night (€56/66, 3½ hours) and a foodie tour of Testaccio (€80, four hours).

Arcult Walking Tour
(☎339 650 31 72; www.arcult.it) Run by architects, Arcult offers excellent customisable group tours focusing on Rome's contemporary architecture. Prices depend on the itinerary.

Roma Cristiana Walking Tour
(☎06 69 89 61; www.operaromanapellegrinaggi.org) Runs various tours, including guided visits to the Vatican Museums (adult/reduced €35/25) and two-hour tours of St Peter's Basilica (€14).

Football

Throughout the football season there's a game most Sundays at the Stadio Olimpico involving one of the city's two Serie A (Italy's premier league) teams: **Roma**, known as the *giallorossi* (yellow and reds; www.asroma.it), or **Lazio**, the *biancazzuri* (white and blues; www.sslazio.it).

The Rome derby is one of the football season's highest-profile games. The rivalry between Roma and Lazio is fierce and little love is lost between the fans. If you go to the stadium, make sure you get it right – Roma fans flock to the Curva Sud (Southern Stand), while Lazio supporters stand in the Curva Nord (Northern Stand). If you want to sit on the fence, head to the Tribuna Tevere or Tribuna Monte Mario.

To get to the Olimpico, take metro line A to Ottaviano–San Pietro and then bus 32.

A new Roma stadium is currently being built at Tor di Valle, due to be completed in time for the 2017/18 season.

A S Roma fans
MARCO IACOBUCCI EPP/SHUTTERSTOCK ©

Top Bike Rental & Tours Bicycle Tour
(☎06 488 28 93; www.topbikerental.com; Via Labicana 49; ⏱10am-7pm) Offers a series of bike tours throughout the city, including a four-hour 16km exploration of the city

Six Nations Rugby

Italy's rugby team, the Azzurri (the Blues), entered the Six Nations tournament in 2000, and has been the competition underdog ever since. However, it has scored some big wins in recent years, with shock wins over France in 2011 and 2013, and Scotland in 2015. But in 2016 normal business was resumed and the side finished with the wooden spoon.

Despite poor results, home games, which are played at the Stadio Olimpico, draw huge crowds of Azurri fans and high-spirited away supporters.

Six Nations rugby game, Stadio Olimpico (p198)

centre (€45) and an all-day 30km ride through Via Appia Antica and environs (€79). Out-of-town tours take in Castel Gandolfo, Civita di Bagnoregio and Orvieto.

Bici & Baci Tour
(Map p255; ☎06 482 84 43; www.bicibaci.com; Via del Viminale 5; ⌚8am-7pm) Bici & Baci runs daily bike tours of central Rome, taking in the historic centre, Campidoglio and the Colosseum, as well as tours on vintage Vespas and in classic Fiat 500 cars. Reckon on €49 for the bike tour, €145 for the Vespa ride, and €290 for the four-hour guided drive.

Rome Boat Experience Boat Tour
(☎06 8956 7745; www.romeboatexperience.com; adult/reduced €18/12) From April to October, hop-on, hop-off boats cruise along the Tiber. From May to October, there are also dinner cruises (€62, two hours) every Friday and Saturday, and a daily wine-bar cruise (€25, 1½ hours) from Monday to Thursday. Embarkation points are at Molo Sant'Angelo and Isola Tiberina.

Open Bus Cristiana Bus Tour
(www.operaromanapellegrinaggi.org; single tour €15, 24/48hr ticket €20/48; ⌚9am-6pm) The Vatican-sponsored Opera Romana Pellegrinaggi runs a hop-on, hop-off bus departing from Via della Conciliazione and Termini. Stops are situated near to main sights including St Peter's Basilica, Piazza Navona, the Trevi Fountain and the Colosseum. Tickets are available on board or at the meeting point just off St Peter's Sq.

Courses

Roman Kitchen Cooking Course
(Map p250; ☎06 678 57 59; www.italiangourmet.com; per day €200) Cookery writer Diane Seed *(The Top One Hundred Pasta Sauces)* runs cooking courses from her kitchen in Palazzo Doria Pamphilj. There are one-day, two-day, three-day and weeklong courses costing €200 per day and €1000 per week.

Vino Roma Course
(☎328 4874497; www.vinoroma.com; Via in Selci 84/G; 2hr tastings per person €50) With beautifully appointed 1000-year-old cellars and a chic tasting studio, Vino Roma guides novices and experts in tasting wine, under the knowledgeable stewardship of sommelier Hande Leimer and his expert team. Tastings are in English, but German, Japanese, Italian and Turkish sessions are available on special request. It also offers a wine-and-cheese dinner (€60), with snacks, cheeses and cold cuts to accompany the wines, as well as bespoke three-hour food tours. Book online.

Città di Gusto Cooking Course
(☎06 5511 2264; www.gamberorosso.it; Via Fermi 161) Demonstrations, workshops, lessons and courses are held at this foodie complex.

Art Studio Café Art Course

(☎06 3260 9104; www.artstudiocafe.it; Via dei Gracchi 187a) This cafe, exhibition space and mosaic school offers a range of classes. Learn how to cut and glaze enamel tesserae, how to mix colours and how best to mount your final composition. One-day sessions start at €50.

Torre di Babele Centro di Lingua e Cultura Italiana Language Course

(☎06 4425 2578; www.torredibabele.com; Via Cosenza 7) As well as language lessons, you can take courses on cooking, art, architecture and several other subjects.

Divulgazione Lingua Italiana Language Course

(☎06 446 25 93; www.dilit.it; Via Marghera 22) School offering a range of language and cultural courses.

Spas

Hotel De Russie Wellness Zone Spa

(Map p256; ☎06 3288 8820; www.hotelderussie.it; Via del Babuino 9; ⏲6.30am-10pm; Ⓜ Flaminio) This glamorous and gorgeous day spa is in one of Rome's best hotels, and allows admission to the gym and steam room. Treatments are also available, including shiatsu and deep-tissue massage (a 50-minute massage costs around €95).

Kami Spa Spa

(Map p255; ☎06 42010039; www.kamispa.com; Via Degli Avignonesi 11-12; massage €120-160; Ⓜ Barberini) A chic, soothing spa not far from the Trevi Fountain, this is a great place to recharge your batteries if you want to fork out for the luxurious prices.

REST YOUR HEAD

Top tips for the best accommodation

Rest Your Head

With a wide range of hotels, pensioni *(pensions) B&Bs, hostels and convents, Rome has accommodation to please everyone, from the fussiest prince to the most impecunious nun. At the top end of the market, opulent five-star hotels occupy stately historic palazzi (mansions) and chic boutique guesthouses boast discreet luxury. Family-run B&Bs and pensioni offer character and a warm welcome while religious houses cater to pilgrims and budget-minded travellers. Hostel goers can choose between party-loving hang-outs or quieter, more restrained digs.*

But while there's plenty of choice, rates are universally high in Rome and you'll need to book early to get the best deal.

In This Section

Tipping

Tipping is not necessary, but up to €5 for porter, housekeeping or room service in a top-end hotel is fine.

ERIC VANDEVILLE/GAMMA-RAPHO/GETTY IMAGES ©

Luxurious hotel room, Trevi

Reservations

- Always try to book ahead, especially for the major religious festivals.
- Ask for a *camera matrimoniale* for a room with a double bed or a *camera doppia* for twin beds.

Checking In & Out

- When you check in you'll need to present your passport or ID card.
- Checkout is usually between 10am and noon. In hostels it's around 9am.
- Some guesthouses and B&Bs require you to arrange a time to check in.

Accommodation Types

Pensioni & Hotels

The bulk of Rome's accommodation consists of *pensioni* (pensions) and *alberghi* (hotels).

A *pensione* is a small, family-run hotel, often in a converted apartment. Rooms are usually fairly simple, though most come with a private bathroom.

Hotels are rated from one to five stars, though this rating relates to facilities only and gives no indication of value, comfort, atmosphere or friendliness. Most hotels in Rome's historic centre tend to be three-star and up. As a rule, a three-star room will come with a hairdryer, a minibar (or fridge), a safe, air-con and wi-fi. Some may also have satellite TV.

A common complaint in Rome is that hotel rooms are small. This is especially true in the *centro storico* and Trastevere, where many hotels are housed in centuries-old *palazzi* (mansions). Similarly, a spacious lift is a rare find, particularly in older *palazzi*, and you'll seldom find one that can accommodate more than one average-sized person with luggage.

Breakfast in cheaper hotels is rarely worth setting the alarm for, so if you have the option, save a few bob and pop into a bar for a coffee and *cornetto* (croissant).

B&Bs & Guesthouses

Alongside traditional B&Bs, Rome has many boutique-style guesthouses offering chic, upmarket accommodation at midrange to top-end prices. Breakfast in a Roman B&B usually consists of bread rolls, croissants, yoghurt, ham and cheese.

Hostels

Rome's hostels cater to everyone from backpackers to budget-minded families. Many offer traditional dorms as well as smart hotel-style rooms (singles, doubles, even family rooms) with private bathrooms. Curfews are a thing of the past and some places even have 24-hour receptions. Many hostels don't accept prior reservations for

Views over Rome on a from a hotel rooftop

dorm beds, so arrive after 10am and it's first come, first served.

Religious Accommodation

Unsurprisingly, Rome is well furnished with religious institutions, many of which offer cheap(-ish) rooms for the night. Bear in mind, though, that many have strict curfews and that the accommodation, while spotlessly clean, tends to be short on frills. Also, while there are a number of centrally located options, many convents are situated out of the centre, typically in the districts north and west of the Vatican. Book well in advance.

Rental Accommodation

For longer stays, renting an apartment will generally work out cheaper than an extended hotel sojourn. Bank on about €900 per month for a studio apartment or one-bedroom flat. For longer stays, you'll probably have to pay bills plus a building maintenance charge.

Seasons & Rates

Rome doesn't have a low season as such but rates are at their lowest from November to March (excluding Christmas and New Year) and from mid-July through August. Expect to pay top whack in spring (April to June) and autumn (September and October) and over the main holiday periods (Christmas, New Year and Easter).

Most midrange and top-end hotels accept credit cards. Budget places might, but it's always best to check in advance.

Hotel Tax

Everyone overnighting in Rome has to pay a room-occupancy tax on top of their regular accommodation bill. This amounts to the following:

- €3 per person per night in one- and two-star hotels.
- €3.50 in B&Bs and room rentals.
- €4/6/7 in three-/four-/five-star hotels.

The tax is applicable for a maximum of 10 consecutive nights. Prices in reviews do not include the tax.

Online Rentals

Holiday rentals are booming in Rome right now thanks to online outfits like AirBnb and VRBO.

These sites offer a vast range of options, from single rooms in private houses to fully equipped apartments. They are often good value and will almost certainly save you money, especially in expensive areas such as the historic centre or Trastevere. Alternatively, they give the chance to get away from the touristy hot spots and see another side to the city – characterful neighbourhoods include Testaccio and Garbatella. Just make sure to research the location when you book and, if necessary, work out how you're going to get there (public transport, lift from the property owner etc).

Always make sure to check the property's reviews for things like noise (an issue in central locations) and privacy. You'll also need to check whether Rome's obligatory hotel tax is included in the rate or has to be paid separately.

Table settings on a hotel rooftop
PAOLO CORDELLI/GETTY IMAGES ©

Getting There

Most tourist areas are a bus ride or metro journey away from Stazione Termini. If you come by car, be warned that much of the city centre is a ZTL (limited traffic zone)

PAOLO CORDELLI/GETTY IMAGES ©

and off limits to unauthorised traffic. Note also that there is a terrible lack of on-site parking facilities in the city centre, although your hotel should be able to direct you to a private garage. Street parking is not recommended.

Useful Websites

Lonely Planet (www.lonelyplanet.com/italy/rome/hotels) Author-reviewed accommodation options and online booking.

060608 (www.060608.it/en/accoglienza/dormire) Official Comune di Roma site with accommodation lists; details not always up to date.

B&B Association of Rome (www.b-b.rm.it) Lists B&Bs and short-term apartment rentals.

B&B Italia (www.bbitalia.it) Rome's longest-established B&B network.

Rome As You Feel (www.romeasyoufeel.com) Apartment rentals, from cheap studio flats to luxury apartments.

Where to Stay

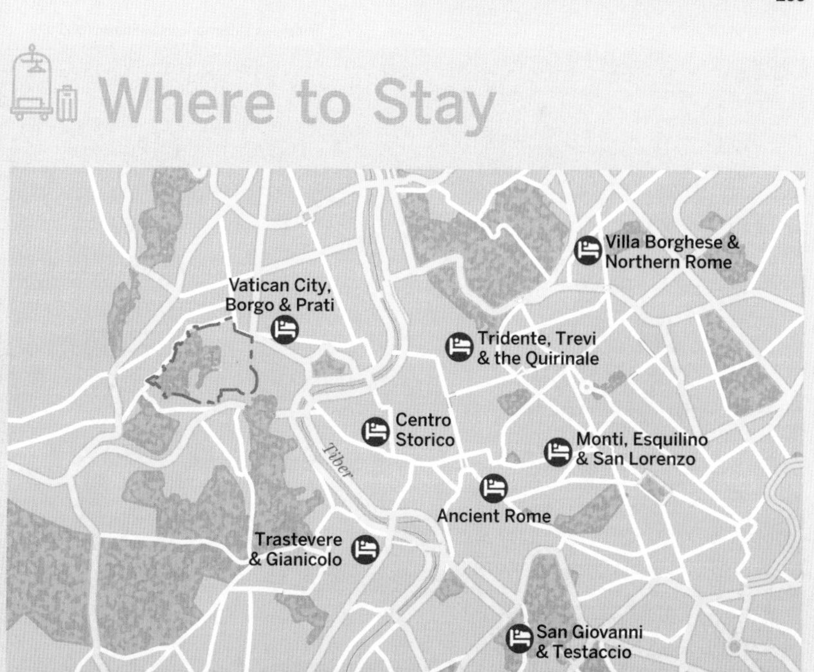

Neighbourhood	Atmosphere
Ancient Rome	Close to major sights such as Colosseum, Roman Forum and Capitoline Museums; quiet at night; not cheap; restaurants are touristy.
Centro Storico	Atmospheric area with everything on your doorstep: Pantheon, Piazza Navona, restaurants, bars, shops; most expensive part of town; can be noisy.
Tridente, Trevi & the Quirinale	Good for Spanish Steps, Trevi Fountain and designer shopping; excellent midrange to top-end options; good transport links; subdued after dark.
Monti, Esquilino & San Lorenzo	Lots of budget accommodation around Stazione Termini; top eating in Monti and good nightlife in San Lorenzo; good transport links; some dodgy streets near Termini.
San Giovanni & Testaccio	Authentic atmosphere with good eating and drinking options; Aventino is a quiet, romantic area; Testaccio is a top food and nightlife district; not many big sights.
Trastevere & Gianicolo	Gorgeous, atmospheric area; party vibe with hundreds of bars, cafes and restaurants; some interesting sights; expensive; noisy, particularly on summer nights.
Vatican City, Borgo & Prati	Near St Peter's Basilica; decent range of accommodation; some excellent shops and restaurants; on the metro; not much nightlife; sells out quickly for religious holidays.
Villa Borghese & Northern Rome	Largely residential area good for the Auditorium and Stadio Olimpico; some top museums; generally quiet after dark.

Tempio di Saturno, Roman Forum (p80)

In Focus

Rome Today

In late 2014, Rome was shocked as details emerged of deep-rooted corruption in city hall and links to organised crime. This, coming after a year of spending cuts and economic uncertainty, was a bitter pill for the city's residents to swallow. On the plus side, restoration work concluded on several high-profile monuments and Pope Francis continues to win over fans in the Vatican.

Above: Colosseum interior WIBOWO RUSLI/GETTY IMAGES ©

Scandal Rocks Rome

Rome is no stranger to controversy, but even hardened observers were shocked by the Mafia Capitale scandal that rocked the city in late 2014. The controversy centred on a criminal gang that had infiltrated city hall and was making millions by milking funds earmarked for immigration centres and camps for the city's Roma population. Thirty-seven people were arrested and up to 100 politicians and public officials, including former mayor Gianni Alemanno, were placed under police investigation. The Romans, not unused to the colourful behaviour of their elected officials, were appalled and a feeling of genuine outrage swept the city.

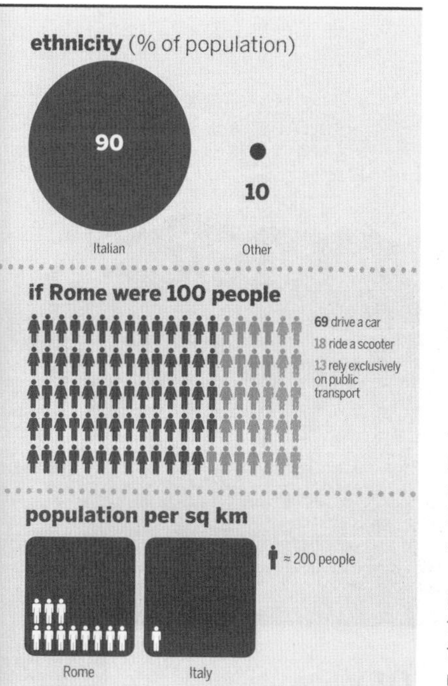

Restoration & Street Art

Rome's recent economic straits have put an enormous strain on the city's capacity to maintain its historical monuments. But walk around town today and you'll see several high-profile monuments gleaming in the wake of recent makeovers.

Despite the very public nature of these restoration projects, most of them were actually financed by private money. In the past few years Rome's cultural administrators have been actively courting private investment to help shore up municipal budgets. And while this has attracted heated debate, it has proved successful: the Colosseum clean-up was sponsored by Tod's; the Fendi fashion house footed the bill for work on the Trevi Fountain; and Bulgari has pledged €1.5 million towards an overhaul of the Spanish Steps.

Away from the spotlight, street art has exploded onto the scene in recent times, and giant murals have become a feature of some of the city's lesser-known neighbourhoods.

Pope Francis & the Jubilee

Over on the west bank of the Tiber, the Vatican continues to capture the world's attention. In April 2014 an estimated 800,000 people flocked to St Peter's to witness the dual canonisation of Popes John Paul II and John XXIII. Overseeing events was Pope Francis, the popular Argentinian pontiff whose papacy has done much to resurrect the Church's image in the wake of Benedict XVI's troubled tenure.

The city celebrated a Jubilee, or Holy Year, in 2016. During a Jubilee, the Church offers a plenary indulgence to Catholics who visit one of Rome's patriarchal basilicas and observe certain religious conditions.

Tourism Thrives

As well as wooing Catholics back to the fold, Pope Francis has had a positive effect on tourism, and in 2014 he was credited with inspiring a 28% increase in the number of Argentinian visitors to the city. Surprisingly, Rome's recent tribulations haven't adversely affected tourism, which continues to go from strength to strength – arrivals in 2014 were up on previous years, confirming the recent upward trend.

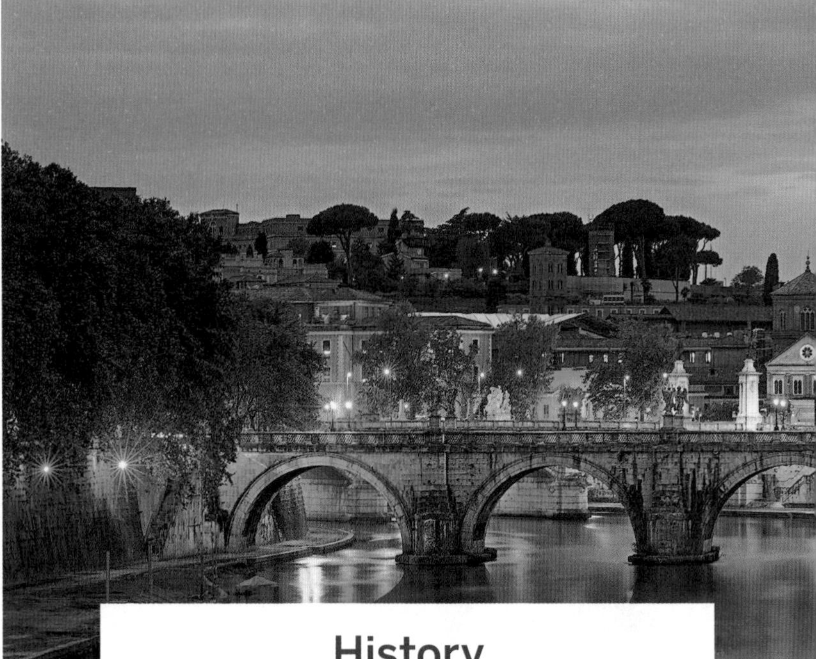

History

Rome's history spans three millennia, from the classical myths of vengeful gods to the follies of Roman emperors, from Renaissance excess to swaggering 20th-century fascism. Emperors, popes and dictators have come and gone, playing out their ambitions and conspiring for their place in history. Everywhere you go in this remarkable city, you're surrounded by the past.

Above: St Peter's Basilica (p46) and the River Tiber

753 BC
According to legend, Romulus kills his twin brother Remus and founds Rome on the Palatino.

509 BC
The Roman Republic is founded, paving the way for Rome's rise to European domination.

15 March 44 BC
On the Ides of March, Julius Caesar is stabbed to death in the Teatro di Pompeo (on modern-day Largo di Torre Argentina).

The Myth of Ancient Rome

As much a mythical construct as a historical reality, ancient Rome's image has been carefully nurtured throughout history.

Rome's original myth makers were the first emperors. Eager to reinforce the city's status as *caput mundi* (capital of the world), they turned to writers such as Virgil, Ovid and Livy to create an official Roman history. These authors, while adept at weaving epic narratives, were less interested in the rigours of historical research and frequently presented myth as reality. In the *Aeneid*, Virgil brazenly draws on Greek legends and stories to tell the tale of Aeneas, a Trojan prince who arrives in Italy and establishes Rome's founding dynasty.

Ancient Rome's rulers were sophisticated masters of spin; under their tutelage, art, architecture and elaborate public ceremony were employed to perpetuate the image of Rome as an invincible and divinely sanctioned power.

AD 67

St Peter and St Paul become martyrs as Nero massacres Rome's Christians in a ploy to win popularity after the great fire of AD 64.

285

Diocletian splits the Roman Empire in two. The eastern half later joins the Byzantine Empire; the western half falls to the barbarians.

476

The fall of Romulus Augustulus marks the end of the Western Empire.

★ Best Historical Sites

Palatino (p60)

Roman Forum (p80)

Via Appia Antica (p76)

Castel Sant'Angelo (p49)

Ruins along Via Appia Antica

Legacy of an Empire

Rising out of the bloodstained remnants of the Roman Republic, the Roman Empire was the Western world's first great superpower. At its zenith under Emperor Trajan (r AD 98–117), it extended from Britannia in the north to North Africa in the south, from Hispania (Spain) in the west to Palestina (Palestine) and Syria in the east. Rome itself had more than 1.5 million inhabitants. Decline eventually set in during the 3rd century and by the latter half of the 5th century Rome was in barbarian hands.

In AD 285 the emperor Diocletian, prompted by widespread disquiet across the empire, split the empire into eastern and western halves – the west centred on Rome and the east on Byzantium (later called Constantinople) – in a move that was to have far-reaching consequences for centuries. In the west, the fall of the Western Roman Empire in AD 476 paved the way for the emergence of the Holy Roman Empire and the Papal States, while in the east, Roman (later Byzantine) rule continued until 1453 when the empire was finally conquered by rampaging Ottoman armies.

Christianity & Papal Power

For much of its history Rome has been ruled by the pope, and today the Vatican still wields immense influence over the city.

Emergence of Christianity

Christianity swept in from the Roman province of Judaea in the 1st century AD. Its early days were marred by persecution, most notably under Nero (r 54–68), but it slowly caught on, thanks to its popular message of heavenly reward.

However, it was the conversion of Emperor Constantine (r 306–37) that really set Christianity on the path to European domination. In 313 Constantine issued the Edict of Milan, officially legalising Christianity, and in 378, Theodosius (r 379–95) made Christianity Rome's state religion. By this time, the Church had developed a sophisticated organisa-

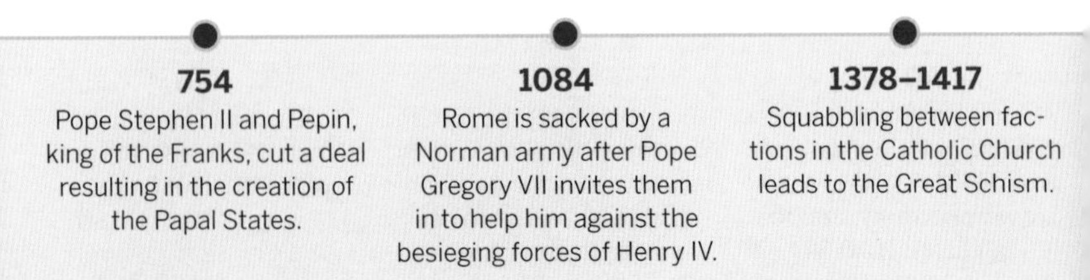

754
Pope Stephen II and Pepin, king of the Franks, cut a deal resulting in the creation of the Papal States.

1084
Rome is sacked by a Norman army after Pope Gregory VII invites them in to help him against the besieging forces of Henry IV.

1378–1417
Squabbling between factions in the Catholic Church leads to the Great Schism.

tional structure based on five major sees: Rome, Constantinople, Alexandria, Antioch and Jerusalem. At the outset, each bishopric carried equal weight, but in subsequent years Rome emerged as the senior party. The reasons for this were partly political – Rome was the wealthy capital of the Roman Empire – and partly religious – early Christian doctrine held that St Peter, founder of the Roman Church, had been sanctioned by Christ to lead the universal Church.

Papal Control

But while Rome had control of Christianity, the Church had yet to conquer Rome. This it did in the dark days that followed the fall of the Roman Empire by skilfully stepping into the power vacuum created by the demise of imperial power. And although no one person can take credit for this, Pope Gregory the Great (r 590–604) did more than most to lay the groundwork. A leader of considerable foresight, he won many friends by supplying free bread to Rome's starving citizens and restoring the city's water supply. He also stood up to the menacing Lombards, who presented a very real threat to the city.

It was this threat that pushed the papacy into an alliance with the Frankish kings, resulting in the creation of the two great powers of medieval Europe: the Papal States and the Holy Roman Empire. In Rome, the battle between these two superpowers translated into endless feuding between the city's baronial families and frequent attempts by the French to claim the papacy for their own. This political and military fighting eventually culminated in the papacy transferring to the French city of Avignon between 1309 and 1377, and the Great Schism (1378–1417), a period in which the Catholic world was headed by two popes, one in Rome and one in Avignon.

As both religious and temporal leaders, Rome's popes wielded influence well beyond their military capacity. For much of the medieval period, the Church held a virtual monopoly on Europe's reading material (mostly religious scripts written in Latin) and was the authority on virtually every aspect of human knowledge.

Romulus & Remus

The most famous of Rome's many legends is the story of Romulus and Remus, the mythical twins who are said to have founded Rome on 21 April 753 BC.

Romulus and Remus were born to the vestal virgin Rhea Silva after she'd been seduced by Mars. At their birth they were sentenced to death by their great-uncle Amulius, who had stolen the throne of Alba Longa from his brother, Rhea Silva's father, Numitor. The sentence was never carried out, and the twins were abandoned in a basket on the banks of the river Tiber. They were saved by a she-wolf and raised by a shepherd, Faustulus. Years later, the twins decided to build a city on the site where they'd originally been saved. They didn't know where this was, so they asked the omens. Remus, on the Aventine Hill, saw six vultures; his brother over on the Palatine saw 12. Romulus began building, which outraged his brother. They argued and Romulus killed Remus, going on to found his city.

1527
Pope Clement VII hides in Castel Sant'Angelo as Rome is sacked by troops loyal to Charles V, king of Spain and Holy Roman Emperor.

1626
St Peter's Basilica is completed after 150 years' construction. The basilica remains the largest church in the world until 1997.

1798
Napoleon marches into Rome. A republic is announced, but it doesn't last long and in 1801 Pope Pius VII returns to Rome.

★ Best Frescoes & Mosaics

Sistine Chapel (p44)

Stanze di Raffaello (p43)

Museo Nazionale Romano: Palazzo Massimo alle Terme (p68)

Basilica di Santa Maria in Trastevere (p106)

Stanze di Raffaello, Vatican Museums

Modern Influence

Almost 1000 years on and the Church is still a major influence on modern Italian life. In recent years, Vatican intervention in political and social debate has provoked fierce divisions within Italy. This relationship between the Church and Italy's modern political establishment is a fact of life that dates to the establishment of the Italian Republic in 1946. For much of the First Republic (1946–94), the Vatican was closely associated with Democrazia Cristiana (DC; Christian Democrat Party), Italy's most powerful party and an ardent opponent of communism. At the same time, the Church, keen to weed communism out of the political landscape, played its part by threatening to excommunicate anyone who voted for Italy's Partito Comunista Italiano (PCI; Communist Party). Today, no one political party has a monopoly on Church favour, and politicians across the spectrum tread warily around Catholic sensibilities.

Renaissance: A New Beginning

Bridging the gap between the Middle Ages and the modern age, the Renaissance (Rinascimento in Italian) was a far-reaching intellectual, artistic and cultural movement. It emerged in 14th-century Florence but quickly spread to Rome, where it gave rise to one of the greatest makeovers the city had ever seen.

Humanism & Rebuilding

The movement's intellectual cornerstone was humanism, a philosophy that focused on the central role of humanity within the universe, a major break from the medieval world view, which placed God at the centre of everything. It was not anti-religious though. One of the most celebrated humanist scholars of the 15th century was Pope Nicholas V (r 1447–84), who is considered the harbinger of the Roman Renaissance.

When Nicholas became pope in 1447, Rome was not in a good state. Centuries of medieval feuding had reduced the city to a semi-deserted battleground. In political terms, the

1870
Nine years after Italian unification, Rome's city walls are breached at Porta Pia and Pope Pius IX cedes the city to Italy.

1922
Some 40,000 fascists march on Rome. King Vittorio Emanuele III invites the 39-year-old Mussolini to form a government.

1929
The Lateran Treaty is signed, creating the state of Vatican City. To celebrate, Via della Conciliazione is bulldozed through the medieval Borgo.

papacy was recovering from the trauma of the Great Schism and attempting to face down Muslim encroachment in the east.

Against this background, Nicholas decided to rebuild Rome as a showcase of Church power, setting off an enormous program that would see the Sistine Chapel and St Peter's built.

Sack of Rome & Protestant Protest

But outside Rome an ill wind was blowing. The main source of trouble was the long-standing conflict between the Holy Roman Empire, led by the Spanish Charles V, and the Italian city-states. This simmering tension came to a head in 1527 when Rome was invaded by Charles' marauding army and ransacked as Pope Clement VII (r 1523–34) hid in Castel Sant'Angelo. The sack of Rome, regarded by most historians as the nail in the coffin of the Roman Renaissance, was a hugely traumatic event. It left the papacy reeling and gave rise to the view that the Church had been greatly weakened by its own moral shortcomings. That the Church was corrupt was well known, and it was with considerable public support that Martin Luther pinned his famous 95 Theses to a church door in Wittenberg in 1517, thus sparking off the Protestant Reformation.

Counter-Reformation

The Catholic reaction to the Reformation was all-out. The Counter-Reformation was marked by a second wave of artistic and architectural activity, as the Church once again turned to bricks and mortar to restore its authority. But in contrast to the Renaissance, the Counter-Reformation was a period of persecution and official intolerance. With the full blessing of Pope Paul III, Ignatius Loyola founded the Jesuits in 1540, and two years later the Holy Office was set up as the Church's final appeals court for trials prosecuted by the Inquisition. In 1559 the Church published the *Index Librorum Prohibitorum* (Index of Prohibited Books) and began to persecute intellectuals and freethinkers.

Despite, or perhaps because of, the Church's policy of zero tolerance, the Counter-Reformation was largely successful in re-establishing papal prestige. From being a rural backwater with a population of around 20,000 in the mid-15th century, Rome had grown to become one of Europe's great 17th-century cities, home to Christendom's most spectacular churches and a population of 100,000 people.

The First Tourists

While Rome has a long past as a pilgrimage site, its history as a modern tourist destination can be traced back to the late 1700s and the fashion for the Grand Tour. The 18th-century version of a gap year, the Tour was considered an educational rite of passage for wealthy young men from northern Europe, and Britain in particular.

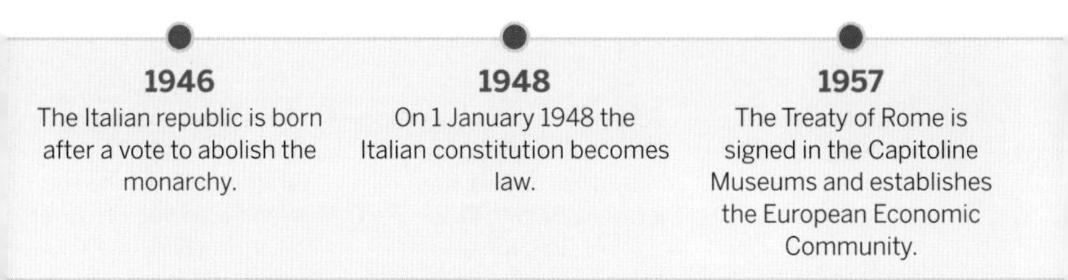
1946
The Italian republic is born after a vote to abolish the monarchy.

1948
On 1 January 1948 the Italian constitution becomes law.

1957
The Treaty of Rome is signed in the Capitoline Museums and establishes the European Economic Community.

Natale di Roma parade

GIORGIO COSULICH/STRINGER/GETTY IMAGES ©

★ Best Historical Celebrations

Carnevale (p7)

Natale di Roma (p9)

Festa de' Noantri (p12)

Re-enactment of Caesar's death (p8)

Festa della Repubblica (p11)

Rome, enjoying a rare period of peace, was perfectly set up for this English invasion. The city was basking in the aftermath of the 17th-century baroque building boom, and a craze for all things classical was sweeping Europe. Rome's papal authorities were also crying out for money after their excesses had left the city coffers bare, reducing much of the population to abject poverty.

Thousands came, including Goethe, who stopped off to write his travelogue *Italian Journey* (1817), as well as Byron, Shelley and Keats, who all fuelled their romantic sensibilities in the city's vibrant streets.

Artistically, rococo was the rage of the moment. The Spanish Steps, built between 1723 and 1726, proved a major hit with tourists, as did the exuberant Trevi Fountain.

Ghosts of Fascism

Rome's fascist history is a deeply sensitive subject, and in recent years historians on both sides of the political spectrum have accused each other of recasting the past to suit their views.

Mussolini

Benito Mussolini was born in 1883 in Forlì, a small town in Emilia-Romagna. As a young man he was an active member of the Italian Socialist Party, rising through the ranks to become editor of the party's official newspaper. However, service in WWI and Italy's subsequent descent into chaos led to a change of heart and in 1919 he founded the Italian Fascist Party.

In 1921 Mussolini was elected to the Chamber of Deputies. His parliamentary support was limited, but on 28 October 1922 he marched on Rome with 40,000 black-shirted followers. The march was largely symbolic but it had the desired effect, and King Vittorio Emanuele III, fearful of a civil war between the fascists and socialists, invited Mussolini to form a government. By the end of 1925 he had seized complete control of Italy. In order to silence the Church he signed the Lateran Treaty in 1929, which made Catholicism the state

1960
Rome hosts the Games of the XVII Olympiad.

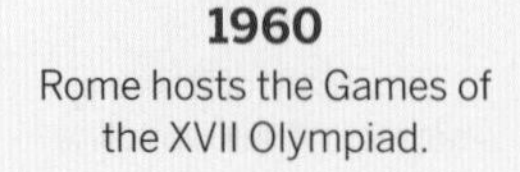

1978
Former prime minister Aldo Moro is kidnapped and shot by a cell of the extreme left-wing Brigate Rosse (Red Brigades).

2005
Pope John Paul II dies after 27 years on the papal throne. He is replaced by his long-standing ally Josef Ratzinger (Benedict XVI).

religion and recognised the sovereignty of the Vatican State.

Abroad, Mussolini invaded Abyssinia (now Ethiopia) in 1935 and sided with Hitler in 1936. In 1940, from the balcony of Palazzo Venezia, he announced Italy's entry into WWII to a vast, cheering crowd. The good humour didn't last, as Rome suffered, first at the hands of its own fascist regime, then, after Mussolini was ousted in 1943, at the hands of the Nazis. Rome was finally liberated from German occupation on 4 June 1944.

Jubilee Years

As seat of the Catholic Church, Rome was already one of the main pilgrim destinations in the Middle Ages, when, in 1300, Pope Boniface VIII proclaimed the first ever Holy Year (Jubilee). Promising full forgiveness for anyone who made the pilgrimage to St Peter's Basilica and the Basilica di San Giovanni in Laterano, his appeal to the faithful proved a resounding success. Hundreds of thousands answered his call and the Church basked in popular glory.

Some 700 years later and the Holy Year tradition is still going strong. A Jubilee was held in 2000, attracting up to 24 million visitors, and another, more low-key affair, in 2016.

Postwar Period

Defeat in WWII didn't kill off Italian fascism, and in 1946 hardline Mussolini supporters founded the Movimento Sociale Italiano (MSI; Italian Social Movement). For close on 50 years this overtly fascist party participated in mainstream Italian politics, while on the other side of the spectrum the Partito Comunista Italiano (PCI; Italian Communist Party) grew into Western Europe's largest communist party. The MSI was finally dissolved in 1994.

Outside the political mainstream, fascism (along with communism) was a driving force of the domestic terrorism that rocked Rome and Italy during the *anni di piombo* (years of lead), between the late 1960s and the early '80s.

2013
Pope Benedict XVI becomes the first pope to resign since 1415. Argentinian cardinal Jorge Mario Bergoglio is elected as Pope Francis.

2014
Ex-mayor Gianni Alemanno and 100 politicians and public officials are investigated as the Mafia Capitale scandal rocks Rome.

2015–16
Pope Francis declares a Jubilee Year to run from 8 December 2015 to 20 November 2016.

The Arts

Rome's turbulent history and magical cityscape have long provided inspiration for painters, sculptors, film makers, writers and musicians. The classical works of Roman antiquity fuelled the imagination of Renaissance artists; Counter-Reformation persecution led to baroque art and street satire; and the trauma of Mussolini and WWII found expression in neorealist cinema.

Above: Sarcofago degli Sposi (Sarcophagus of the Betrothed; p58)

Painting & Sculpture

Home to some of the Western world's most recognisable art, Rome is a visual feast. Its churches alone contain more masterpieces than many small countries, and the city's galleries are laden with works by world-famous artists.

Etruscan Groundwork

The Etruscans placed great importance on their funerary rites and they developed sepulchral decoration into a highly sophisticated art form. An important example is the *Sarcofago degli Sposi* (Sarcophagus of the Betrothed) in the Museo Nazionale Etrusco di

Villa Giulia. They were also noted for their bronze work, an example of which is the iconic *Lupa Capitolina* in the Capitoline Museums.

Roman Developments

In terms of decorative art, the Roman use of mosaics and wall paintings was derived from Etruscan funerary decoration. Typical themes included landscapes, still lifes, geometric patterns and depictions of gods. In the Museo Nazionale Romano: Palazzo Massimo alle Terme, you'll find some spectacular wall mosaics and 1st century BC frescoes.

Sculpture was an important element of Roman art, and was largely influenced by Greek styles. Early Roman sculptures were often made by Greek artists or were copies of Greek works. They were largely concerned with visions of male beauty – classic examples are the *Apollo Belvedere* and the *Laocoön* in the Museo Pio-Clementino of the Vatican Museums.

From the time of Augustus (r 27 BC–AD 14), Roman art became increasingly propagandistic. A new style of narrative art developed which often took the form of relief decoration – the Ara Pacis is a stunning example.

Early Christian Art

Some of Rome's earliest preserved Christian artworks are the faint biblical frescoes in the Catacombe di San Sebastiano on Via Appia Antica.

With the legalisation of Christianity in the 4th century, these images began to move into the public arena, appearing in churches such as the Basilica di Santa Maria Maggiore.

Eastern influences became much more pronounced between the 7th and 9th centuries, when Byzantine styles swept in from the East – you can see such brighter, golden works in the Basilica di Santa Maria in Trastevere.

The Renaissance

The Renaissance arrived in Rome in the latter half of the 15th century, and was to have a profound impact on the city, as the top artists of the day were summoned to decorate the many new buildings going up around town.

Rome's most celebrated works of Renaissance art are Michelangelo's paintings in the Sistine Chapel – his cinematic ceiling frescoes, painted between 1508 and 1512, and the *Giudizio Universale* (Last Judgment), which he worked on between 1536 and 1541.

Central to the Last Judgment and much Renaissance art was the human form. This led artists to develop a far greater appreciation of perspective. But while early Renaissance painters made great strides in formulating rules of perspective, they still struggled to paint harmonious arrangements of figures. And it was this that Raffaello Sanzio (Raphael; 1483–1520) tackled in his great masterpiece *La Scuola di Atene* (The School of Athens; 1510–11) in the Vatican Museums.

Counter-Reformation & the Baroque

The baroque burst onto Rome's art scene in the early 17th century. Combining a dramatic sense of dynamism with highly charged emotion, it was enthusiastically appropriated by the Catholic Church, which used it as a propaganda tool in its persecution of Counter-Reformation heresy. The powerful popes of the day eagerly championed the likes of Caravaggio, Gian Lorenzo Bernini, Domenichino, Pietro da Cortona and Alessandro Algardi.

Not surprisingly, much baroque art has a religious theme and you'll often find depictions of martyrdoms, ecstasies and miracles.

★ Best Rome Films

Roma città aperta (1945)

Roman Holiday (1953)

La dolce vita (1960)

The Talented Mr Ripley (1999)

La grande belleza (2013)

La dolce vita poster

One of its premier exponents was Milan-born Caravaggio (1573–1610), whose realistic interpretations of religious subjects often outraged his patrons. In contrast, the exquisite sculptural works of Gian Lorenzo Bernini (1598–1680) proved an instant hit.

Literature

Rome has a rich literary tradition, encompassing everything from ancient satires to contemporary thrillers.

Classics

Famous for his blistering oratory, Marcus Tullius Cicero (106–43 BC) was the Roman Republic's preeminent author. His contemporary, Catullus (c 84–54 BC) cut a very different figure with his epigrams and erotic verse.

On becoming emperor, Augustus (aka Octavian) encouraged the arts, and Virgil (70–19 BC), Ovid, Horace and Tibullus all enjoyed freedom to write.

Rome As Inspiration

Rome has provided inspiration for legions of foreign authors.

In the 18th century, historians and grand tourists poured into Rome from northern Europe. The German author Goethe captures the elation of discovering ancient Rome in his travelogue *Italian Journey* (1817). The city was also favoured by the Romantic poets: John Keats, Lord Byron, Percy Bysshe Shelley, Mary Shelley and other writers all spent time here.

In more recent fiction, Rome has provided a setting for many a blockbuster, including Dan Brown's thriller *Angels and Demons* (2001).

Writing Today

Born in Rome in 1966, Niccolò Ammaniti is the best known of the city's crop of contemporary authors. In 2007 he won the Premio Strega, Italy's top literary prize, for his novel *Come Dio comanda* (As God Commands), although he's probably best known for *Io non ho paura* (I'm Not Scared; 2001)

Cinema

Rome has a long cinematic tradition, spanning the works of the postwar neorealists and film-makers as diverse as Federico Fellini, Sergio Leone, Nanni Moretti, and Paolo Sorrentino, the Oscar-winning director of *La grande belleza* (The Great Beauty).

The 1940s was Roman cinema's Golden Age, when Roberto Rossellini (1906–77) produced a trio of neorealist masterpieces, most notably *Roma città aperta* (Rome Open City; 1945). Also important was Vittorio de Sica's 1948 *Ladri di biciclette* (Bicycle Thieves).

Federico Fellini (1920–94) took the creative baton from the neorealists, producing his era-defining hit *La dolce vita* in 1960. The films of Pier Paolo Pasolini (1922–75) are similarly demanding in their depiction of Rome's gritty underbelly in the postwar period.

Idiosyncratic and whimsical, Nanni Moretti continues to make films that fall into no mainstream tradition, such as *Habemus Papam*, his 2011 portrayal of a pope having a crisis of faith.

Recently, the big news in cinema circles has been the return of international film-making to Rome. In 2015 Daniel Craig charged around town filming the latest 007 outing, *Spectre,* while Ben Stiller was camping it up for *Zoolander 2*. Down in the city's southern reaches, a remake of *Ben-Hur* was filmed at the Cinecittà film studios, the very same place where the original sword-and-sandal epic was shot in 1959.

Neoclassicism

Emerging in the late 18th and early 19th centuries, neoclassicism signalled a departure from the emotional abandon of the baroque and a return to the clean, sober lines of classical art. Its major exponent was the sculptor Antonio Canova (1757–1822), whose study of Paolina Bonaparte Borghese as *Venere Vincitrice* (Venus Victrix) in the Museo e Galleria Borghese is typical of the mildly erotic style for which he became known.

Music

Despite years of austerity-led cutbacks, Rome's music scene is bearing up well. International orchestras perform to sell-out audiences, jazz greats jam in steamy clubs, and rappers rage in underground venues.

Jazz has long been a mainstay of Rome's music scene, while recent years have seen the emergence of a vibrant rap and hip-hop culture. Opera is served up at the Teatro dell'Opera and, in summer, at the spectacular Terme di Caracalla.

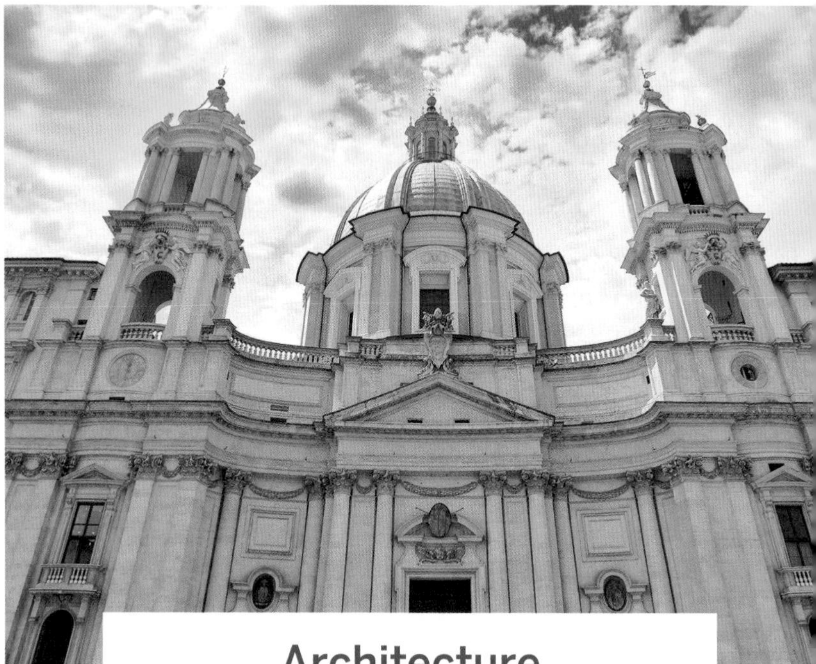

Architecture

From ancient ruins and Renaissance basilicas to baroque churches and hulking fascist palazzi *(palaces), Rome's architectural legacy is unparalleled. Michelangelo, Bramante, Borromini and Bernini, as well as more recently contemporary stars Renzo Piano and Zaha Hadid, are among the architects who have stamped their genius on Rome's remarkable cityscape.*

Chiesa di Sant'Agnese in Agone (p66) KAGENMI/GETTY IMAGES ©

The Ancients

Architecture was central to the success of the ancient Romans. In building their great capital, they were among the first people to use architecture to tackle problems of infrastructure, urban management and communication. For the first time, architects and engineers designed houses, roads, aqueducts and shopping centres alongside temples, tombs and imperial palaces. To do this, the Romans advanced methods devised by the Etruscans and Greeks, developing construction techniques and building materials that allowed them to build on a massive and hitherto unseen scale.

Etruscan Roots

By the 7th century BC the Etruscans were the dominant force on the Italian peninsula, with important centres at Tarquinia, Caere (Cerveteri) and Veii (Veio). But they built with wood and brick, which didn't last, and much of what we now know about the Etruscans derives from their impressive cemeteries. These were constructed outside the city walls and harboured richly decorated stone vaults covered by mounds of earth.

Roman Developments

When Rome was founded in 753 BC (or earlier if recent archaeological findings are to be believed), the Etruscans were at the height of their power and Greek colonists were establishing control over southern Italy. Against this background, Roman architects borrowed heavily from Greek and Etruscan traditions, gradually developing their own styles and techniques.

Ancient Roman architecture was monumental in form and often propagandistic in nature. Huge amphitheatres, aqueducts and temples joined muscular and awe-inspiring basilicas, arches and thermal baths in trumpeting the skill and vision of the city's early rulers and the nameless architects who worked for them.

Temples

Early Republican-era temples were based on Etruscan designs, but over time the Romans turned to the Greeks for their inspiration. But whereas Greek temples had steps and colonnades on all sides, the classic Roman temple had a high podium with steps leading up to a deep porch.

The Roman use of columns was also Greek in origin, even if the Romans favoured the more slender Ionic and Corinthian columns over the plain Doric pillars – to see how these differ study the exterior of the Colosseum, which incorporates all three styles.

Aqueducts & Sewers

One of the Romans' crowning architectural achievements was the development of a water supply infrastructure.

To meet the city's water demand, the Romans constructed a complex system of aqueducts to bring water in from the hills of central Italy and distribute it around town. The first aqueduct to serve Rome was the 16.5km Aqua Appia, which became fully operational in 312 BC. Over the next 700 years or so, up to 800km of aqueducts were built in the city, a network capable of supplying up to 1 million cu metres of water a day.

At the other end of the water cycle, waste water was drained away via an underground sewerage system known as the Cloaca Maxima (Great Sewer) and emptied downstream into the river Tiber.

Residential Housing

While Rome's emperors and aristocrats lived in luxury on the Palatino (Palatine Hill), the city's poor huddled together in large residential blocks called *insulae*. These poorly built structures were sometimes up to six or seven storeys high, accommodating hundreds of people in dark, unhealthy conditions. Near the foot of the steps that lead up to the Chiesa di Santa Maria in Aracoeli, you can still see a section of what was once a typical city-centre *insula*.

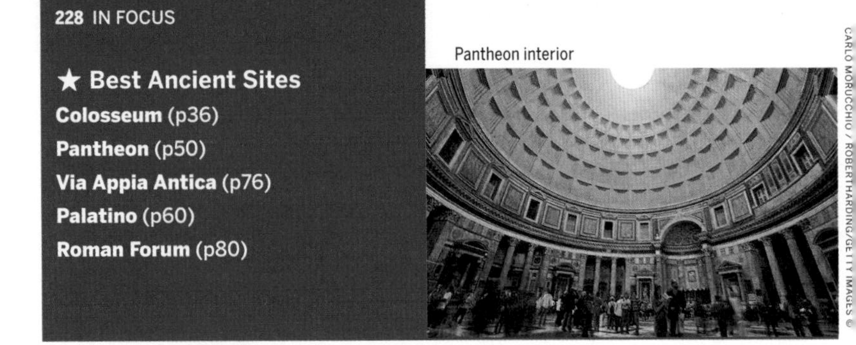

★ **Best Ancient Sites**

Colosseum (p36)
Pantheon (p50)
Via Appia Antica (p76)
Palatino (p60)
Roman Forum (p80)

Pantheon interior

CARLO MORUCCHIO / ROBERTHARDING/GETTY IMAGES ©

Concrete & Monumental Architecture

Grandiose structures such as the Colosseum, Pantheon and the Forums are not only reminders of the sophistication and scale of ancient Rome – just as they were originally designed to be – but also monuments to the vision of the city's ancient architects.

One of the key breakthroughs the Romans made was the invention of concrete in the 1st century BC. Made by mixing volcanic ash with lime and an aggregate, often tufa rock or brick rubble, concrete was quick to make, easy to use and cheap. It allowed the Romans to develop vaulted roofing, which they used to span the Pantheon's ceiling and the huge vaults at the Terme di Caracalla.

Early Christian

The most startling reminders of early Christian activity are the catacombs, a series of subterranean burial grounds built under Rome's ancient roads. Christian belief in the resurrection meant that the Christians could not cremate their dead, as was the custom in Roman times, and with burial forbidden inside the city walls they were forced to go outside the city.

The Christians began to abandon the catacombs in the 4th century and increasingly opted to be buried in the churches being built in the city. The most notable of the many churches that Constantine commissioned is the Basilica di San Giovanni in Laterano, the model on which many subsequent basilicas were based. Other period showstoppers include the Basilica di Santa Maria in Trastevere and the Basilica di Santa Maria Maggiore.

A second wave of church-building hit Rome in the period between the 8th and 12th centuries. As the early papacy battled for survival against the threatening Lombards, its leaders took to construction to leave some sort of historic imprint, resulting in churches such as the Chiesa di Santa Maria in Cosmedin, home of the Bocca della Verità (Mouth of Truth).

The Renaissance

Many claim it was the election of Pope Nicholas V in 1447 that sparked the Renaissance in Rome. Nicholas believed that as head of the Christian world Rome had a duty to impress, a theory that was endorsed by his successors, and it was at the behest of the great papal dynasties – the Barberini, Farnese and Pamphilj – that the leading artists of the day were summoned to Rome.

Bramante & the High Renaissance

It was under Julius II (1503–13) that the Roman Renaissance reached its peak, thanks largely to a classically minded architect from Milan, Donato Bramante (1444–1514).

Considered the high priest of Renaissance architecture, Bramante arrived in Rome in 1499 and developed a hugely influential classical style. His 1502 Tempietto is a masterpiece of elegance. In 1506 Julius commissioned him to start work on the job that would finally finish him off – the rebuilding of St Peter's Basilica. The fall of Constantinople's Aya Sofya (Church of the Hagia Sofia) to Islam in the mid-14th century had pricked Nicholas V into ordering an earlier revamp, but the work had never been completed and it wasn't until Julius took the bull by the horns that progress was made. However, Bramante died in 1514 and he never got to see how his original Greek-cross design was developed.

St Peter's Basilica occupied most of the other notable architects of the High Renaissance, including Michelangelo (1475–1564) who took over in 1547 and created the basilica's crowning dome. Modelled on Brunelleschi's cupola for the Duomo in Florence, this is considered the artist's finest architectural achievement and one of the most important works of the Roman Renaissance.

Rococo Frills

In the early days of the 18th century, as baroque fashions began to fade and neoclassicism waited to make its 19th-century entrance, the rococo burst into theatrical life. Drawing on the excesses of the baroque, it was a short-lived fad but one that left a memorable mark.

The Spanish Steps (p100), built between 1723 and 1726 by Francesco de Sanctis, provided a focal point for the many Grand Tourists who were busy discovering Rome's classical past. A short walk to the southwest, Piazza Sant'Ignazio was designed by Filippo Raguzzini (1680–1771) to provide a suitably melodramatic setting for the Chiesa di Sant'Ignazio di Loyola, Rome's second most important Jesuit church.

Most spectacular of all, however, was the Trevi Fountain (p86), one of the city's most exuberant and enduringly popular monuments. It was designed in 1732 by Nicola Salvi (1697–1751) and completed three decades later.

The Baroque

The Catholic Church became increasingly powerful in the 16th century. But with power came corruption and calls for reform. These culminated in the far-reaching Protestant Reformation, which prompted the Counter-Reformation (1560–1648), a vicious campaign to get people back into the Catholic fold. Art and architecture emerged as an effective form of propaganda. Stylistically, baroque architecture aims for a dramatic sense of dynamism, an effect that it often achieves by combining spatial complexity with clever lighting and a flamboyant use of decorative painting and sculpture.

The end of the 16th century and the papacy of Sixtus V (1585–90) marked the beginning of major urban-planning schemes that saw Domenico Fontana (1543–1607) and other architects create a network of major thoroughfares to connect parts of the sprawling medieval city.

Bernini vs Borromini

No two people did more to fashion the face of Rome than the two great figures of the Roman baroque – Gian Lorenzo Bernini (1598–1680) and Francesco Borromini (1599–1667). Naples-born Bernini, confident and suave, is best known for his work in the Vatican where he designed St Peter's Square and was chief architect at St Peter's Basilica from 1629.

Under the patronage of the Barberini pope Urban VIII, Bernini was given free rein to transform the city, and his churches, *palazzi,* piazzas and fountains remain landmarks to

★ Best Baroque

St Peter's Square (p49)
Fontana dei Quattro Fiumi (p66)
Chiesa di Sant'Agnese in Agone (p66)

Fontana dei Quattro Fiumi, Piazza Navona

this day. However, his fortunes nose-dived when the pope died in 1644. Urban's successor, Innocent X, wanted as little contact as possible with the favourites of his hated predecessor, and instead turned to Borromini.

Borromini, a solitary, peculiar man from Lombardy, created buildings involving complex shapes and exotic geometry such as the Chiesa di Sant'Agnese in Agone on Piazza Navona.

Rationalism & Fascism

Rome entered the 20th century in good shape. During the late 19th century it had been treated to one of its periodic makeovers, after being made capital in 1870. Piazzas were built, including Piazza Vittorio Emanuele II and neoclassical Piazza della Repubblica. To celebrate unification and pander to the ruling Savoy family, the ostentatious Vittoriano monument was built (1885–1911).

The 1920s saw the emergence of architectural rationalism. Its main Italian proponents, Gruppo Sette, combined classicism with modernism, which tied in perfectly with Mussolini's vision of fascism as the modern bearer of ancient Rome's imperialist ambitions. Mussolini's most famous architectural legacy is Rome's southern EUR district, built for the Esposizione Universale di Roma in 1942.

Modern Rome

The 21st century has witnessed a flurry of architectural activity in Rome as a clutch of starchitects have worked on projects in the city. Renzo Piano worked on the acclaimed Auditorium; American Richard Meier built the controversial pavilion for the 1st century AD Ara Pacis; and Anglo-Iraqi Zaha Hadid won plaudits for the Museo Nazionale delle Arti del XXI Secolo (MAXXI). Out in EUR, work continues on a striking conference centre, known as the Nuvola (Cloud), designed by Massimiliano Fuksas.

The Roman Way of Life

As a visitor, it's often difficult to see beyond Rome's spectacular veneer to the large, modern city that lies beneath. But how do the Romans live in their city? Where do they work? Who do they live with?

Vespa riders SIMON MONTGOMERY/ROBERTHARDING/GETTY IMAGES ©

Day in the Life

Rome's Mr Average, Signor Rossi, lives in a small, two-bedroom apartment and works in a government ministry.

His morning routine is somewhat the same as city dwellers the world over: a quick espresso followed by a short bus ride to the nearest metro station. On the way he'll stop at an *edicola* (kiosk) to pick up his daily newspaper *(Il Messaggero)* – a quick scan of the headlines reveals few surprises: Matteo Renzi promoting his latest reforms; the usual political shenanigans in city hall; Roma and Lazio match reports.

His work, like many in the swollen state bureaucracy, is secure and with a much sought-after *contratto a tempo indeterminato* (permanent contract) he doesn't have to worry about losing it. In contrast, younger colleagues work in constant fear that their temporary contracts won't be renewed.

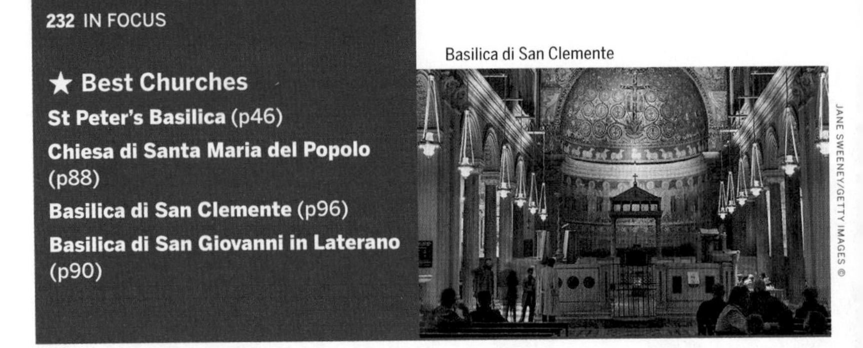

★ Best Churches

St Peter's Basilica (p46)

Chiesa di Santa Maria del Popolo (p88)

Basilica di San Clemente (p96)

Basilica di San Giovanni in Laterano (p90)

Basilica di San Clemente

JANE SWEENEY/GETTY IMAGES ©

Lunch, usually a snack or *pizza al taglio* (pizza by the slice) from a nearby takeaway, is followed by a quick coffee in the usual bar.

Clocking-off time in most ministries is typically from 5pm onwards and by about 7pm the evening rush hour is in full swing.

Home Life & the Family

Romans, like most Italians, live in apartments. These are often small – 75 sq metres to 100 sq metres is typical – and expensive. House prices in central Rome are among the highest in the country and many first-time buyers are forced to move to distant suburbs.

It's still the rule rather than the exception for young Romans to stay at home until they marry, which they typically do at around 30.

Religion

Rome is packed with churches. And with the Vatican in the centre of town, the Church is an important point of reference. Yet the role of religion in modern Roman society is an ambiguous one. On the one hand, most people consider themselves Catholic, but on the other, church attendance is in freefall, particularly among the young.

St Peter's Basilica interior

Survival Guide

Directory A–Z

Customs Regulations

If you're arriving from a non-EU country you can import, duty free, 200 cigarettes, 1L of spirits (or 2L fortified wine), 4L wine, 60ml perfume, 16L beer, and goods, including electronic devices, up to a value of €300/430 (travelling by land/air or sea); anything over this value must be declared on arrival and the duty paid.

Non-EU residents can reclaim value-added tax (VAT) on expensive purchases on leaving the EU.

Electricity

230V/50Hz

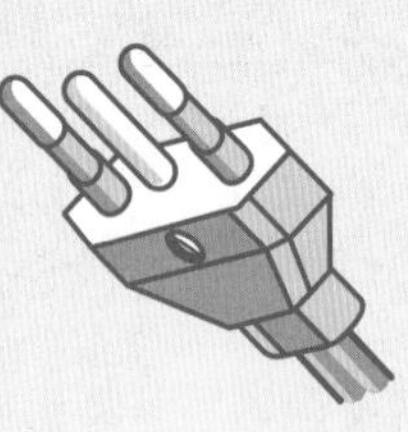

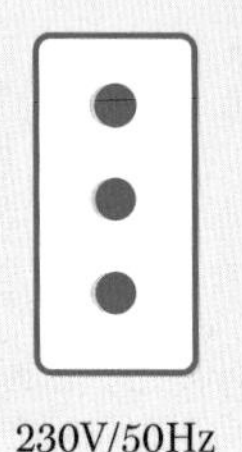

230V/50Hz

Emergency

Ambulance	☎118
Fire	☎115
Police	☎112, ☎113

Health

Nationals of the EU are entitled to reduced-cost, sometimes free, medical care with a European Health Insurance Card (EHIC), available from your home health authority; non-EU citizens should take out medical insurance.

For emergency treatment, you can go to the *pronto soccorso* (casualty) section of an *ospedale* (public hospital). For less serious ailments call the **Guardia Medica** (☎06 8840113; Via Mantova 44; 24hr).

More convenient, if you have insurance and can afford to pay up front, would be to call a private doctor for a home visit. Try the **International Medical Centre** (☎06 488 23 71; Via Firenze 47; GP call-out & treatment fee €140, 8pm-9am & weekends €200; 24hr).

Pharmacies

Marked by a green cross, *farmacie* (pharmacies) open from 8.30am to 1pm and 4pm to 7.30pm Monday to Friday and on Saturday mornings. Outside these hours they open on a rotational basis, and all are legally required to post a list of places open in the vicinity.

Internet Access

You'll find several internet cafes in the area around Termini station. Most hotels have wi-fi these days, though with signals of varying quality.

Wi-fi access is available in much of central Rome courtesy of Roma Wireless (www.romawireless.com). It's free (for two hours a day) but you'll need to register the first time you use it, and to do that you'll need an Italian mobile-phone number. An easier option is to head to a cafe or bar offering free wi-fi.

Legal Matters

The most likely reason for a brush with the law is to report a theft. If you do have something stolen and you want to claim it on insurance, you must make a statement to the police, as insurance companies won't pay up without official proof of a crime.

The Italian police is divided into three main bodies: the *polizia*, who wear navy-blue jackets; the *carabinieri*, in a black uniform with a red stripe; and the grey-clad *guardia di finanza* (fiscal police). If you run into trouble, you're most likely to end up dealing with the *polizia* or *carabinieri*.

Roma Pass

A comprehensive discount card, the Roma Pass comes in two forms:

Classic Roma Pass (€36; valid for three days) Provides free admission to two museums or sites, as well as reduced entry to extra sites, unlimited city transport, and discounted entry to other exhibitions and events.

48-Hour Roma Pass (€28; valid for 48 hours) Gives free admission to one museum or site and then as per the classic pass.

LGBT Travellers

Hardly San Fran on the Med, Rome nevertheless has a thriving, if low-key, gay scene. Close to the Colosseum, San Giovanni in Laterano has several bars dotted along it, including **Coming Out** (p181) and **My Bar** (p181).

Homosexuality is legal (over the age of 16) and even widely accepted, but Italy is notably conservative in its attitudes. In early 2016, however, the Italian parliament voted to recognise same-sex civil unions.

The main gay cultural and political organisation, **Circolo Mario Mieli di Cultura Omosessuale** (800 110611; www.mariomieli.org; Via Efeso 2a), organises debates, cultural events and social functions, including Gay Pride.

The national organisation for lesbians, **Coordinamento Lesbiche Italiano** (www.clrbp.it; Via San Francesco di Sales 1b), holds regular conferences and literary evenings.

Money

Italy's currency is the euro. The seven euro notes come in denominations of €500, €200, €100, €50, €20, €10 and €5. The euro coins are in denominations of €2 and €1, and 50, 20, 10, five, two and one cents.

ATMs

ATMs (*bancomat* in Italian) are widely available and most will accept cards tied into the Visa, MasterCard, Cirrus and Maestro systems. The daily limit for cash withdrawal is €250. Let your bank know when you are going abroad, in case they block your card when payments from unusual locations appear.

Remember that every time you withdraw cash, your home bank charges you a foreign exchange fee (usually around 1% to 3%) as well as a transaction charge of around 1%.

Changing Money

You can change money in banks, at post offices or at a *cambio* (exchange office). There are exchange booths at Stazione Termini and at Fiumicino and Ciampino airports. In the centre, there are numerous bureaux de change, including **American Express** (06 6 76 41; Piazza di Spagna 38; 9am-5.30pm Mon-Fri, 9am-12.30pm Sat).

Credit Cards

Credit cards are widely accepted but it's still a good idea to carry a cash backup. Virtually all midrange and top-end hotels accept cards, as do most restaurants and large shops. You can also use them to obtain cash advances at some banks. Some of the cheaper

Watch Your Valuables

Pickpockets follow the tourists, so watch out around the Colosseum, Piazza di Spagna, St Peter's Square and Stazione Termini. Be particularly vigilant around the bus stops on Via Marsala where thieves prey on travellers fresh in from Ciampino Airport. Crowded public transport is another hot spot – the 64 Vatican bus is notorious.

pensioni (guesthouses), trattorias and pizzerias accept nothing but cash.

Major cards such as Visa, MasterCard, Eurocard, Cirrus and Eurocheques are widely accepted. Amex is also recognised, although it's less common than Visa or MasterCard.

Opening Hours

Banks 8.30am to 1.30pm and 2.45pm to 4.30pm Monday to Friday

Bars & Cafes 7.30am to 8pm, sometimes until 1am or 2am

Clubs 10pm to 4am

Restaurants noon to 3pm and 7.30pm to 11pm (later in summer)

Shops 9am to 7.30pm or 10am to 8pm Monday to Saturday, some 11am to 7pm Sunday; smaller shops 9am to 1pm and 3.30pm to 7.30pm (or 4pm to 8pm) Monday to Saturday

Public Holidays

Most Romans take their annual holiday in August. Many businesses and shops close for at least part of the month, particularly around Ferragosto on 15 August.

Public holidays include the following:

Capodanno (New Year's Day) 1 January

Epifania (Epiphany) 6 January

Pasquetta (Easter Monday) March/April

Giorno della Liberazione (Liberation Day) 25 April

Festa del Lavoro (Labour Day) 1 May

Festa della Repubblica (Republic Day) 2 June

Festa dei Santi Pietro e Paolo (Feast of St Peter & St Paul) 29 June

Ferragosto (Feast of the Assumption) 15 August

Festa di Ognisanti (All Saints' Day) 1 November

Festa dell'Immacolata Concezione (Feast of the Immaculate Conception) 8 December

Natale (Christmas Day) 25 December

Festa di Santo Stefano (Boxing Day) 26 December

Safe Travel

Rome is not a dangerous city, but petty crime is a problem. Road safety is also an issue, so don't take it for granted that cars and scooters will stop at pedestrian crossings, or even at red lights.

Telephone

Domestic Calls

Rome's area code is ☎06. Area codes are an integral part of all Italian phone numbers and must be dialled even when calling locally. Mobile-phone numbers are nine or 10 digits long and begin with a three-digit prefix starting with a 3.

International Calls

To call abroad from Italy dial ☎00, then the country and area codes, followed by the telephone number.

You can Skype from most internet cafes.

Mobile Phones

Italian mobile phones operate on the GSM 900/1800 network, which is compatible with the rest of Europe and Australia but not always with the North American GSM or CDMA systems – check with your service provider.

If you can unlock your phone, it can cost as little as €10 to activate a *prepagato* (prepaid) SIM card in Italy. TIM (Telecom Italia Mobile; www.tim.it), Wind (www.wind.it) and Vodafone (www.vodafone.it) all offer SIM cards and have retail outlets across town. Note that by Italian law all SIM cards must be registered, so make sure you have a passport or

ID card with you when you buy one.

Time

Italy is in a single time zone, one hour ahead of GMT. Daylight-saving time, when clocks move forward one hour, starts on the last Sunday in March. Clocks are put back an hour on the last Sunday in October.

Italy operates on a 24-hour clock.

Toilets

There are toilets at the Colosseum, St Peter's Square, Castel Sant'Angelo and Stazione Termini (€1). Alternatively, nip into a cafe or bar, all of which are required by law to have a toilet.

Tourist Information

For phone enquiries, the Comune di Roma runs a free multilingual tourist information line (06 06 08).

There are tourist information points at **Fiumicino** (Terminal 3, International Arrivals; 8am-7.30pm) and **Ciampino** (International Arrivals, baggage claim area; 9am-6.30pm) airports, and at locations across the city:

Piazza delle Cinque Lune (9.30am-7.15pm)

Stazione Termini (8am-7.45pm)

Fori Imperiali (Via dei Fori Imperiali; 9.30am-7pm)

Via Marco Minghetti (Via Marco Minghetti; 9.30am-7.15pm)

Via Nazionale (Via Nazionale; 9.30am-7.15pm)

For information about the Vatican, contact the **Centro Servizi Pellegrini e Turisti** (06 6988 1662; St Peter's Sq; 8.30am-6pm Mon-Sat).

Useful websites include:

060608 (www.060608.it) Comprehensive information on sites, upcoming events, transport etc.

Roma Turismo (www.turismoroma.it) Rome's official tourist website with listings and up-to-date information.

Travellers with Disabilities

Cobbled streets, blocked pavements, tiny lifts and relentless traffic make Rome a difficult city for travellers with disabilities.

On metro line B all stations have wheelchair access except for Circo Massimo, Colosseo, Cavour and EUR Magliana; on line A only Cipro–Musei Vaticani and Valle Aurelia have lifts. Bus 590 covers the same route as metro line A and is wheelchair accessible. Newer buses and trams have disabled access; it's indicated on bus stops which routes are wheelchair accessible.

Contact ADR Assistance (www.adrassistance.it) for assistance at Fiumicino or Ciampino airports.

Some taxis are equipped to carry passengers in wheelchairs; ask for a taxi for a *sedia a rotelle* (wheelchair).

Visas

EU citizens do not need a visa to enter Italy. Nationals of some other countries, including Australia, Canada, Israel, Japan, New Zealand,

Practicalities

Media Vatican Radio (www.radiovaticana.org; 93.3 FM and 105 FM in Rome) is broadcast in Italian, English and other languages; RAI-1, RAI-2 and RAI-3 (www.rai.it) is the national broadcaster, running state TV and radio; and Canale 5 (www.mediaset.it/canale5), Italia 1 (www.mediaset.it/italia1), Rete 4 (www.mediaset.it/rete4) and La 7 (www.la7.it) are the main commercial stations.

Smoking Banned in enclosed public spaces, which includes restaurants, bars, shops and public transport. It's also been recently banned in Villa Borghese and all other public parks.

Weights and measures Italy uses the metric system.

Switzerland and the USA, do not need a visa for stays of up to 90 days.

Italy is one of the 15 signatories of the Schengen Convention, an agreement whereby participating countries abolished customs checks at common borders. The standard tourist visa for non-European visitors to a Schengen country is valid for 90 days; for more information see www.schengenvisainfo.com/tourist-schengen-visa.

Women Travellers

Sexual harassment can be an issue in Rome; if you get groped, a loud *'che schifo!'* (how disgusting!) will draw attention to the incident. Otherwise, women should take the usual precautions as they would in any large city, and, as in most places, avoid wandering around alone late at night, especially in the area around Termini.

Transport

Arriving in Rome

Most people arrive in Rome by plane, landing at one of its two airports: Leonardo da Vinci, better known as Fiumicino; or Ciampino, the hub for European low-cost carrier Ryanair. Flights from New York take around nine hours; from London 2¾ hours; from Sydney at least 22 hours.

As an alternative to short-haul flights, trains serve Rome's main station, Stazione Termini, from a number of European destinations, including Paris (about 15 hours), as well as cities across Italy.

Ferries serve Civitavecchia, some 80km north of the city, from a number of Mediterranean ports.

Flights, cars and tours can be booked online at lonelyplanet.com/bookings.

Leonardo da Vinci Airport

Rome's main international airport, **Leonardo da Vinci** (Fiumicino; ☎06 6 59 51; www.adr.it/fiumicino), is 30km west of the city. It's divided into four terminals: Terminals 1, 2 and 3 are for domestic and international flights; Terminal 5 is for American and Israeli airlines flying to the US and Israel.

Terminals 1, 2 and 3 are within easy walking distance of each other in the main airport building; Terminal 5 is accessible by shuttle bus from Terminal 3.

The easiest way to get into town is by train, but there are also buses and private shuttle services.

Train

Leonardo Express (one way €14) Runs to/from Stazione Termini. Departures from the airport every 30 minutes between 6.23am and 11.23pm; from Termini between 5.35am and 10.35pm. Journey time is 30 minutes.

FL1 (one way €8) Connects to Trastevere, Ostiense and Tiburtina stations, but not Termini. Departures from the airport every 15 minutes (half-hourly on Sunday and public holidays) between 5.57am and 10.42pm; from Tiburtina every 15 minutes between 5.46am and 7.31pm, then half-hourly until 10.02pm.

Bus

SIT (☎06 591 68 26; www.sitbusshuttle.it; one way €6) Regular departures from the airport to Stazione Termini (Via Marsala) from 8.30am to 11.50pm; from Termini between 5am and 8.30pm. All buses stop at the Vatican en route. Tickets are available on the bus. Journey time is approximately one hour.

Cotral (www.cotralspa.it; one way €5, if bought on the bus €7) Runs to/from Fiumicino from Stazione Tiburtina via Termini. Eight daily departures including night services from the airport at 1.15am, 2.15am, 3.30am and 5am, and from Tiburtina at 12.30am, 1.15am, 2.30am and 3.45am. Journey time is one hour.

Terravision (www.terravision.eu; one way €6, online €4) Regular services from the airport to Stazione Termini (Via Marsala) between 5.35am and 11pm; from Termini between 4.40am and 9.50pm. Allow about an hour for the journey.

Private Shuttle

Airport Connection Services (☎06 2111 6248; www.airport-connection.it) Transfers to/from the city centre start at €35 per person.

Airport Shuttle (www.airport-shuttle.it) Transfers to/from your hotel for €13 per person for up to four passengers, then €5 for each additional passenger up to a maximum of eight.

Taxi

The set fare to/from the city centre is €48, which is valid for up to four passengers including luggage. Journey time is approximately 45 minutes to an hour depending on traffic.

Climate Change & Travel

Every form of transport that relies on carbon-based fuel generates CO², the main cause of human-induced climate change. Modern travel is dependent on aeroplanes, which might use less fuel per kilometre per person than most cars but travel much greater distances. The altitude at which aircraft emit gases (including CO²) and particles also contributes to their climate change impact. Many websites offer 'carbon calculators' that allow people to estimate the carbon emissions generated by their journey and, for those who wish to do so, to offset the impact of the greenhouse gases emitted with contributions to portfolios of climate-friendly initiatives throughout the world. Lonely Planet offsets the carbon footprint of all staff and author travel.

Ciampino Airport

Ciampino (☎06 6 59 51; www.adr.it/ciampino), 15km southeast of the city centre, is used by Ryanair. It's not a big airport but there's a steady flow of traffic and at peak times it can get extremely busy.

To get into town, the best option is to take one of the dedicated bus services.

Bus

Terravision (www.terravision.eu; one way €6, online €4) Twice hourly departures to/from Stazione Termini (Via Marsala). From the airport, services run between 8.15am and 12.15am; from Via Marsala, between 4.30am and 9.20pm. Buy tickets at Terracafè in front of the Via Marsala bus stop. Journey time is 40 minutes.

SIT (☎06 591 68 26; www.sit-busshuttle.com; from/to airport €4/6) Regular departures from the airport to Stazione Termini (Via Marsala) between 7.45am and 11.15pm; from Termini between 4.30am and 9.30pm. Get tickets on the bus. Journey time is 45 minutes.

Atral (www.atral-lazio.com) Runs buses to/from Anagnina metro station (€1.20) and Ciampino train station (€1.20), where you can get a train to Termini (€1.30).

Private Shuttle

Airport Shuttle (www.airportshuttle.it) Transfers to/from your hotel for €25 for one person, then €5 for each additional passenger up to a maximum of eight.

Taxi

The set rate to/from the airport is €30. Journey time is approximately 30 minutes depending on traffic.

Termini Train Station

Almost all trains arrive at and depart from **Stazione Termini** (Piazza dei Cinquecento; M Termini), Rome's main train station and principal transport hub. There are regular connections to other European countries, all major Italian cities, and many smaller towns.

Train information is available from the Customer Service area on the main concourse to the left of the ticket desks. Alternatively, check www.trenitalia.com or phone ☎892021.

From Termini, you can connect with the metro or take a bus from Piazza dei Cinquecento out front. Taxis are outside the main entrance/exit.

Civitavecchia Port

The nearest port to Rome is at Civitavecchia, about 80km north of town. Ferries sail here from destinations across the Mediterranean, including Sicily and Sardinia. Check www.traghettiweb.it for route details and prices, and to book.

From Civitavecchia there are half-hourly trains to

Stazione Termini (€5 to €15, 40 minutes to 1¼ hours). Civitavecchia's station is about 700m from the entrance to the port.

Getting Around

Rome is a sprawling city, but the historic centre is relatively compact and it's quite possible to explore much of it on foot. The city's public transport system includes buses, trams, a metro and a suburban train system. Tickets, which come in various forms, are valid for all forms of transport.

Metro

- Rome has two main metro lines, A (orange) and B (blue), which cross at Stazione Termini. A branch line, 'B1', serves the northern suburbs, and a line C runs through the southeastern outskirts, but you're unlikely to use these.
- Trains run between 5.30am and 11.30pm (to 1.30am on Friday and Saturday).
- Take line A for the Trevi Fountain (Barberini), Spanish Steps (Spagna) and St Peter's (Ottaviano–San Pietro).
- Take line B for the Colosseum (Colosseo).

Bus & Tram

The main bus station is in front of Stazione Termini on Piazza dei Cinquecento, where there's an **information booth** (7.30am-8pm). Other important hubs are at Largo di Torre Argentina and Piazza Venezia.

Buses generally run from about 5.30am until midnight, with limited services throughout the night.

Rome's night bus service comprises more than 25 lines, many of which pass Termini and/or Piazza Venezia. Buses are marked with an 'n' before the number and bus stops have a blue owl symbol. Departures are usually every 15 to 30 minutes between about 1am and 5am, but can be much slower.

The most useful routes:

n1 Follows the route of metro line A.

n2 Follows the route of metro line B.

n7 Piazzale Clodio, Piazza Cavour, Via Zanardelli, Corso del Rinascimento, Corso Vittorio Emanuele II, Largo di Torre Argentina, Piazza Venezia, Via Nazionale and Stazione Termini.

Taxi

- Official licensed taxis are white with an ID number and 'Roma Capitale' on the sides.
- Always go with the metered fare, never an arranged price (the set fares to/from the airports are exceptions).
- Official rates are posted in taxis.
- You can hail a taxi, but it's often easier to wait at a rank or phone for one. There are taxi ranks at the airports, Stazione Termini, Piazza della Repubblica, Piazza Barberini, Piazza di Spagna, the Pantheon, the Colosseum, Largo di Torre Argentina, Piazza Belli, Piazza Pio XII and Piazza del Risorgimento.
- Note that when you call for a cab, the meter is switched on straight away

Public Transport Tickets

Public transport tickets are valid on all of Rome's bus, tram and metro lines, except for routes to Leonardo da Vinci (Fiumicino) airport. Children under 10 travel free.

Buy tickets at *tabacchaio*, newsstands and from vending machines at main bus stops and metro stations. Tickets must be purchased before you start your journey and validated in the machines on buses, at the entrance gates to the metro, or at train stations. If you travel without a ticket you risk an on-the-spot €50 fine.

Tickets come in various forms:

BIT (*biglietto integrato a tempo;* a single ticket valid for 100 minutes and one metro ride) €1.50

Roma 24h (valid for 24 hours) €7

Roma 48h (valid for 48 hours) €12.50

Roma 72h (valid for 72 hours) €18

CIS (*carta integrata settimanale;* a weekly ticket) €24 and you pay for the cost of the journey from wherever the driver receives the call.

La Capitale (☎06 49 94)

Pronto Taxi (☎06 66 45)

Radio 3570 (☎06 35 70; www.3570.it)

Car & Motorcycle

Driving around Rome is not recommended. Riding a scooter or motorbike is faster and makes parking easier, but Rome is no place for learners, so if you're not an experienced rider give it a miss. Hiring a car for a day trip out of town is worth considering.

Most of Rome's historic centre is closed to unauthorised traffic from 6.30am to 6pm Monday to Friday, from 2pm to 6pm (10am to 7pm in some places) Saturday, and from 11pm to 3am Friday and Saturday. Evening restrictions also apply in Trastevere, San Lorenzo, Monti and Testaccio, typically from 9.30pm or 11pm to 3am on Friday and Saturday.

All streets accessing the 'Limited Traffic Zone' (ZTL) are monitored by electronic-access detection devices. If you're staying in this zone, contact your hotel. For further information, check www.agenziamobilita.roma.it.

Driving Licence

All EU driving licences are recognised in Italy. Holders of non-EU licences should get an International Driving Permit (IDP) to accompany their national licence. Apply to your national motoring association.

A licence is required to ride a scooter – a car licence will do for bikes up to 125cc; for anything over 125cc you'll need a motorcycle licence.

A good source of information is the **Automobile Club d'Italia** (ACI; www.aci.it), Italy's national motoring organisation.

Hire

To hire a car you'll require a driving licence (plus IDP if necessary) and credit card. Age restrictions vary but generally you'll need to be 21 or over.

Car hire is available at both Rome's airports and Stazione Termini.

Avis (☎199 100 133; www.avisautonoleggio.it)

Europcar (☎199 30 70 30; www.europcar.it)

Hertz (☎02 6943 0019; www.hertz.it)

Maggiore National (☎199 151 120; www.maggiore.it)

To hire a scooter, prices range from about €30 to €120 depending on the size of the vehicle. Reliable operators:

Bici & Baci (☎06 482 84 43; www.bicibaci.com; Via del Viminale 5; ⏰8am-7pm)

Eco Move Rent (☎06 4470 4518; www.ecomoverent.com; Via Varese 48-50; ⏰8.30am-7.30pm)

Parking

Blue lines denote pay-and-display parking – get tickets from meters (coins only) and *tabacchaio* (tobacconist's shops).

Expect to pay up to €1.20 per hour between 8am and 8pm (11pm in some places). After 8pm (or 11pm) parking is free until 8am the next morning.

Traffic wardens are vigilant and fines are not uncommon. If your car gets towed away, call ☎06 6769 2303.

Useful car parks:

Piazzale dei Partigiani (per hr €0.77; ⏰7am-11pm)

Stazione Termini (Piazza dei Cinquecento; per hr/day €2.20/18; ⏰6am-1am)

Villa Borghese (Viale del Galoppatoio 33; per hr/day €2.20/18; ⏰24hr)

Train

Apart from connections to Leonardo da Vinci airport, you'll probably only need the overground rail network if you head out of town.

- Train information is available from the customer service area on the main concourse. Alternatively, check www.trenitalia.com or phone ☎892021.
- Buy tickets on the main station concourse, from automated ticket machines, or from an authorised travel agency – look for an FS or *biglietti treni* sign in the window.
- Rome's second train station is Stazione Tiburtina, four stops from Termini on metro line B. Of the capital's eight other train stations, the most important are Stazione Roma-Ostiense and Stazione Trastevere.

Language

Italian pronunciation isn't difficult as most sounds are also found in English. The pronunciation of some consonants depends on which vowel follows, but if you read our pronunciation guides below as if they were English, you'll be understood just fine. Just remember to pronounce double consonants as a longer, more forceful sound than single ones. The stressed syllables in words are in italics in our pronunciation guides.

To enhance your trip with a phrasebook, visit **lonelyplanet.com**. Lonely Planet iPhone phrasebooks are available through the Apple App store.

Basics

Hello.
Buongiorno./Ciao. (pol/inf) bwon·*jor*·no/chow

How are you?
Come sta? *ko*·me sta

I'm fine, thanks.
Bene, grazie. *be*·ne *gra*·tsye

Excuse me.
Mi scusi. mee *skoo*·zee

Yes./No.
Sì./No. see/no

Please. (when asking)
Per favore. per fa·*vo*·re

Thank you.
Grazie. *gra*·tsye

Goodbye.
Arrivederci./Ciao. (pol/inf) a·ree·ve·*der*·chee/chow

Do you speak English?
Parla inglese? *par*·la een·*gle*·ze

I don't understand.
Non capisco. non ka·*pee*·sko

How much is this?
Quanto costa? *kwan*·to *ko*·sta

Accommodation

I'd like to book a room.
Vorrei prenotare una camera. vo·*ray* pre·no·*ta*·re *oo*·na *ka*·me·ra

How much is it per night?
Quanto costa per una notte? *kwan*·to *kos*·ta per *oo*·na *no*·te

Eating & Drinking

I'd like ..., please.
Vorrei ..., per favore. vo·*ray* ... per fa·*vo*·re

What would you recommend?
Cosa mi consiglia? *ko*·za mee kon·*see*·lya

That was delicious!
Era squisito! *e*·ra skwee·*zee*·to

Bring the bill/check, please.
Mi porta il conto, per favore. mee *por*·ta eel *kon*·to per fa·*vo*·re

I'm allergic (to peanuts).
Sono allergico/a (alle arachidi). (m/f) *so*·no a·*ler*·jee·ko/a (*a*·le a·*ra*·kee·dee)

I don't eat ...
Non mangio ... non *man*·jo ...

fish	*pesce*	*pe*·she
meat	*carne*	*kar*·ne
poultry	*pollame*	po·*la*·me

Emergencies

I'm ill.
Mi sento male. mee *sen*·to *ma*·le

Help!
Aiuto! a·*yoo*·to

Call a doctor!
Chiami un medico! *kya*·mee oon *me*·dee·ko

Call the police!
Chiami la polizia! *kya*·mee la po·lee·*tsee*·a

Directions

I'm looking for (a/the) ...
Cerco ... *cher*·ko ...

bank	*la banca*	la *ban*·ka
... embassy	*la ambasciata de ...*	la am·ba·*sha*·ta de ...
market	*il mercato*	eel mer·*ka*·to
museum	*il museo*	eel moo·*ze*·o
restaurant	*un ristorante*	oon rees·to·*ran*·te
toilet	*un gabinetto*	oon ga·bee·*ne*·to
tourist office	*l'ufficio del turismo*	loo·*fee*·cho del too·*reez*·mo

Behind the Scenes

Acknowledgements

Climate map data adapted from Peel MC, Finlayson BL & McMahon TA (2007) 'Updated World Map of the Koppen-Geiger Climate Classification', *Hydrology and Earth System Sciences*, 11, 163344

Illustrations pp84–5 by Javier Martinez Zarracina

This Book

This book was curated by Duncan Garwood and researched and written by Duncan and Abigail Blasi.
This guidebook was commissioned in Lonely Planet's Melbourne office, and produced by the following:
Destination Editor Anna Tyler
Series Designer Katherine Marsh
Cartographic Series Designer Wayne Murphy
Associate Product Director Liz Heynes
Senior Product Editor Catherine Naghten
Product Editor Kathryn Rowan
Senior Cartographer Anthony Phelan
Book Designers Virginia Moreno, Wibowo Rusli
Assisting Editors Victoria Harrison, Gabrielle Innes, Charlotte Orr, Gabrielle Stefanos
Cover Researchers Campbell McKenzie and Naomi Parker
Thanks to Indra Kilfoyle, Anne Mason, Kate Mathews, Jenna Myers, Kirsten Rawlings, Alison Ridgway, Dianne Schallmeiner, Luna Soo, Angela Tinson

Send Us Your Feedback

We love to hear from travellers – your comments keep us on our toes and help make our books better. Our well-travelled team reads every word on what you loved or loathed about this book. Although we cannot reply individually to postal submissions, we always guarantee that your feedback goes straight to the appropriate authors, in time for the next edition. Each person who sends us information is thanked in the next edition, the most useful submissions are rewarded with a selection of digital PDF chapters.

Visit lonelyplanet.com/contact to submit your updates and suggestions or to ask for help. Our award-winning website also features inspirational travel stories, news and discussions.

Note: We may edit, reproduce and incorporate your comments in Lonely Planet products such as guidebooks, websites and digital products, so let us know if you don't want your comments reproduced or your name acknowledged. For a copy of our privacy policy visit lonelyplanet.com/privacy.

A–Z

Index

000 Map pages

STACIA_S/SHUTTERSTOCK ©

Rome Maps

Centro Storico

Sights

1 Area Archeologica del Teatro di Marcello e del Portico d'Ottavia ... E8
2 Basilica di Santa Maria Sopra Minerva ... E4
3 Campo de' Fiori ... B5
4 Chiesa del Gesù ... E5
5 Chiesa di San Luigi dei Francesi ... C3
6 Chiesa di Sant'Agnese in Agone ... B3
7 Chiesa di Sant'Ignazio di Loyola ... E3
8 Colonna di Marco Aurelio ... F2
9 Fontana dei Quattro Fiumi ... B3
10 Fontana del Moro ... B4
11 Fontana del Nettuno ... B3
12 Fontana delle Tartarughe ... D7
13 Galleria Doria Pamphilj ... F4
14 Museo Ebraico di Roma ... E8
15 Museo Nazionale del Palazzo Venezia ... F5
16 Museo Nazionale Romano: Crypta Balbi ... E6
17 Museo Nazionale Romano: Palazzo Altemps ... B2
18 Palazzo Chigi ... F2
19 Palazzo di Montecitorio ... E2
20 Palazzo Farnese ... A6
21 Palazzo Pamphilj ... B4
22 Palazzo Venezia ... F5
23 Pantheon ... D3
24 Piazza Colonna ... F2
25 Piazza Navona ... B3
26 Stadio di Domiziano ... B2
27 Teatro di Marcello ... E8

Eating

28 Antico Forno Urbani ... D7
29 Armando al Pantheon ... D3
30 Baguetteria del Fico ... A3
31 Campo de' Fiori ... B5
32 Casa Bleve ... C4
33 Casa Coppelle ... D2
34 Chiostro del Bramante Caffè ... B2
35 Cremeria Romana ... D7
36 Ditirambo ... B5
37 Forno di Campo de' Fiori ... B5
38 Forno Roscioli ... C6
39 Gelateria del Teatro ... A2
40 Grappolo D'Oro ... B5
41 La Ciambella ... D5
42 La Rosetta ... D3
43 Nonna Betta ... E7
44 Osteria dell'Ingegno ... E2
45 Pasticceria De Bellis ... C5
46 Piperno ... D7
47 Renato e Luisa ... D6
48 Salumeria Roscioli ... C6

Shopping

49 A S Roma Store ... F2
50 Ai Monasteri ... C2
51 Alberta Gloves ... E5
52 Aldo Fefè ... D1
53 Arsenale ... A5
54 Bartolucci ... E3
55 Borini ... B7
56 Bottega Pio La Torre ... D1
57 Confetteria Moriondo & Gariglio ... E4
58 daDADA 52 ... B6
59 De Sanctis ... E3
60 Galleria Alberto Sordi ... F2
61 Ibiz – Artigianato in Cuoio ... C6
62 Le Artigiane ... D4
63 Le Tele di Carlotta ... B2
64 Leone Limentani ... E7
65 Loco ... B5
66 Luna & L'Altra ... B4
67 Nardecchia ... B3
68 Officina Profumo Farmaceutica di Santa Maria Novella ... C3
69 Rachele ... A5
70 SBU ... B4
71 Tartarughe ... E4
72 Tempi Moderni ... A4

Drinking & Nightlife

73 Barnum Cafe ... A4
74 Caffè Sant'Eustachio ... C4
75 Circus ... A2
76 Etablì ... A3
77 Freni e Frizioni ... A8
78 La Casa del Caffè Tazza d'Oro ... D3
79 La Mescita ... A8
80 L'Angolo Divino ... B6
81 No.Au ... B2
82 Open Baladin ... C7
83 Salotto 42 ... E2

Entertainment

84 Teatro Argentina ... D5

Activities, Courses & Tours

85 Roman Kitchen ... F4

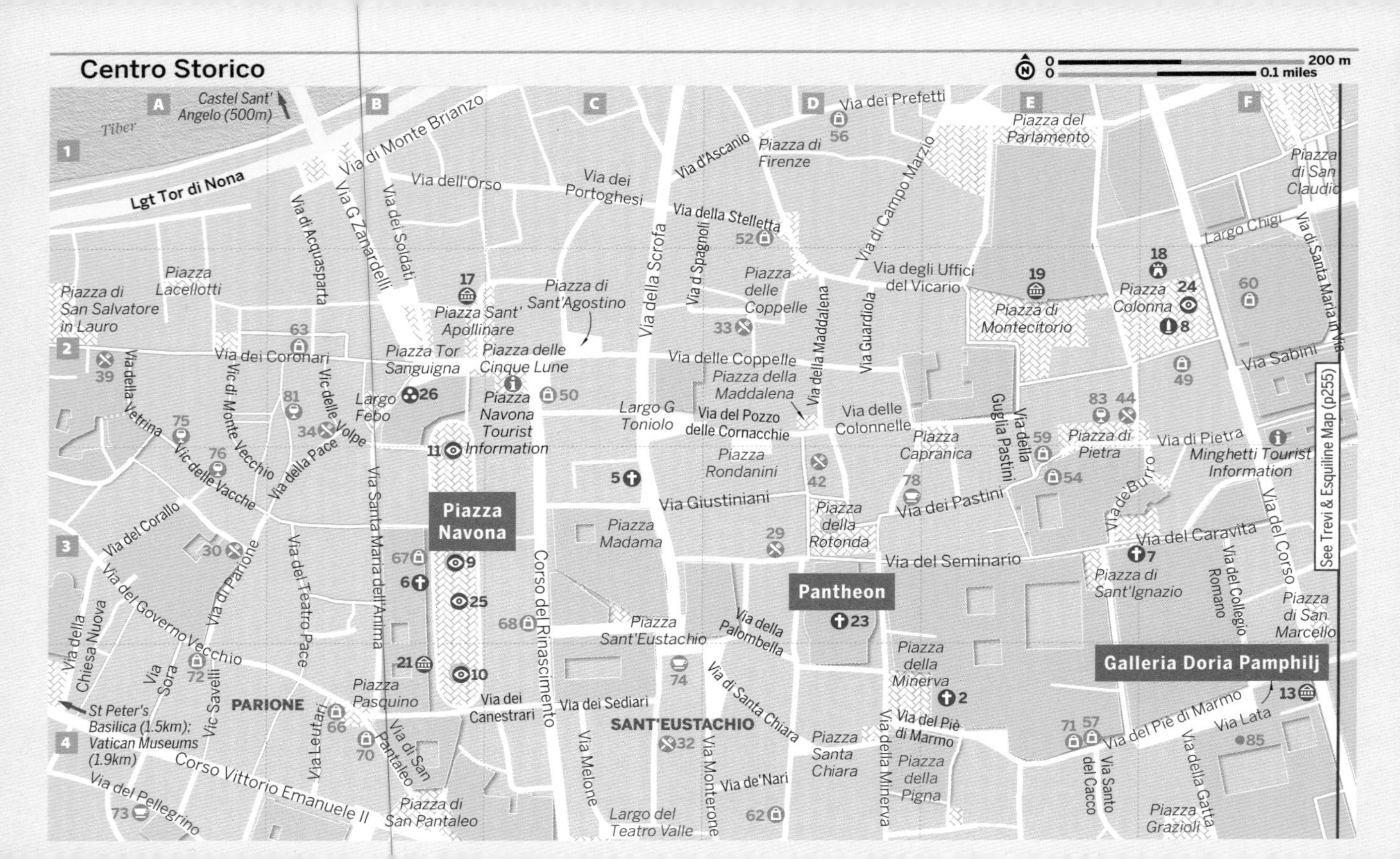
Centro Storico
200 m
0.1 miles
See Trevi & Esquiline Map (p255)
Castel Sant' Angelo (500m)
St Peter's Basilica (1.5km); Vatican Museums (1.9km)
Tiber
Lgt Tor di Nona
Piazza Navona
Pantheon
Galleria Doria Pamphilj
SANT'EUSTACHIO
PARIONE
Piazza Navona Tourist Information
Minghetti Tourist Information
Via di Monte Brianzo
Via dell'Orso
Via dei Portoghesi
Via d'Ascanio
Via dei Prefetti
Piazza di Firenze
Via di Campo Marzio
Piazza del Parlamento
Piazza di San Claudio
Largo Chigi
Via di Santa Maria in Via
Via della Stelletta
Via di Acquasparta
Via G Zanardelli
Via dei Soldati
Via della Scrofa
Via d Spagnoli
Piazza delle Coppelle
Via degli Uffici del Vicario
Piazza di Montecitorio
Piazza Colonna
Piazza di San Salvatore in Lauro
Piazza Lancellotti
Piazza Sant' Apollinare
Piazza di Sant'Agostino
Via della Maddalena
Via Guardiola
Via dei Coronari
Piazza Tor Sanguigna
Piazza delle Cinque Lune
Via delle Coppelle
Piazza della Maddalena
Via Sabini
Via della Vetrina
Vic di Monte Vecchio
Vic delle Volpe
Largo Febo
Largo G Toniolo
Via del Pozzo delle Cornacchie
Via delle Colonnelle
Via della Guglia Pastini
Piazza di Pietra
Via di Pietra
Vic delle Vacche
Via della Pace
Via Santa Maria dell'Anima
Piazza Capranica
Piazza Rondanini
Via Giustiniani
Via dei Pastini
Via de'Burrò
Via del Corallo
Piazza Madama
Piazza della Rotonda
Via del Caravita
Via del Corso
Via di Parione
Via del Teatro Pace
Corso del Rinascimento
Via del Seminario
Piazza di Sant'Ignazio
Via del Collegio Romano
Via della Chiesa Nuova
Via del Governo Vecchio
Piazza Sant'Eustachio
Via della Palombella
Piazza di San Marcello
Via Sora
Vic Savelli
Piazza della Minerva
Piazza Pasquino
Via dei Canestrari
Via dei Sediari
Via di Santa Chiara
Via Leutari
Via di San Pantaleo
Via del Piè di Marmo
Via Lata
Corso Vittorio Emanuele II
Via Melone
Via Monterone
Via de'Nari
Piazza Santa Chiara
Via della Minerva
Piazza della Pigna
Via Santo del Cacco
Via della Gatta
Via del Pellegrino
Piazza di San Pantaleo
Largo del Teatro Valle
Piazza Grazioli
A B C D E F
1 2 3 4

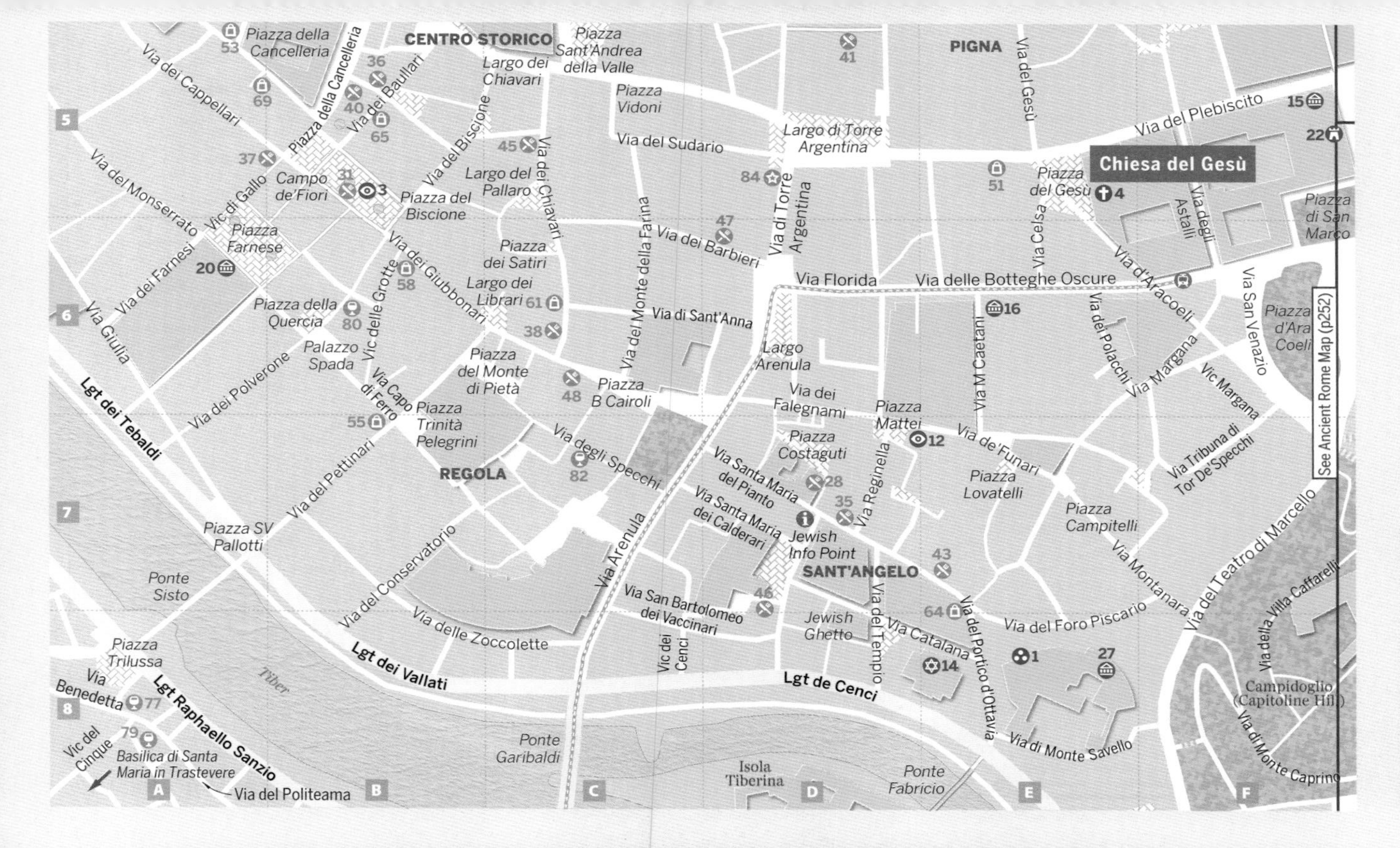

CENTRO STORICO
PIGNA
REGOLA
SANT'ANGELO
Chiesa del Gesù
See Ancient Rome Map (p252)
Piazza della Cancelleria
Piazza Sant'Andrea della Valle
Largo dei Chiavari
Via dei Cappellari
Piazza della Cancelleria
Via dei Baullari
Piazza Vidoni
Via del Gesù
Via del Plebiscito
Via del Biscione
Via del Sudario
Largo di Torre Argentina
Campo de' Fiori
Largo del Pallaro
Via dei Chiavari
Piazza del Gesù
Piazza del Biscione
Via Celsa
Via degli Astalli
Piazza di San Marco
Via del Monserrato
Vic di Gallo
Piazza Farnese
Via di Torre Argentina
Via dei Barbieri
Via del Monte della Farina
Piazza dei Satiri
Via dei Farnesi
Via dei Giubbonari
Via delle Grotte
Largo dei Librari
Via Florida
Via delle Botteghe Oscure
Via d'Aracoeli
Via San Venazio
Piazza d'Ara Coeli
Piazza della Quercia
Via di Sant'Anna
Via dei Polacchi
Via Giulia
Palazzo Spada
Largo Arenula
Via M Caetani
Via Margana
Vic Margana
Piazza del Monte di Pietà
Via Capo di Ferro
Piazza B Cairoli
Via dei Falegnami
Piazza Mattei
Lgt dei Tebaldi
Via dei Polverone
Piazza Trinità Pelegrini
Via de' Funari
Via Tribuna di Tor De'Specchi
Piazza Costaguti
Via degli Specchi
Via Santa Maria del Pianto
Via Reginella
Piazza Lovatelli
Via del Pettinari
Via Santa Maria dei Calderari
Jewish Info Point
Piazza Campitelli
Piazza SV Pallotti
Via del Conservatorio
Via Arenula
Via Montanara
Via del Teatro di Marcello
Ponte Sisto
Via San Bartolomeo dei Vaccinari
Via della Villa Caffarelli
Jewish Ghetto
Via del Tempio
Via del Portico d'Ottavia
Via del Foro Piscario
Via delle Zoccolette
Via Catalana
Piazza Trilussa
Tiber
Lgt dei Vallati
Vic dei Cenci
Lgt de Cenci
Campidoglio (Capitoline Hill)
Via Benedetta
Lgt Raphaello Sanzio
Vic del Cinque
Basilica di Santa Maria in Trastevere
Ponte Garibaldi
Via di Monte Savello
Via di Monte Caprino
Isola Tiberina
Ponte Fabricio
Via del Politeama
A
B
C
D
E
F
5
6
7
8

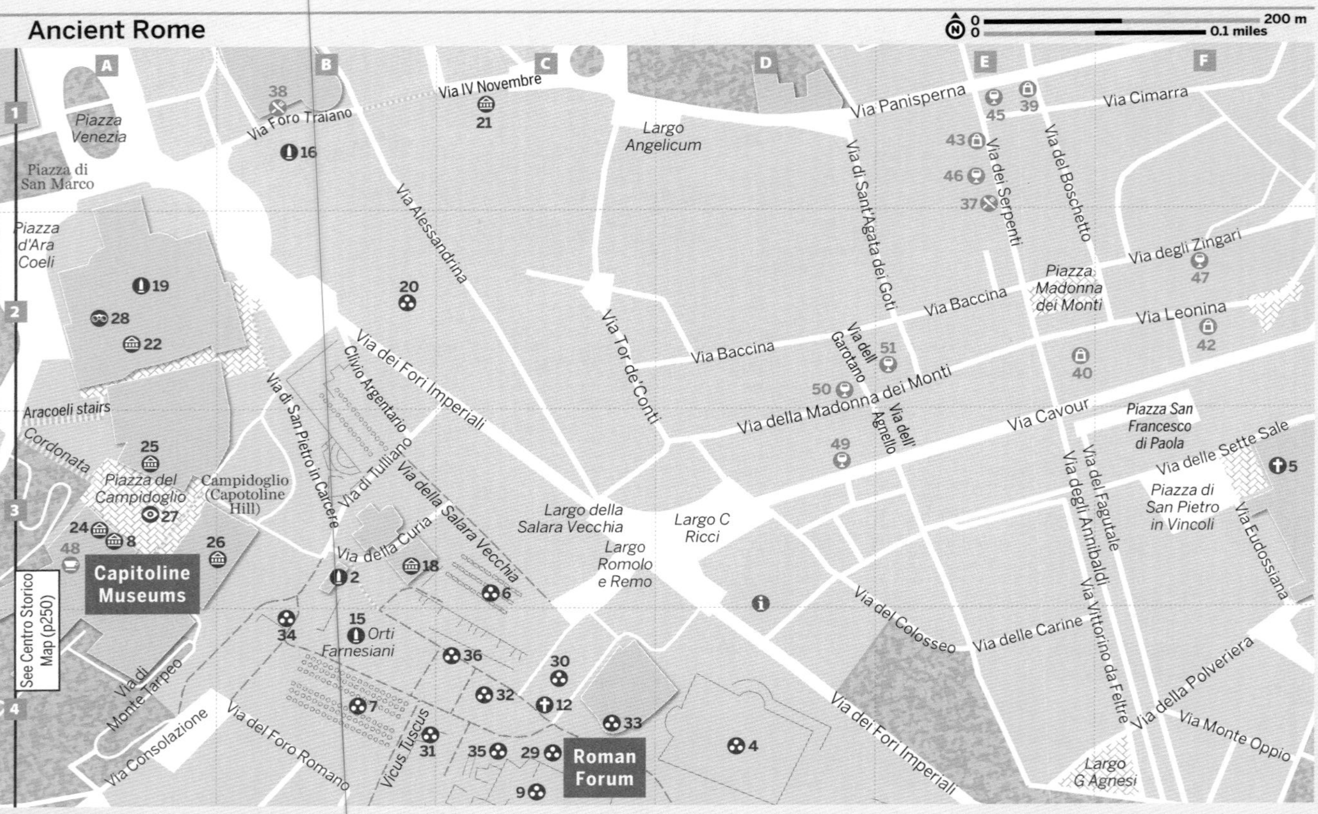

Ancient Rome
200 m
0.1 miles
Piazza Venezia
Piazza di San Marco
Piazza d'Ara Coeli
Aracoeli stairs
Cordonata
Piazza del Campidoglio
Campidoglio (Capotoline Hill)
Capitoline Museums
See Centro Storico Map (p250)
Via di Monte Tarpeo
Via Consolazione
Via del Foro Romano
Vicus Tuscus
Orti Farnesiani
Roman Forum
Via di San Pietro in Carcere
Clivio Argentario
Via di Tulliano
Via della Curia
Via della Salara Vecchia
Via dei Fori Imperiali
Via Foro Traiano
Via IV Novembre
Via Alessandrina
Largo Angelicum
Via Tor de'Conti
Largo della Salara Vecchia
Largo Romolo e Remo
Largo C Ricci
Via Panisperna
Via di Sant'Agata dei Goti
Via dei Serpenti
Via del Boschetto
Via Cimarra
Via degli Zingari
Piazza Madonna dei Monti
Via Baccina
Via Leonina
Via dell Garotano
Via della Madonna dei Monti
Via dell' Agnello
Via Cavour
Piazza San Francesco di Paola
Via delle Sette Sale
Piazza di San Pietro in Vincoli
Via Eudossiana
Via degli Annibaldi
Via del Fagutale
Via Vittorino da Feltre
Via del Colosseo
Via delle Carine
Via della Polveriera
Via Monte Oppio
Largo G Agnesi

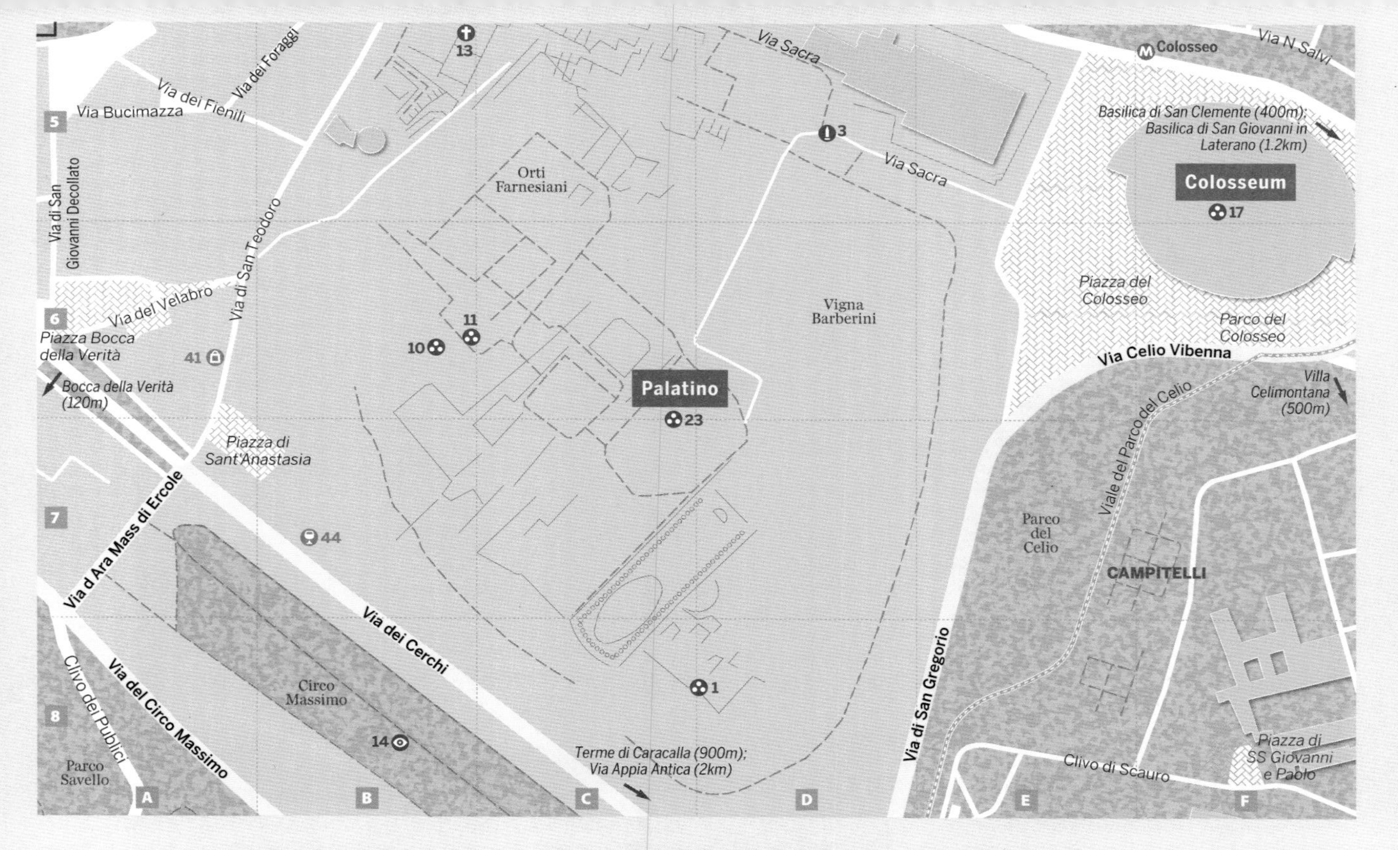

Via Sacra
13
Via dei Foraggi
Via dei Fienili
Via Bucimazza
Via di San Giovanni Decollato
Colosseo
Via N Salvi
Basilica di San Clemente (400m); Basilica di San Giovanni in Laterano (1.2km)
Colosseum
17
3
Via Sacra
Orti Farnesiani
Via di San Teodoro
Via del Velabro
Piazza Bocca della Verità
41
11
10
Vigna Barberini
Piazza del Colosseo
Parco del Colosseo
Via Celio Vibenna
Villa Celimontana (500m)
Bocca della Verità (120m)
Palatino
23
Piazza di Sant'Anastasia
Viale del Parco del Celio
Via d Ara Mass di Ercole
44
Parco del Celio
CAMPITELLI
Via dei Cerchi
Via del Circo Massimo
Clivo dei Publici
Circo Massimo
1
Via di San Gregorio
14
Parco Savello
Terme di Caracalla (900m); Via Appia Antica (2km)
Clivo di Scauro
Piazza di SS Giovanni e Paolo
A
B
C
D
E
F
5
6
7
8

Ancient Rome

Sights

1 Arcate Severiane D8
2 Arco di Settimio Severo B3
3 Arco di Tito D5
4 Basilica di Massenzio D4
5 Basilica di San Pietro in Vincoli F3
6 Basilica Fulvia Aemilia C3
7 Basilica Giulia B4
8 Capitoline Museums A3
9 Casa delle Vestali C4
10 Casa di Augusto B6
11 Casa di Livia B6
12 Chiesa di San Lorenzo in Miranda C4
13 Chiesa di Santa Maria Antiqua B5
14 Circo Massimo B8
15 Colonna di Foca B4
16 Colonna di Traiano B1
17 Colosseum F5
18 Curia B3
19 Il Vittoriano A2
20 Imperial Forums B2
21 Mercati di Traiano Museo dei Fori Imperiali C1
22 Museo Centrale del Risorgimento A2
23 Palatino C6
24 Palazzo dei Conservatori A3
25 Palazzo Nuovo A3
26 Palazzo Senatorio B3
27 Piazza del Campidoglio A3
28 Roma dal Cielo A2
29 Roman Forum C4
30 Tempio di Antonino e Faustina C4
31 Tempio di Castore e Polluce B4
32 Tempio di Giulio Cesare C4
33 Tempio di Romolo C4
34 Tempio di Saturno B4
35 Tempio di Vesta C4
36 Via Sacra C4

Eating

37 Temakinho E1
38 Terre e Domus B1

Shopping

39 Fabio Piccioni E1
40 La Bottega del Cioccolato E2
41 Mercato di Circo Massimo A6
42 Mercato Monti Urban Market F2
43 Podere Vecciano E1

Drinking & Nightlife

44 0,75 .... B7
45 Ai Tre Scalini E1
46 Al Vino al Vino E1
47 Bohemien F2
48 Caffè Capitolino A3
49 Cavour 313 .... D3
50 Fafiuché D2
51 Ice Club E2

Trevi & Esquiline

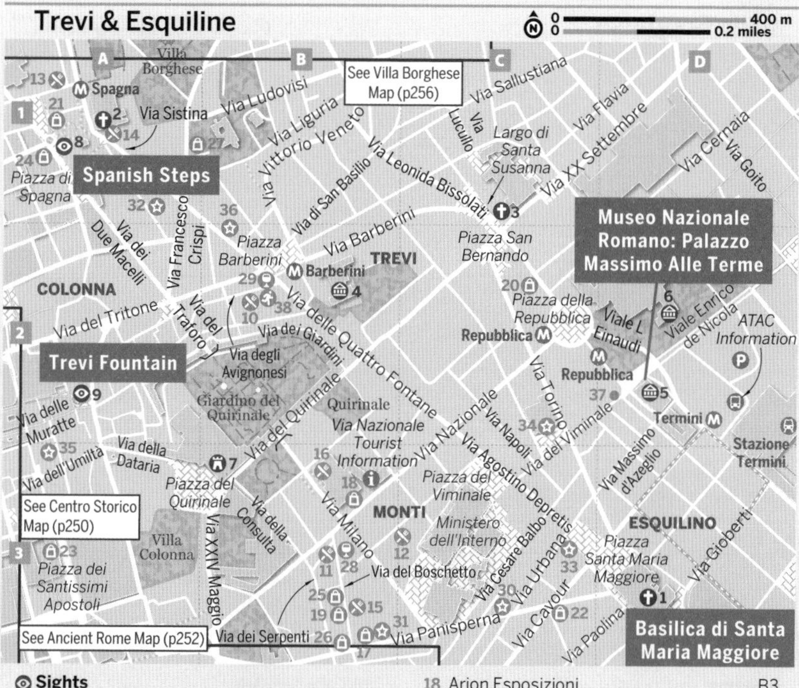

Sights

1 Basilica di Santa Maria Maggiore D3
2 Chiesa della Trinità dei Monti A1
3 Chiesa di Santa Maria della Vittoria C1
4 Galleria Nazionale d'Arte Antica: Palazzo Barberini B2
Keats-Shelley House .... (see 8)
5 Museo Nazionale Romano: Palazzo Massimo alle Terme D2
6 Museo Nazionale Romano: Terme di Diocleziano D2
7 Palazzo del Quirinale B3
8 Piazza di Spagna & the Spanish Steps A1
9 Trevi Fountain A2

Eating

10 Colline Emiliane B2
11 Da Valentino B3
12 Doozo C3
13 Gina A1
14 Imàgo A1
15 L'Asino d'Oro B3
16 Open Colonna B3

Shopping

17 Abito B3
18 Arion Esposizioni B3
19 Creje B3
20 Feltrinelli International C2
21 Furla A1
22 Giacomo Santini C3
23 Lucia Odescalchi A3
24 Sermoneta A1
25 Spot B3
26 Tina Sondergaard B3
27 Underground A1

Drinking & Nightlife

28 La Barrique B3
29 Micca Club B2

Entertainment

30 Blackmarket C3
31 Charity Café B3
32 Gregory's A1
33 Orbis C3
34 Teatro dell'Opera di Roma C2
35 Teatro Quirino A3
36 Teatro Sistina B2

Activities, Courses & Tours

37 Bici & Baci D2
38 Kami Spa B2

Villa Borghese

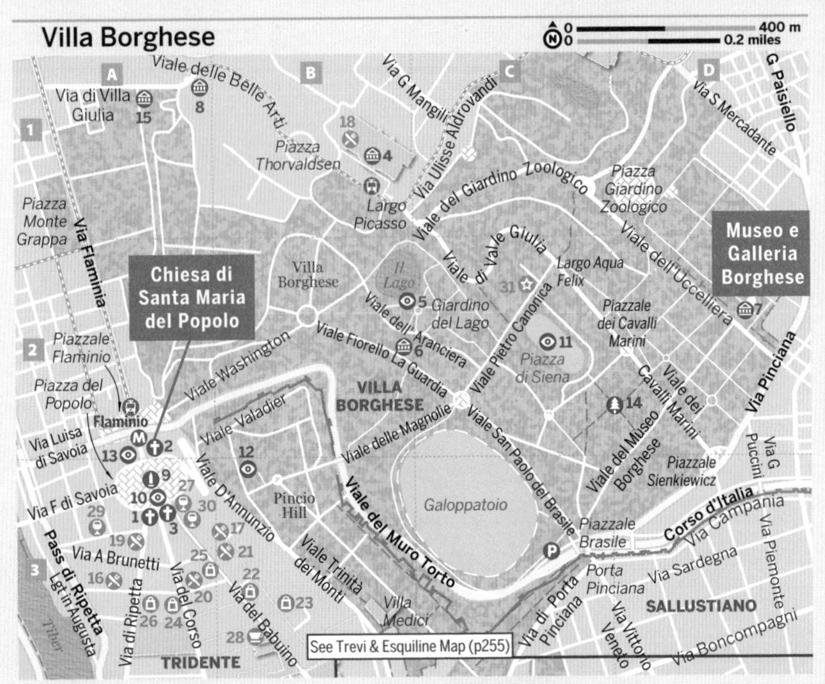

Sights
1 Chiesa di Santa Maria dei Miracoli A3
2 Chiesa di Santa Maria del Popolo A2
3 Chiesa di Santa Maria in Montesanto A3
4 Galleria Nazionale d'Arte Moderna e Contemporanea B1
5 Giardino del Lago B2
6 Museo Carlo Bilotti B2
7 Museo e Galleria Borghese D2
8 Museo Nazionale Etrusco di Villa Giulia A1
9 Obelisk A3
10 Piazza del Popolo A3
11 Piazza di Siena C2
12 Pincio Hill Gardens B3
13 Porta del Popolo A2
14 Villa Borghese D2
15 Villa Poniatowski A1

Eating
Al Gran Sasso .... (see 26)
16 All'Oro A3
17 Babette B3
18 Caffè delle Arti B1
19 Dei Gracchi A3
20 Fatamorgana A3
21 Il Margutta RistorArte B3

Shopping
22 Barrilà Boutique B3
23 Bottega del Marmoraro B3
24 DaDADA A3
25 Fabriano A3
26 L'Olfattorio A3

Drinking & Nightlife
27 Canova A3
28 Canova Tadolini B3
29 La Scena A3
30 Stravinskij Bar – Hotel de Russie A3

Entertainment
31 Silvano Toti Globe Theatre C2

Activities, Courses & Tours
Hotel De Russie Wellness Zone .... (see 30)

Symbols & Map Key

Look for these symbols to quickly identify listings:

- Sights
- Activities
- Courses
- Tours
- Festivals & Events
- Eating
- Drinking
- Entertainment
- Shopping
- Information & Transport

These symbols and abbreviations give vital information for each listing:

Sustainable or green recommendation

FREE No payment required

- Telephone number
- Opening hours
- Parking
- Nonsmoking
- Air-conditioning
- Internet access
- Wi-fi access
- Swimming pool
- Bus
- Ferry
- Tram
- Train
- English-language menu
- Vegetarian selection
- Family-friendly

Find your best experiences with these Great For... icons.

- Budget
- Food & Drink
- Drinking
- Cycling
- Shopping
- Sport
- Art & Culture
- Events
- Photo Op
- Scenery
- Family Travel
- Short Trip
- Detour
- Walking
- Local Life
- History
- Entertainment
- Beaches
- Winter Travel
- Cafe/Coffee
- Nature & Wildlife

Sights

- Beach
- Bird Sanctuary
- Buddhist
- Castle/Palace
- Christian
- Confucian
- Hindu
- Islamic
- Jain
- Jewish
- Monument
- Museum/Gallery/Historic Building
- Ruin
- Shinto
- Sikh
- Taoist
- Winery/Vineyard
- Zoo/Wildlife Sanctuary
- Other Sight

Points of Interest

- Bodysurfing
- Camping
- Cafe
- Canoeing/Kayaking
- Course/Tour
- Diving
- Drinking & Nightlife
- Eating
- Entertainment
- Sento Hot Baths/Onsen
- Shopping
- Skiing
- Sleeping
- Snorkelling
- Surfing
- Swimming/Pool
- Walking
- Windsurfing
- Other Activity

Information

- Bank
- Embassy/Consulate
- Hospital/Medical
- Internet
- Police
- Post Office
- Telephone
- Toilet
- Tourist Information
- Other Information

Geographic

- Beach
- Gate
- Hut/Shelter
- Lighthouse
- Lookout
- Mountain/Volcano
- Oasis
- Park
- Pass
- Picnic Area
- Waterfall

Transport

- Airport
- BART station
- Border crossing
- Boston T station
- Bus
- Cable car/Funicular
- Cycling
- Ferry
- Metro/MRT station
- Monorail
- Parking
- Petrol station
- Subway/S-Bahn/Skytrain station
- Taxi
- Train station/Railway
- Tram
- Tube Station
- Underground/U-Bahn station
- Other Transport

Our Story

A beat-up old car, a few dollars in the pocket and a sense of adventure. In 1972 that's all Tony and Maureen Wheeler needed for the trip of a lifetime – across Europe and Asia overland to Australia. It took several months, and at the end – broke but inspired – they sat at their kitchen table writing and stapling together their first travel guide, *Across Asia on the Cheap*. Within a week they'd sold 1500 copies. Lonely Planet was born. Today, Lonely Planet has offices in Dublin, Melbourne, London, Oakland, Franklin, Delhi and Beijing, with more than 600 staff and writers. We share Tony's belief that 'a great guidebook should do three things: inform, educate and amuse'.

Our Writers

Duncan Garwood

A Brit travel writer based in the Castelli Romani hills just outside Rome, Duncan moved to the Italian capital just in time to see the new millennium in at the Colosseum. He has since clocked up endless kilometres walking around his adopted hometown and exploring the far-flung reaches of the surrounding Lazio region. He has worked on the past six editions of the *Rome* city guide as well as previous editions of the *Pocket Rome* guide and a whole host of LP Italy publications. He has also written on Italy for newspapers and magazines.

Abigail Blasi

Abigail moved to Rome in 2003 and lived there for three years, got married alongside Lago Bracciano and her first son was born in Rome. Nowadays she divides her time between Rome, Puglia and London. She has worked on four editions of Lonely Planet's Italy and Rome guides, and co-wrote the first edition of *Puglia & Basilicata*. She also regularly writes on Italy for various publications, including the *Independent*, the *Guardian*, and *Lonely Planet Traveller*.

STAY IN TOUCH LONELYPLANET.COM/CONTACT

EUROPE Unit E, Digital Court, The Digital Hub, Rainsford St, Dublin 8, Ireland

AUSTRALIA Levels 2 & 3 551 Swanston St, Carlton, Victoria 3053 ☎ 03 8379 8000, fax 03 8379 8111

USA 150 Linden Street, Oakland, CA 94607 ☎ 510 250 6400, toll free 800 275 8555, fax 510 893 8572

UK 240 Blackfriars Road, London SE1 8NW ☎ 020 3771 5100, fax 020 3771 5101

 twitter.com/lonelyplanet

 facebook.com/lonelyplanet

 instagram.com/lonelyplanet

youtube.com/lonelyplanet

 lonelyplanet.com/newsletter